HVAC/R

PROFESSIONAL REFERENCE

MASTER EDITION

Paul Rosenberg

Created exclusively
for DeWALT by:

www.palpublications.com
1-800-246-2175

Titles Available From DeWALT

DeWALT Trade Reference Series

Construction Professional Reference

Construction Estimating Professional Reference

Blueprint Reading Professional Reference

Electric Motor Professional Reference

Electrical Estimating Professional Reference

Electrical Professional Reference

Lighting & Maintenance Professional Reference

Plumbing Professional Reference

Residential Remodeling & Repair Professional Reference

Spanish/English Construction Dictionary – Illustrated

Wiring Diagrams Professional Reference

DeWALT Exam and Certification Series

Construction Licensing Exam Guide

Electrical Licensing Exam Guide

HVAC Technician Certification Exam Guide

Plumbing Licensing Exam Guide

For a complete list of The DeWALT Professional Trade Reference Series visit **www.palpublications.com**.

This Book Belongs To:

Name:_____

Company: _____

Title: _____

Department: _____

Company Address: _____

_____

_____

Company Phone: _____

Home Phone: _____

Pal Publications, Inc.
800 Heritage Drive, Suite 810
Pottstown, PA 19464-3810

Copyright © 2006 by Pal Publications, Inc.
First edition published 2006

ISBN 0-9770003-8-9

10   09   08   07   06   5   4   3   2   1

Printed in the United States of America

A Note To Our Customers

We have manufactured this book to the highest quality standards possible. The cover is made of a flexible, durable and water-resistant material able to withstand the toughest on-the-job conditions. We also utilize the Otabind process which allows this book to lay flatter than traditional paperback books that tend to snap shut while in use.

Preface

HVAC/R is a huge trade and requires the combination of various specialized skills. HVAC/R involves the combustion of several types of fuels, various refrigeration systems, piping, ductwork, motors, compressors, fans, pumps, and a fair amount of electrical work.

We previously published a reference book, but found it was too small to include everything that an HVAC/R technician requires. The book you are now holding contains twice the amount of information as the previous edition. I have included much more regarding the core HVAC/R technologies, as well as service, troubleshooting and mechanical piping. This new version gives you the additional reference material that you will need on the job.

Chapter One contains essential HVAC/R formulas along with necessary design data that may be required by a contractor. Chapters Two, Three and Four cover the central technologies of heating, ventilation, air conditioning, and refrigeration. Chapter Five covers mechanical piping systems. In addition to providing needed information, Chapters Six and Seven also contain useful maintenance forms for a wide variety of service work as well as a section including troubleshooting charts.

Chapter Eight is loaded with information regarding electricity and motors. Chapters Nine, Ten, Eleven, and Twelve contain conversion factors, tools and materials, plan symbols and abbreviations, as well as a detailed glossary.

In all, I think this is the most complete book of its kind, and that you will find it useful and necessary to keep close at hand on a daily basis.

Naturally, there may be some aspects of the HVAC/R trade that are not covered in sufficient depth for some readers. I will update this book on a continual basis and attempt to include material suggested by readers as well as keep pace with developments in the trade.

Best wishes,
Paul Rosenberg

CONTENTS

CHAPTER 2 – *Heating* 2-1

CHAPTER 7 – *Troubleshooting*. 7-1

CHAPTER 1
Formulas and Design Data

HEAT TRANSFER

This equation is used to determine the amount of heat required to heat or cool a given amount of any specific substance.

BTU = Specific heat $\times$ lbs. $\times$ ΔT

BTU = British Thermal Unit
ΔT = Temperature Difference (°F)

The specific heat of a substance is the amount of heat in Btu that it takes to change the temperature of one pound of that substance by one degree Fahrenheit.

For example, the specific heat of water is 1.0 Btu/lb.°F. Therefore, to raise the temperature of 1 pound of water by 1 degree F takes 1 Btu. Specific heats of other materials vary.

HEATING AND COOLING CAPACITY

BTU/h = 1.08 $\times$ CFM $\times$ ΔT

BTU/h = Btu/h per hour, a quantity of heat
CFM = Cubic feet per minute of air flow

(for sensible heat only)

TOTAL HEAT CAPACITY

$$\text{BTU/h} = 4.5 \times \text{CFM} \times \Delta H$$

ΔH = Difference between return & supply air, in Btu/lb.

(all systems, including both sensible & latent heats)

COOLING AND HEATING

$$\text{Sensible heat} = 1.08 \times \text{CFM} \times \Delta T$$
$$\text{Sensible heat} = 1.1 \times \text{CFM} \times \Delta T$$
$$\text{Latent heat} = 0.68 \times \text{CFM} \times \Delta W_{GR.}$$
$$\text{Latent heat} = 4840 \times \text{CFM} \times \Delta W_{LB.}$$
$$\text{Total heat} = 4.5 \times \text{CFM} \times \Delta H$$
$$\text{Total heat} = H_S + H_L$$
$$\text{Total heat} = U \times A \times \Delta T$$

$$\text{SHR} = \frac{H_S}{H_T} = \frac{H_S}{H_S + H_L}$$

H_S = Sensible heat (Btu/h)
H_L = Latent heat (Btu/h)
H_T = Total heat (Btu/h)
ΔT = Temperature difference (°F)
$\Delta W_{GR.}$ = Humidity ratio difference (Gr.H_2O/Lb.DA)
$\Delta W_{LB.}$ = Humidity ratio difference (Lb.H_2O/Lb.DA)
ΔH = Enthalpy difference (Btu/Lb.DA)
CFM = Air flow rate (cubic feet per minute)
U = U-value (Btu/h/sq. ft./°F)
A = Area (sq. ft.)
SHR = Sensible heat ratio
H_{FG} = Latent heat of vaporization at design pressure

R-VALUES/U-VALUES

$$R = \frac{1}{C} = \frac{1}{K} \times \text{Thickness}$$

$$U = \frac{1}{\Sigma R}$$

R = R-value (hr. sq. ft. °F/Btu)
U = U-value (Btu/h sq. ft. °F)
C = Conductance (Btu/h sq. ft.°F)
K = Conductivity (Btu in./hr. sq. ft. °F)
ΣR = Sum of the individual R-values

OBTAINING CFM FROM CAPACITY

$$\text{CFM} = \frac{\text{Btu/h}}{1.08 \times \Delta T}$$

CFM OF AN ELECTRIC FURNACE

$$\text{CFM} = \frac{\text{Watts} \times 3.42 \ (\text{Btu/Watt})}{1.08 \times \Delta T}$$

CHANGE IN TEMPERATURE

$$\Delta T = \frac{\text{Btu/h}}{1.08 \times \text{CFM}}$$

OBTAINING RPM OF EXISTING BLOWER

$$\text{Blower RPM} = \frac{\text{Motor pulley diameter} \times \text{RPM of motor}}{\text{Diameter of blower pulley}}$$

CHANGE IN PULLEY, NEW BELT LENGTH

Belt length = $2 \times C + 1.57 \times D + d + \dfrac{(D - d)^2}{(4 \times C)}$

C = Distance, center to center, between pulleys

D = Larger pulley diameter

d = Smaller pulley diameter

PULLEY LAWS

RPM of driven = $\dfrac{\text{Diameter of driver} \times \text{RPM}}{\text{Diameter of driven}}$

RPM of driver = $\dfrac{\text{Diameter of driven} \times \text{RPM}}{\text{Diameter of driver}}$

Diameter of driven = $\dfrac{\text{Diameter of driver} \times \text{RPM}}{\text{RPM of driven}}$

Diameter of the driver = $\dfrac{\text{Diameter of driven} \times \text{RPM}}{\text{RPM of driver}}$

COMPRESSION RATIO

Compression ratio = $\dfrac{\text{Absolute head pressure in psia}}{\text{Absolute suction pressure in psia}}$

CONDENSER TONNAGE

$$\text{Condenser tons} = \frac{\text{GPM} \times 1.25 \times \Delta T \text{ conditioned water}}{24}$$

GPM = Gallons per minute of flow

ΔT = Temperature Difference (°F)

CHILLER TONNAGE

$$\text{Chiller tons} = \frac{\text{GPM} \times \Delta T \text{ chiller water}}{30}$$

GPM = Gallons per minute of flow

ΔT = Temperature Difference (°F)

WATER SYSTEM

$$H = 500 \times \text{GPM} \times \Delta T$$

$$\text{GPM}_{\text{EVAP.}} = \frac{\text{Tons} \times 24}{\Delta T}$$

$$\text{GPM}_{\text{COND.}} = \frac{\text{Tons} \times 30}{\Delta T}$$

H = Total heat (BTU/h)

GPM = Water flow rate (gallons per minute)

ΔT = Temperature difference (°F)

Tons = Air conditioning load (tons)

$\text{GPM}_{\text{EVAP.}}$ = Evaporator water flow rate (gallons per minute)

$\text{GPM}_{\text{COND.}}$ = Condenser water flow rate (gallons per minute)

COOLING TOWERS AND HEAT EXCHANGERS

$APPROACH_{CT} = LWT - AWB$

$APPROACH_{HE} = EWT_{HS} - LWT_{CS}$

$RANGE = EWT - LWT$

EWT = Entering water temperature (°F)

LWT = Leaving water temperature (°F)

AWB = Ambient wet bulb temperature (design WB, °F)

CT = Cooling Tower

HE = Heat Exchange

HS = Hot side

CS = Cold side

PUMP NET POSITIVE SUCTION HEAD (NPSH)

$NPSH_{AVAIL} > NPSH_{REQD}$

$NPSH_{AVAIL} = H_A \pm H_S - H_F - H_{VP}$

$NPSH_{AVAIL}$ = Net positive suction available at pump (ft.)

$NPSH_{REQD}$ = Net positive suction required at pump (ft.)

H_A = Pressure at liquid surface (ft. — 34 ft. for water at atmospheric pressure)

H_S = Height of liquid surface above (+) or below (−) pump (ft.)

H_F = Friction loss between pump and source (ft.)

H_{VP} = Absolute pressure of water vapor at liquid temperature

PUMP LAWS

$$\frac{GPM_2}{GPM_1} = \frac{RPM_2}{RPM_1}$$

$$\frac{HD_2}{HD_1} = \left[\frac{GPM_2}{GPM_1}\right]^2 = \left[\frac{RPM_2}{RPM_1}\right]^2$$

$$\frac{BHP_2}{BHP_1} = \left(\frac{GPM_2}{GPM_1}\right)^3 = \left(\frac{RPM_2}{RPM_1}\right)^3 = \left(\frac{HD_2}{HD_1}\right)^{1.5}$$

GPM   = Gallons per minute

RPM   = Revolutions per minute

HD   = Feet of water

BHP   = Break horsepower

Pump size   = Constant

Water density   = Constant

PUMP LAWS *(cont.)*

$$BHP = \frac{GPM \times HD \times SP.GR.}{3960 \times PUMP_{EFF}}$$

$$MHP = \frac{BHP}{M/D_{EFF}}$$

$$VH = \frac{V^2}{2g}$$

$$HD = \frac{P \times 2.31}{SP.GR.}$$

SP.GR.	=	Specific gravity of liquid with respect to water
SP.GR. (water)	=	1.0
PUMP_{EFF}	=	60–80%
M/D_{EFF}	=	85–95%
M/D	=	Motor/drive
MHP	=	Motor horsepower
P	=	Pressure in psi
VH	=	Velocity head in ft.
V	=	Velocity in ft./sec.
g	=	Acceleration due to gravity (32.16 ft./sec.²)

REFRIGERATION FORMULAS

Net refrigerating effect (Btu/lb.)	= Enthalpy of vapor leaving evaporator (Btu/lb.) − Enthalpy of liquid entering evaporator (Btu/lb.)	× Refrigerant circulated (lb./min.)
Compression work (Btu min.)	= Heat of compression (Btu/lb.)	× Refrigerant circulated (lb./min.)
Compression HP	= Compression work (Btu/min.)/42.4	
Compression HP	= Capacity (Btu/min.)/42.4 × COP	
Compression HP (per ton)	= 4.715/COP	

1-9

REFRIGERATION FORMULAS (cont.)

Power (Watts) = Compression HP per ton × 745.7

$$COP = \frac{\text{Net refrigerating effect (Btu/lb.)}}{\text{Heat of compression (Btu/lb.)}}$$

Capacity (Btu/min.) = Refrigerant circulated (lb./min.) × Net refrigerating effect (Btu/lb.)

$$\text{Compressor displacement (ft.}^3\text{/min.)} = \frac{\text{Capacity (Btu/min.)}}{\text{Net refrigerating effect (Btu/lb.)}} \times \text{Volume of gas entering compressor (ft.}^3\text{/lb.)}$$

Heat of compression (Btu/lb.) = Enthalpy of vapor leaving compressor (Btu/lb.) − Enthalpy of vapor entering compressor (Btu/lb.)

REFRIGERATION FORMULAS (cont.)

$$\text{Volumetric efficiency} = 100 \times \left(\frac{\text{Actual weight of refrigerant}}{\text{Theoretical weight of refrigerant}} \right)$$

$$\text{Compression ratio} = \frac{\text{Head pressure, psia (absolute)}}{\text{Suction pressure, psia (absolute)}}$$

$$\text{Refrigerant circulated,} = \frac{200}{\text{Refrigerating effect}}$$
$$\text{lb./(min.)/(ton)}$$

42.4 = heat flow, Btu/(min.) (hp); 200 = Btu/(min.) (ton); COP = coefficient of performance

AIR FLOW

CFM = FPM × area (ft.²)

FPM = CFM/area (ft.²)

Area = CFM/FPM

AIR CHANGE RATE

$$AC/HR = \frac{CFM \times 60}{VOLUME}$$

$$CFM = \frac{AC/HR \times VOLUME}{60}$$

AC/HR = Air change rate per hour
CFM = Air flow rate (cubic feet per minute)
VOLUME = Space volume (cubic feet)

AIR BALANCE EQUATIONS

SA = RA + OA = RA + EA + RFA
If minimum OA (ventilation air) is greater than EA,
OA = EA + RFA
If EA is greater than minimum OA (ventilation air),
OA = EA   RFA = 0
For economizer cycle: OA = SA = EA + RFA   RA = 0

SA = Supply air
RA = Return air
OA = Outside air
EA = Exhaust air
RFA = Relief air

FAN LAWS

$$\frac{CFM_2}{CFM_1} = \frac{RPM_2}{RPM_1}$$

$$CFM_2 = CFM_1 \times \frac{RPM_2}{RPM_1}$$

$$RPM_2 = RPM_1 \times \frac{CFM_2}{CFM_1}$$

$$\frac{SP_2}{SP_1} = \left(\frac{CFM_2}{CFM_1}\right)^2 = \left(\frac{RPM_2}{RPM_1}\right)^2$$

$$SP_2 = SP_1 \times \left(\frac{RPM_2}{RPM_1}\right)^2$$

CFM	=	Cubic feet per minute
RPM	=	Revolutions per minute
SP	=	In. W.G.
BHP	=	Break horsepower
Fan size	=	Constant
Air density	=	Constant

FAN LAWS (cont.)

$$\frac{BHP_2}{BHP_1} = \left(\frac{CFM_2}{CFM_1}\right)^3 = \left(\frac{RPM_2}{RPM_1}\right)^3 = \left(\frac{SP_2}{SP_1}\right)^{1.5}$$

$$BHP_2 = BHP_1 \times \left(\frac{RPM_2}{RPM_1}\right)^3$$

$$BHP = \frac{CFM \times SP \times SP.GR.}{6356 \times FAN_{EFF}}$$

$$MHP = \frac{BHP}{M/D_{EFF}}$$

CFM	=	Cubic feet per minute
RPM	=	Revolutions per minute
SP	=	In. W.G.
BHP	=	Break horsepower
Fan size	=	Constant
Air density	=	Constant
SP.GR. (air)	=	1.0
FAN_{EFF}	=	65–85%
M/D_{EFF}	=	80–95%
M/D	=	Motor/drive
MHP	=	Motor horsepower

MIXED AIR TEMPERATURE

$$T_{MA} = \left(T_{ROOM} \times \frac{CFM_{RA}}{CFM_{SA}} \right) + \left(T_{OA} \times \frac{CFM_{OA}}{CFM_{SA}} \right)$$

$$T_{MA} = \left(T_{RA} \times \frac{CFM_{RA}}{CFM_{SA}} \right) + \left(T_{OA} \times \frac{CFM_{OA}}{CFM_{SA}} \right)$$

CFM_{SA} = Supply air (CFM)

CFM_{RA} = Return air (CFM)

CFM_{OA} = Outside air (CFM)

T_{MA} = Mixed air temperature (°F)

T_{ROOM} = Room design temperature (°F)

T_{RA} = Return air temperature (°F)

T_{OA} = Outside air temperature (°F)

AIR CONDITIONING CONDENSATE

$$GPM_{AC\ COND} = \frac{CFM \times \Delta W_{LB.}}{SpV \times 8.33}$$

$$GPM_{AC\ COND} = \frac{CFM \times \Delta W_{GR.}}{SpV \times 8.33 \times 7000}$$

$GPM_{AC\ COND}$ = Air conditioning condensate flow (gallons/min.)

CFM = Air flow rate (cu. ft./min.)

SpV = Specific volume of air (cu. ft./lb.DA)

$\Delta W_{LB.}$ = Specific humidity (lb.H_2O/lb.DA)

$\Delta W_{GR.}$ = Specific humidity (gr.H_2O/lb.DA)

MOISTURE CONDENSATION ON GLASS

$$T_{GLASS} = T_{ROOM} - \left[\frac{R_{IA}}{R_{GLASS}} \times (T_{ROOM} - T_{OA}) \right]$$

$$T_{GLASS} = T_{ROOM} - \left[\frac{U_{GLASS}}{U_{IA}} \times (T_{ROOM} - T_{OA}) \right]$$

(if $T_{GLASS} < DP_{ROOM}$ condensation occurs)

T = Temperature (°F)
R = R-value (hr. sq. ft. °F/Btu)
U = U-value (Btu/h sq. ft. °F)
IA = Inside airfilm
OA = Design outside air temperature
DP = Dew point

HUMIDIFIER SENSIBLE HEAT GAIN

$$H_S = (0.244 \times Q \times \Delta T) + (L \times 380)$$

H_S = Sensible heat gain (Btu/h)
Q   = Steam flow (lb. steam/hr.)
ΔT = Steam temperature – supply air temperature (°F)
L   = Length of humidifier manifold (ft.)

HUMIDIFICATION

$$\text{GRAINS}_{\text{REQD}} = \left(\frac{W_{\text{GR.}}}{\text{SpV}}\right)_{\text{ROOM AIR}} - \left(\frac{W_{\text{GR.}}}{\text{SpV}}\right)_{\text{SUPPLY AIR}}$$

$$\text{POUNDS}_{\text{REQD}} = \left(\frac{W_{\text{GR.}}}{\text{SpV}}\right)_{\text{ROOM AIR}} - \left(\frac{W_{\text{GR.}}}{\text{SpV}}\right)_{\text{SUPPLY AIR}}$$

$$\text{Lb. STM/hr.} = \frac{\text{CFM} \times \text{GRAINS}_{\text{REQD}} \times 60}{7000}$$

$$= \text{CFM} \times \text{POUNDS}_{\text{REQD}} \times 60$$

$\text{GRAINS}_{\text{REQD}}$ = Grains of moisture required (gr.H_2O/cu. ft.)

$\text{POUNDS}_{\text{REQD}}$ = Pounds of moisture required (lb.H_2O/ cu. ft.)

CFM = Air flow rate (cu. ft./min.)

SpV = Specific volume of air (cu. ft./lb.DA)

$W_{\text{GR.}}$ = Specific humidity (gr.H_2O/lb.DA)

$W_{\text{LB.}}$ = Specific humidity (lb.H_2O/lb.DA)

EXPANSION TANKS

Closed $V_T = V_S \times \dfrac{[(v_2/v_1) - 1] - 3\alpha\Delta T}{[(P_A/P_1) - (P_A/P_2)]}$

Open $V_T = 2 \times \{(V_S \times [(v_2/v_1) - 1]) - 3\alpha\Delta T\}$

Diaphragm $V_T = V_S \times \dfrac{[(v_2/v_1) - 1] - 3\alpha\Delta T}{1 - (P_1/P_2)}$

V_T = Volume of expansion tank (gal.)

V_S = Volume of water in piping system (gal.)

$\Delta T = T_2 - T_1$ (°F)

T_1 = Lower system temperature (°F)

Heating water	T_1 = 45–50°F temperature at fill condition
Chilled water	T_1 = Supply water temperature
Dual temperature	T_1 = Chilled water supply temperature

T_2 = Higher system temperature (°F)

Heating water	T_2 = Supply water temperature
Chilled water	T_2 = 95°F ambient temperature (design weather data)
Dual temperature	T_2 = Heating water supply temperature

1-18

EXPANSION TANKS *(cont.)*

P_A = Atmospheric pressure (14.7 psia)

P_1 = System fill pressure/minimum system pressure (psia)

P_2 = System operating pressure/maximum operating pressure (psia)

V_1 = SpV of H_2O at T_1 (cu. ft./lb.H_2O)
V_2 = SpV of H_2O at T_2 (cu. ft./lb.H_2O)

α = Linear coefficient of expansion
$\alpha_{STEEL} = 6.5 \times 10^{-6}$
$\alpha_{COPPER} = 9.5 \times 10^{-6}$

System volume estimate:
12 gal./ton
35 gal./BHP

System fill pressure/minimum system pressure estimate:
To height of system add 5 to 10 psi

System operating pressure/maximum operating pressure estimate:

| 150-lb. systems | 45–125 psi |
| 250-lb. systems | 125–225 psi |

EFFICIENCIES

EER = Energy efficiency ratio

$$EER = \frac{Btu\ output}{Watts\ input}$$

COP = Coefficient of Performance

$$COP = \frac{Btu\ output}{Btu\ input} = \frac{EER}{3.413}$$

or

$$COP = \frac{Btu\ output}{Input\ watts \times 3.413\ (Btu\ per\ watt)}$$

Turndown ratio = Maximum firing rate: Minimum firing rate (5:1, etc.)

$$Overall\ thermal\ eff. = \frac{Gross\ Btu\ output}{Gross\ Btu\ input} \times 100\%$$

$$Combustion\ eff. = \frac{Btu\ input - BTU\ stack\ loss}{Btu\ input} \times 100\%$$

Overall thermal efficiency range 75%–90%

Combustion efficiency range 85%–95%

DUCTWORK HVAC CONVERSIONS

Unit of Measure	Multiply	By	To Obtain	Symbol
Ductwork air flow	CFM	0.0004719	cubic meters/sec.	m^3/s
	FPM	0.00508	meters/sec.	m/s
	FPM	0.508	centimeters/sec.	cm/s
	FPS	0.3048	meters/sec.	m/s
	MPH	0.447	meters/sec.	m/s
Ductwork pressure	in. H_2O	0.25	kilopascals	kPa
	in. H_2O	249 (use 250)	pascals	Pa
	in. $H_2O/100°$	8.176	pascals (N/m^2)	Pa
	in. Hg	3.386	kilopascals (kN/m^2)	kPa
Ductwork length & area	in.	25.4	millimeters	mm
	in.	2.54	centimeters	cm
	in.	0.0254	meters	m

FAN AND PUMP DUTY HVAC CONVERSIONS

Unit of Measure	Multiply	By	To Obtain	Symbol
Ductwork length & area	in.^{2}	6.452	centimeters squared	cm^2
	ft.	0.3048	meters	m
	ft.^{2}	0.0929	meters squared	m^2
	ft.^{3}	0.02832	meters cubed	m^3
Fan duty	CFM	0.4719	liters/sec.	l/s
	in. H$_2$O	249 (use 250)	pascals	Pa
	Hp	0.7460	kilowatts	kW
	RPM	0.10472	radians/sec.	rad/s
	RPM	60	revolutions/sec.	rev/s
Pump duty	Psf	47.88 (use 50)	pascals (N/m^2)	Pa
	Psi	6895 (use 7000)	pascals (N/m^2)	Pa
	Psi	6.895	kilopascals (kN/m^2)	kPa

OFFICE DESIGN CRITERIA

General

Total heat	300–400 sq. ft./ton (range 230–520)
Total heat	30–40 Btu/h/sq. ft. (range 23–52)
Room sens. heat	25–28 Btu/h/sq. ft. (range 19–37)
SHR	0.75–0.93
Perimeter spaces	1.0–3.0 CFM/sq. ft.
Interior spaces	0.5–1.5 CFM/sq. ft.
Building block CFM	1.0–1.5 CFM/sq. ft.
Air change rate	4–10 AC/hr.

Large, Perimeter

Total heat	225–275 sq. ft./ton
Total heat	43–53 Btu/h/sq. ft.

Large, Interior

Total heat	300–350 sq. ft./ton
Total heat	34–40 Btu/h/sq. ft.

Small

Total heat	325–375 sq. ft./ton
Total heat	32–37 Btu/h/sq. ft.

GENERAL DESIGN VALUES

	Total Heat (Ton)	Total Heat BTU/h/sq. ft.	Room Sens. Heat BTU/h/sq. ft.	SHR	Spaces	Air Flow	Air Change Rate
Banks, Court Houses, Municipal Buildings	200–250 sq. ft./ton (range 160–340)	48–60 BTU/h/sq. ft. (range 35–75)	28–38 BTU/h/sq. ft. (range 21–48)	0.75–0.90	–	–	4–10 AC/hr.
Police Stations, Fire Stations, Post Offices	250–350 sq. ft./ton (range 200–400)	34–48 BTU/h/sq. ft. (range 30–60)	25–35 BTU/h/sq. ft. (range 20–40)	0.75–0.90	–	–	4–10 AC/hr.
Precision Manufacturing	50–300 sq. ft./ton	40–240 BTU/h/sq. ft.	32–228 BTU/h/sq. ft.	0.80–0.95	–	–	10–50 AC/hr.
Computer Rooms	50–150 sq. ft./ton	80–240 BTU/h/sq. ft.	64–228 BTU/h/sq. ft.	0.80–0.95	–	2.0–4.0 CFM/sq. ft.	15–20 AC/hr.
Restaurants	100–250 sq. ft./ton (range 75–300)	48–120 BTU/h/sq. ft. (range 40–155)	21–62 BTU/h/sq. ft. (range 20–80)	0.65–0.80	–	1.5–4.0 CFM/sq. ft.	8–12 AC/hr.
Kitchens	150–350 sq. ft./ton (at 85°F space)	34–80 BTU/h/sq. ft. (at 85°F space)	20–56 BTU/h/sq. ft. (at 85°F space)	0.60–0.70	–	1.5–2.5 CFM/sq. ft.	12–15 AC/hr.

	Total Heat (Ton)	Total Heat BTU/h/sq. ft.	Room Sens. Heat BTU/h/sq. ft.	SHR	Spaces	Air Flow	Air Change Rate
Cocktail Lounges, Bars, Taverns, Clubhouses, Nightclubs	150–200 sq. ft./ton (range 75–300)	60–80 BTU/h/sq. ft. (range 40–155)	27–40 BTU/h/sq. ft. (range 20–80)	0.65–0.80	1.5–4.0 CFM/sq. ft.	—	15–20 AC/hr. (cocktail lounges, bars) 20–30 AC/hr. night clubs
Hospital Patient Rooms, Nursing Home Patient Rooms	250–300 sq. ft./ton (range 200–400)	40–48 BTU/h/sq. ft. (range 30–60)	32–46 BTU/h/sq. ft. (range 25–50)	0.75–0.85	—	—	—
Medical/Dental Centers, Clinics, and Offices	250–300 sq. ft./ton (range 200–400)	40–48 BTU/r/sq. ft. (range 30–60)	32–46 BTU/h/sq. ft. (range 25–50)	0.75–0.85	—	—	8–12 AC/hr.
Residences	500–700 sq. ft./ton	17–24 BTU/h/sq. ft.	12–20 BTU/h/sq. ft.	0.80–0.95	—	—	—
Apartments	350–450 sq. ft./ton (range 300–500)	27–34 BTU/h/sq. ft. (range 24–40)	22–30 BTU/h/sq. ft. (range 20–35)	0.80–0.95	—	—	—

Total heat includes ventilation. Room sensible heat does not include ventilation.

GENERAL DESIGN VALUES (cont.)

	Total Heat (Ton)	Total Heat BTU/h/sq. ft.	Room Sens. Heat BTU/h/sq. ft.	SHR	Spaces	Air Flow	Air Change Rate
Motel and Hotel Public Areas	250–300 sq. ft./ton (range160–375)	40–48 BTU/h/sq. ft. (range 32–74)	32–46 BTU/h/sq. ft. (range 25–60)	0.75–0.90	–	–	–
Motel and Hotel Guest Rooms	400–500 sq. ft./ton (range 300–600)	24–30 BTU/h/sq. ft. (range 20–40)	20–25 BTU/h/sq. ft. (range 15–35)	0.80–0.95	–	–	–
School Classrooms	225–275 sq. ft./ton (range 150–350)	43–53 BTU/h/sq. ft (range 35–80)	25–42 BTU/h/sq. ft. (range 20–65)	0.65–0.80	–	–	4–12 AC/hr.
Dining Halls	100–250 sq. ft./ton (range 75–300)	48–120 BTU/h/sq. ft (range 40–155)	21–62 BTU/h/sq. ft. (range 20–80)	0.65–0.80	1.5–4.0 CFM/ sq. ft.	–	12–15 AC/hr.
Libraries, Museums	250–350 sq. ft./ton (range 160–400)	34–48 BTU/h/sq. ft. (range 30–75)	22–32 BTU/h/sq. ft. (range 20–50)	0.80–0.90	–	–	8–12 AC/hr.
Retail, Department Stores	200–300 sq. ft./ton (range 200–500)	40–60 BTU/h/sq. ft. (range 24–60)	32–43 BTU/h/sq. ft. (range 16–43)	0.65–0.90	–	–	6–10 AC/hr.
Other Shops	175–225 sq. ft./ton (range 100–350)	53–69 BTU/h/sq. ft. (range 35–115)	23–54 BTU/h/sq. ft. (range 15–90)	0.65–0.90	–	–	6–10 AC/hr.

	Total Heat (Ton)	Total Heat BTU/h/sq. ft.	Room Sens. Heat BTU/h/sq. ft.	SHR	Spaces	Air Flow	Air Change Rate
Supermarkets	250–350 sq. ft./ton (range 150–400)	34–48 BTU/h/sq. ft. (range 30–80)	25–40 BTU/h/sq. ft. (range 22–67)	0.65–0.85	—	—	4–10 AC/hr.
Malls, Shopping Centers	150–350 sq. ft./ton (range 150–400)	34–80 BTU/h/sq. ft. (range 30–80)	25–67 BTU/h/sq. ft. (range 22–67)	0.65–0.85	—	—	6–10 AC/hr.
Jails	350–450 sq. ft./ton (range 300–500)	27–34 BTU/h/sq. ft. (range 24–40)	22–30 BTU/h/sq. ft. (range 20–35)	0.80–0.95	—	—	—
Auditoriums, Theaters	0.05–0.07 tons/seat	600–840 BTU/h/seat	325–385 BTU/h/seat	0.65–0.75	—	15–30 CFM/seat	8–15 AC/hr.
Churches	0.04–0.06 tons/seat	480–720 BTU/h/seat	260–330 BTU/h/seat	0.65–0.75	—	15–30 CFM/seat	8–15 AC/hr.
Bowling Alleys	1.5–2.5 tons/alley	18,000–30,000 BTU/h/alley	—	—	—	—	10–15 AC/hr.
All Spaces	300–500 CFM/ton @ 20°F ΔT	400 CFM/ton ± 20% @ 2°F ΔT			—	—	4 AC/hr. minimum

Total heat includes ventilation. Room sensible heat does not include ventilation.

EQUIPMENT HEAT GAIN

Offices and Commercial Spaces
0.5–5.0 Watts/sq. ft.

Computer Rooms, Data Centers
2.0–300.0 Watts/sq. ft.

Telecom Rooms
50.0–120.0 Watts/sq. ft.

Electrical Equipment
Transformers

150 kVA and smaller	50 Watts/kVA
151–500 kVA	30 Watts/kVA
501–1000 kVA	25 Watts/kVA
1001–2500 kVA	20 Watts/kVA
Larger than 2500 kVA	15 Watts/kVA

Switchgear

Low voltage breaker 0–40 Amps	10 Watts
Low voltage breaker 50–100 Amps	20 Watts
Low voltage breaker 225 Amps	60 Watts
Low voltage breaker 400 Amps	100 Watts
Low voltage breaker 600 Amps	130 Watts
Low voltage breaker 800 Amps	170 Watts
Low voltage breaker 1,600 Amps	460 Watts
Low voltage breaker 2,000 Amps	600 Watts
Low voltage breaker 3,000 Amps	1,100 Watts
Low voltage breaker 4,000 Amps	1,500 Watts
Medium voltage breaker/switch 600 Amps	1,000 Watts
Medium voltage breaker/switch 1,200 Amps	1,500 Watts
Medium voltage breaker/switch 2,000 Amps	2,000 Watts
Medium voltage breaker/switch 2,500 Amps	2,500 Watts

EQUIPMENT HEAT GAIN (cont.)

Panelboards

2 Watts per circuit

Motor Control Centers

500 Watts per section	
Low voltage starters size 00	50 Watts
Low voltage starters size 0	50 Watts
Low voltage starters size 1	50 Watts
Low voltage starters size 2	100 Watts
Low voltage starters size 3	130 Watts
Low voltage starters size 4	200 Watts
Low voltage starters size 5	300 Watts
Low voltage starters size 6	650 Watts
Medium voltage starters size 200 Amp	400 Watts
Medium voltage starters size 400 Amp	1,300 Watts
Medium voltage starters size 700 Amp	1,700 Watts

Variable Frequency Drive

2 to 6 percent of the kVA rating
Bus duct 0.015 Watts/ft./Amp
Capacitors 2 Watts/KVAR

Motors

Motors Only

Motors 0 to 2 Hp	190 Watts/Hp
Motors 3–20 Hp	110 Watts/Hp
Motors 25–200 Hp	75 Watts/Hp
Motors 250 Hp and larger	60 Watts/Hp

HEAT GAIN FROM MOTOR-DRIVEN EQUIPMENT

Motor Horsepower	Location of Motor or Equipment in Conditioned Air Space		
	Motor and Equipment in Space (Btu/h)	Equipment Only in Space (Btu/h)	Motor Only in Space (Btu/h)
$1/20$	360	130	240
$1/12$	580	200	380
$1/8$	900	320	590
$1/6$	1,160	400	760
$1/4$	1,180	640	540
$1/3$	1,500	840	660
$1/2$	2,120	1,270	850
$3/4$	2,650	1,900	740
1	3,390	2,550	850
1.5	4,960	3,820	1,140
2	6,440	5,090	1,350
3	9,430	7,640	1,790
5	15,500	12,700	2,790
7.5	22,700	19,100	3,640
10	29,900	24,500	4,490
15	44,400	38,200	6,210
20	58,500	50,900	7,610
25	72,300	63,600	8,680
30	85,700	76,300	9,440
40	114,000	102,000	12,600
50	143,000	127,000	15,700
60	172,000	153,000	18,900
75	212,000	191,000	21,200
100	283,000	255,000	28,300
125	353,000	318,000	35,300
150	420,000	382,000	37,800
200	569,000	509,000	50,300
250	699,000	636,000	62,900

HEAT REMOVED IN COOLING AIR TO STORAGE ROOM CONDITIONS (BTU PER CU. FT.)

Storage Room Temp. (°F)	Temperature of Outside Air (°F)							
	85		90		95		100	
	Relative Humidity (%)							
	50	60	50	60	50	60	50	60
65	0.32	0.52	0.58	0.81	0.85	1.12	1.15	1.46
60	0.58	0.78	0.83	1.06	1.10	1.37	1.39	1.70
55	0.80	1.00	1.05	1.28	1.32	1.59	1.61	1.92
50	1.01	1.21	1.26	1.49	1.53	1.79	1.82	2.13
45	1.20	1.40	1.45	1.68	1.71	1.98	2.00	2.31
40	1.37	1.57	1.62	1.85	1.88	2.15	2.17	2.48
35	1.54	1.74	1.78	2.01	2.04	2.31	2.33	2.64
30	1.78	2.01	2.05	2.31	2.33	2.64	2.65	3.00

COOLING LOAD PEAK PERIODS

Month of Peak Room Cooling Load for Various Exposures

Probable Month of Peak Room Cooling Load per Exposure

Window Characteristics			North	South	East	West	NE	SE	SW	NW
% Glass	Shade Coef.	Over-hang								
25	0.4	0	July	Sept.	July	July	July	Sept.	Sept.	July
25	0.4	1:2	July	Oct.	July	Aug.	July	Sept.	Sept.	July
25	0.4	1:1	July	Oct.	July	July	July	Sept.	Oct.	July
25	0.6	0	July	Sept.	July	July	July	Sept.	Sept.	July
25	0.6	1:2	July	Oct.	July	Aug.	July	Sept.	Sept.	July
25	0.6	1:1	July	Dec.	July	Sept.	July	Sept.	Oct.	July
50	0.4	0	July	Sept.	July	July	July	Sept.	Sept.	July
50	0.4	1:2	July	Oct.	July	Aug.	July	Sept.	Sept.	July
50	0.4	1:1	July	Dec.	July	Sept.	July	Sept.	Oct.	July
50	0.6	0	July	Oct.	July	July	July	Sept.	Sept.	July
50	0.6	1:2	July	Dec.	July	Aug.	July	Sept.	Oct.	July
50	0.6	1:1	July	Dec.	July	Sept.	July	Sept.	Dec.	July

Notes:
1. % glass is percentage of gross wall area.
2. Shading coefficient of 0.4 is approximately equal to a double-pane glass window combined with blinds.

COOLING LOAD FACTORS

Lighting Load Factors

Fluorescent lights	$1.25 \times$ bulb Watts
Incandescent lights	$1.00 \times$ bulb Watts
HID lighting	$1.25 \times$ bulb Watts

HEATING INFILTRATION

Air Change Rate Calculation

Range 0–10 AC/hr.
Commercial buildings

a. 1.0 AC/hr. 1 exterior wall
b. 1.5 AC/hr. 2 exterior walls
c. 2.0 AC/hr. 3 or 4 exterior walls

Vestibules 3.0 AC/hr.

CFM/Sq. Ft. of Wall Calculation

Range 0–1.0 CFM/sq. ft.

Tight buildings	0.1 CFM/sq. ft.
Average buildings	0.3 CFM/sq. ft.
Leaky buildings	0.6 CFM/sq. ft.

Crack Calculation

Range 0.12–2.8 CFM/ft. of crack
Average 1.0 CFM/ft. of crack

(15 mph wind is assumed)

MOISTURE CONDENSATION ON GLASS WINDOWS

Inside Temp. (°F)	Outside Temp. (°F)	Glass—Single Pane		Glass—Double Pane		Glass—Triple Pane	
		$T_{GLASS}/T_{DEWPOINT}$	% RH	$T_{GLASS}/T_{DEWPOINT}$	% RH	$T_{GLASS}/T_{DEWPOINT}$	% RH
	−30	−6.1	4.5	29.5	25.9	39.2	38.5
	−25	−2.3	5.6	31.3	27.9	40.5	40.5
	−20	1.4	6.9	33.2	30.2	41.9	42.8
	−15	5.2	8.4	35.1	32.6	43.2	45.0
	−5	12.6	12.1	38.8	37.9	46.0	50.1
65	0	16.4	14.5	40.7	40.8	47.3	52.7
	5	20.1	17.2	42.6	44.0	48.7	55.5
	10	23.9	20.3	44.4	47.1	50.0	58.3
	20	31.3	27.9	48.2	54.5	52.8	64.7
	25	35.1	32.6	50.0	58.3	54.1	67.9
	30	38.8	37.9	51.9	62.6	55.5	71.4
	35	42.6	44.0	53.8	67.1	56.8	74.9
	−30	−5.8	4.4	30.1	25.6	39.9	38.2
	−25	−2.1	5.5	32.0	27.7	41.2	40.2
	−20	1.7	6.7	33.8	29.9	42.6	42.5
	−15	5.4	8.2	35.7	32.3	44.0	44.8
	−5	12.9	11.8	39.4	37.4	46.7	49.7
66	0	16.6	14.1	41.3	40.4	48.0	52.2
	5	20.4	16.8	43.2	43.5	49.4	55.1
	10	24.1	19.8	45.1	46.8	50.8	58.0
	20	31.6	27.3	48.8	53.8	53.5	64.1
	25	35.3	31.8	50.7	57.8	54.8	67.2
	30	39.1	37.0	52.5	61.8	56.2	70.8
	35	42.8	42.8	54.4	66.3	57.6	74.4
	−30	−5.6	4.3	30.7	25.4	40.6	37.9
	−25	−1.8	5.4	32.6	27.5	42.0	40.1
	−20	1.9	6.6	34.5	29.7	43.3	42.2
	−15	5.7	8.0	36.3	32.0	44.7	44.5
	−5	13.1	11.6	40.1	37.2	47.4	49.3
67	0	16.9	13.8	41.9	39.9	48.8	52.0
	5	20.6	16.4	43.8	43.0	50.1	54.6
	10	24.4	19.4	45.7	46.2	51.5	57.5
	20	31.8	26.6	49.4	53.2	54.2	63.5
	25	35.6	31.1	51.3	57.1	55.6	66.9
	30	39.3	36.0	53.2	61.3	56.9	70.1
	35	43.1	41.8	55.0	65.4	58.3	73.7

MOISTURE CONDENSATION ON GLASS WINDOWS (cont.)

Inside Temp. (°F)	Outside Temp. (°F)	Glass—Single Pane $T_{GLASS}/T_{DEWPOINT}$	% RH	Glass—Double Pane $T_{GLASS}/T_{DEWPOINT}$	% RH	Glass—Triple Pane $T_{GLASS}/T_{DEWPOINT}$	% RH
68	−30	−5.3	4.3	31.3	25.1	41.3	37.7
	−25	−1.6	5.3	33.2	27.2	42.7	39.8
	−20	2.2	6.5	35.1	29.4	44.1	42.0
	−15	5.9	7.8	37.0	31.8	45.4	44.2
	−5	13.4	11.3	40.7	36.8	48.1	48.9
	0	17.1	13.5	42.6	39.6	49.5	51.6
	5	20.9	16.0	44.4	42.5	50.9	54.4
	10	24.6	18.9	46.3	45.7	52.2	57.0
	20	32.1	26.0	50.0	52.6	54.9	63.0
	25	35.8	30.3	51.9	56.4	56.3	66.3
	30	39.6	35.2	53.8	60.5	57.7	69.7
	35	43.3	40.7	55.7	64.8	59.0	73.0
69	−30	−5.1	4.2	32.0	25.0	42.1	37.6
	−25	−1.3	5.2	33.8	26.9	43.4	39.5
	−20	2.4	6.3	35.7	29.1	44.8	41.7
	−15	6.2	7.7	37.6	31.4	46.2	44.0
	−5	13.6	11.1	41.3	36.4	48.9	48.7
	0	17.4	13.2	43.2	39.2	50.2	51.2
	5	21.1	15.6	45.1	42.2	51.6	53.9
	10	24.9	18.5	46.9	45.2	53.0	56.8
	20	32.3	25.3	50.7	52.1	55.7	62.7
	25	36.1	29.6	52.5	55.7	57.0	65.7
	30	39.8	34.3	54.4	59.8	58.4	69.1
	35	43.6	39.8	56.3	64.0	59.8	72.0
70	−30	−4.8	4.1	32.6	24.8	42.8	37.3
	−25	−1.1	5.0	34.5	26.8	44.2	39.4
	−20	2.7	6.2	36.3	28.8	45.5	41.4
	−15	6.4	7.5	38.2	31.1	46.9	43.7
	−5	13.9	10.8	41.9	36.0	49.6	48.3
	0	17.6	12.9	43.8	38.8	51.0	51.0
	5	21.4	15.3	45.7	41.7	52.3	53.5
	10	25.1	18.0	47.6	44.8	53.7	56.3
	20	32.6	24.8	51.3	51.5	56.4	62.1
	25	36.3	28.8	53.2	55.3	57.8	65.3
	30	40.1	33.6	55.0	59.0	59.1	68.4
	35	43.8	38.8	56.9	63.2	60.5	71.9

MOISTURE CONDENSATION ON GLASS WINDOWS *(cont.)*

Inside Temp. (°F)	Outside Temp. (°F)	Glass—Single Pane		Glass—Double Pane		Glass—Triple Pane	
		$T_{GLASS}/$ $T_{DEWPOINT}$	% RH	$T_{GLASS}/$ $T_{DEWPOINT}$	% RH	$T_{GLASS}/$ $T_{DEWPOINT}$	% RH
71	−30	−4.6	4.0	33.2	23.6	43.5	37.0
	−25	−0.8	5.0	35.1	26.5	44.9	39.1
	−20	2.9	6.0	37.0	28.7	46.2	41.1
	−15	6.7	7.4	38.8	30.8	47.6	43.3
	−5	14.1	10.6	42.6	35.8	50.3	48.0
	0	17.9	12.6	44.4	38.4	51.7	50.5
	5	21.6	14.9	46.3	41.3	53.0	53.0
	10	25.4	17.6	48.2	44.3	54.4	55.8
	20	32.8	24.1	51.9	50.9	57.1	61.6
	25	36.6	28.2	53.8	54.6	58.5	64.7
	30	40.3	32.7	55.7	58.5	59.8	67.8
	35	44.1	37.9	57.5	62.5	61.2	71.3
72	−30	−4.3	4.0	33.8	24.3	44.3	36.9
	−25	−0.6	4.8	35.7	26.3	45.6	38.8
	−20	3.2	5.9	37.6	28.4	47.0	41.0
	−15	6.9	7.2	39.5	30.6	48.3	43.0
	−5	14.4	10.4	43.2	35.4	51.1	47.8
	0	18.1	12.3	45.1	38.1	52.4	50.1
	5	21.9	14.6	46.9	40.8	53.8	52.8
	10	25.6	17.2	48.8	43.8	55.1	55.3
	20	33.1	23.6	52.6	50.5	57.9	61.2
	25	36.8	27.5	54.4	54.0	59.2	64.2
	30	40.6	32.0	56.3	57.8	60.6	67.4
	35	44.3	36.9	58.2	61.9	61.9	70.6
73	−30	−4.1	3.8	34.5	24.2	45.0	36.7
	−25	−0.3	4.8	36.3	26.0	46.3	38.6
	−20	3.4	5.8	38.2	28.1	47.7	40.7
	−15	7.2	7.1	40.1	30.3	49.1	42.9
	−5	14.7	10.2	43.8	35.0	51.8	47.4
	0	18.4	12.1	45.7	37.7	53.1	49.7
	5	22.1	14.3	47.6	40.5	54.5	52.4
	10	25.9	16.9	49.4	43.3	55.9	55.1
	20	33.4	23.1	53.2	49.9	58.6	60.7
	25	37.1	26.9	55.0	53.3	59.9	63.6
	30	40.8	31.2	56.9	57.1	61.3	66.8
	35	44.6	36.1	58.8	61.2	62.7	70.2

MOISTURE CONDENSATION ON GLASS WINDOWS *(cont.)*

Inside Temp. (°F)	Outside Temp. (°F)	Glass—Single Pane		Glass—Double Pane		Glass—Triple Pane	
		$T_{GLASS}/T_{DEWPOINT}$	% RH	$T_{GLASS}/T_{DEWPOINT}$	% RH	$T_{GLASS}/T_{DEWPOINT}$	% RH
	-30	-3.8	3.8	35.1	24.0	45.7	36.4
	-25	-0.1	4.7	37.0	25.9	47.1	38.4
	-20	3.7	5.7	38.8	27.8	48.4	40.4
	-15	7.4	6.9	40.7	30.0	49.8	42.5
	-5	14.9	9.9	44.5	34.8	52.5	47.0
74	0	18.6	11.8	46.3	37.3	53.9	49.5
	5	22.4	14.0	48.2	40.1	55.2	51.9
	10	26.1	16.4	50.1	43.0	56.6	54.6
	20	33.6	22.6	53.8	49.3	59.3	60.2
	25	37.3	26.2	55.7	52.9	60.7	63.3
	30	41.1	30.5	57.5	56.4	62.0	66.2
	35	44.8	35.2	59.4	60.4	63.4	69.6
	-30	-3.5	3.7	35.7	23.8	46.4	36.2
	-25	0.2	4.6	37.6	25.6	47.8	38.2
	-20	3.9	5.6	39.5	27.7	49.2	40.3
	-15	7.7	6.8	41.3	29.7	50.5	42.2
	-5	15.2	9.7	45.1	34.4	53.2	46.7
75	0	18.9	11.6	46.9	36.9	54.6	49.1
	5	22.6	13.6	48.8	39.6	56.0	51.7
	10	26.4	16.1	50.7	42.6	57.3	54.2
	20	33.9	22.1	54.4	48.8	60.0	59.7
	25	37.6	25.7	56.3	52.3	61.4	62.7
	30	41.3	20.7	58.2	56.0	62.8	65.9
	35	45.1	34.4	60.0	59.7	64.1	09.0
	-30	-3.3	3.6	36.4	23.6	47.2	36.1
	-25	0.4	4.5	38.2	25.4	48.5	37.9
	-20	4.2	5.5	40.1	27.4	49.9	39.9
	-15	7.9	6.6	42.0	29.5	51.2	41.9
	-5	15.4	9.5	45.7	34.1	54.0	46.5
76	0	19.1	11.3	47.6	36.6	55.3	48.8
	5	22.9	13.4	49.4	39.2	56.7	51.3
	10	26.6	15.7	51.3	42.1	58.0	53.8
	20	34.1	21.5	55.1	48.4	60.8	59.4
	25	37.8	25.0	56.9	51.7	62.1	62.2
	30	41.6	29.1	58.8	55.3	63.5	65.3
	35	45.3	33.6	60.7	59.2	64.8	68.3

MOISTURE CONDENSATION ON GLASS WINDOWS *(cont.)*

Inside Temp. (°F)	Outside Temp. (°F)	Glass—Single Pane $T_{GLASS}/T_{DEWPOINT}$	% RH	Glass—Double Pane $T_{GLASS}/T_{DEWPOINT}$	% RH	Glass—Triple Pane $T_{GLASS}/T_{DEWPOINT}$	% RH
	−30	−3.0	3.6	37.0	24.4	47.9	35.8
	−25	0.7	4.4	38.8	25.2	49.3	37.8
	−20	4.4	5.3	40.7	27.2	50.6	39.7
	−15	8.2	6.5	42.6	29.3	52.0	41.8
	−5	15.7	9.3	46.3	33.7	54.7	46.1
77	0	19.4	11.1	48.2	36.3	56.1	48.6
	5	23.1	13.0	50.1	38.9	57.4	50.9
	10	26.9	15.4	51.9	41.6	58.8	53.5
	20	34.4	21.1	55.7	47.9	61.5	58.9
	25	38.1	24.5	57.6	51.3	62.9	61.9
	30	41.8	28.4	59.4	54.7	64.2	64.7
	35	45.6	32.8	61.3	58.5	65.6	68.0
	−30	−2.8	3.5	37.6	23.2	48.6	35.6
	−25	0.9	4.3	39.5	25.1	50.0	37.5
	−20	4.7	5.3	41.3	26.9	51.3	39.4
	−15	8.4	6.3	43.2	29.0	52.7	41.5
	−5	15.9	9.1	47.0	33.5	55.4	45.8
78	0	19.7	10.8	48.8	35.9	56.8	48.2
	5	23.4	12.8	50.7	38.5	58.1	50.5
	10	27.1	15.0	52.6	41.3	59.5	53.1
	20	34.6	20.6	56.3	47.3	62.2	58.4
	25	38.4	24.0	58.2	50.7	63.6	61.3
	30	42.1	27.8	60.0	54.0	64.9	64.2
	35	45.8	32.0	61.9	57.8	66.3	67.4
	−30	−2.5	3.5	38.2	23.0	49.4	35.5
	−25	1.2	4.2	40.1	24.8	50.7	37.3
	−20	4.9	5.1	42.0	26.8	52.1	39.3
	−15	8.7	6.2	43.8	28.7	53.4	41.2
	−5	16.2	8.9	47.6	33.2	56.2	45.6
79	0	19.9	10.6	49.5	35.6	57.5	47.8
	5	23.6	12.5	51.3	38.1	58.9	50.3
	10	27.4	14.7	53.2	40.9	60.2	52.7
	20	34.9	20.2	56.9	46.8	63.0	58.1
	25	38.6	23.4	58.8	50.1	64.3	60.8
	30	42.3	27.1	60.7	53.6	65.7	63.9
	35	46.1	31.3	62.5	57.1	67.0	66.8

MOISTURE CONDENSATION ON GLASS WINDOWS *(cont.)*

Inside Temp. (°F)	Outside Temp. (°F)	Glass—Single Pane $T_{GLASS}/T_{DEWPOINT}$	% RH	Glass—Double Pane $T_{GLASS}/T_{DEWPOINT}$	% RH	Glass—Triple Pane $T_{GLASS}/T_{DEWPOINT}$	% RH
80	−30	−2.3	3.4	38.9	22.9	50.1	35.3
	−25	1.5	4.2	40.7	24.6	51.4	37.0
	−20	5.2	5.0	42.6	26.5	52.8	39.0
	−15	8.9	6.1	44.5	28.5	54.2	41.0
	−5	16.4	8.7	48.2	32.8	56.9	45.3
	0	20.2	10.4	50.1	35.3	58.2	47.4
	5	23.9	12.2	51.9	37.7	59.6	49.9
	10	27.6	14.4	53.8	40.5	61.0	52.4
	20	35.1	19.7	57.6	46.4	63.7	57.6
	25	38.9	22.9	59.4	49.5	65.0	60.3
	30	42.6	26.5	61.3	53.0	66.4	63.3
	35	46.3	30.6	63.2	56.6	67.8	66.4

HUMIDITY AND HEALTH

Health Condition	Optimum Relative Humidity Range
Bacteria	20–70%
Viruses	40–78%
Fungi	0–70%
Mites	0–60%
Respiratory infections	40–50%
Allergic rhinitis and asthma	40–60%
Chemical interactions	0–40%
Ozone production	75–100%
Combined health conditions	40–60%

RESIDENTIAL HEAT LOAD DESIGN DATA— SUMMER HEAT GAINS

Room Vol. (ft.³)	CFM	Cooling Btu at Supply Temp.		
		60°F	55°F	50°F
200	14	300	375	450
300	20	430	540	650
400	27	580	730	875
500	34	735	920	1,100
600	40	865	1,080	1,290
700	47	1,015	1,270	1,530
800	55	1,190	1,485	1,780
900	60	1,295	1,620	1,940
1,000	65	1,400	1,755	2,100
1,100	75	1,620	2,000	2,430
1,200	80	1,730	2,160	2,600
1,300	87	1,880	2,350	2,820
1,400	95	2,050	2,560	3,080
1,500	100	2,160	2,700	3,240
1,600	107	2,310	2,890	3,460
1,700	113	2,440	3,050	3,660
1,800	120	2,590	3,240	3,880
1,900	125	2,700	3,370	4,050
2,000	135	2,920	3,645	4,370

Max. Air on One Outlet

Based on 80°F Return Air ◄────────►

1-40

RESIDENTIAL HEAT LOAD DESIGN DATA— SUMMER HEAT GAINS (cont.)

Room Vol. (ft.^{3})	CFM	Cooling Btu at Supply Temp.		
		60°F	55°F	50°F
3,000	200	4,320	5,400	6,480
4,000	265	5,720	7,150	8,580
5,000	335	7,240	9,040	10,850
6,000	400	8,640	10,800	12,960
7,000	465	10,040	12,550	15,060
8,000	535	10,760	14,440	17,830
10,000	670	14,470	18,090	21,700
12,000	800	17,280	21,600	25,920
14,000	935	20,200	25,240	30,290
16,000	1,065	23,000	28,750	34,500
18,000	1,200	25,920	32,400	38,780
20,000	1,335	28,840	36,040	43,250
25,000	1,670	36,070	45,090	54,100
		Based on 80°F Return Air		

RESIDENTIAL HEAT LOAD DESIGN DATA—WINTER HEAT LOSSES

Heating Btu at Supply Temp.

Room Vol. (ft.³)	CFM (Max. Air on One Outlet)	120°F	130°F	140°F	150°F	160°F	170°F
200	14	750	910	1,060	1,210	1,360	1,510
300	20	1,080	1,300	1,510	1,730	1,945	2,160
400	27	1,460	1,750	2,040	2,330	2,625	2,915
500	34	1,830	2,200	2,570	2,940	3,300	3,670
600	40	2,160	2,590	3,025	3,455	3,890	4,320
700	47	2,540	3,025	3,550	4,060	4,565	5,080
800	55	2,970	3,565	4,155	4,750	5,345	5,940
900	60	3,240	3,880	4,535	5,185	5,830	6,480
1,000	65	3,500	4,210	4,915	5,615	6,320	7,000
1,100	75	4,050	4,860	5,670	6,480	7,290	8,100
1,200	80	4,320	5,200	6,040	6,910	7,780	8,640
1,300	87	4,700	5,635	6,570	7,520	8,455	9,400
1,400	95	5,130	6,155	7,180	8,200	9,235	10,260
1,500	100	5,400	6,480	7,560	8,640	9,720	10,800
1,600	107	5,775	6,930	8,090	9,245	10,400	11,550
1,700	113	6,100	7,320	8,540	9,760	10,950	12,200

RESIDENTIAL HEAT LOAD DESIGN DATA—WINTER HEAT LOSSES (cont.)

Heating Btu at Supply Temp.

Room Vol. (ft.³)	CFM	120°F	130°F	140°F	150°F	160°F	170°F
1,800	120	6,480	7,775	9,070	10,360	11,665	12,960
1,900	125	6,750	8,100	9,450	10,800	12,150	13,500
2,000	135	7,300	8,750	10,200	11,660	13,120	14,600
3,000	200	10,800	12,960	15,120	17,280	19,440	21,600
4,000	265	14,300	17,170	20,035	22,890	25,750	28,600
5,000	335	18,100	21,700	25,325	28,940	32,560	36,200
6,000	400	21,600	25,920	30,240	34,560	38,880	43,200
7,000	465	25,100	30,130	35,150	40,170	45,200	50,200
8,000	535	26,900	34,670	40,450	46,220	52,000	53,800
10,000	670	36,180	43,410	50,650	57,890	65,125	72,360
12,000	800	43,200	51,840	60,480	69,120	77,760	86,400
14,000	935	50,500	60,590	70,680	80,780	90,880	101,000
16,000	1,065	57,500	69,000	80,510	92,010	103,520	115,000
18,000	1,200	64,800	77,760	90,720	103,680	116,640	129,600
20,000	1,335	72,100	86,500	100,920	115,340	129,760	144,200
25,000	1,670	90,200	108,215	126,250	144,290	162,320	180,360

Based on 70°F Return Air

HEATING CONVERSIONS

Electric Furnace to Hydronic Baseboard
KWH × 1.0 = KWH for electric boiler
KWH × 0.028 = Gal. for oil-fired boiler
KWH × 0.038 = Therms for gas-fired boiler

Ceiling Cable to Hydronic Baseboard
KWH × 1.06 = KWH for electric boiler
KWH × 0.03 = Gal. for oil-fired boiler
KWH × 0.041 = Therms for gas-fired boiler

Heat Pump to Hydronic Baseboard
KWH × 1.88 = KWH for electric boiler
KWH × 0.052 = Gal. for oil-fired boiler
KWH × 0.073 = Therms for gas-fired boiler

Electric Baseboard to Warm Air Furnace
KWH × 1.19 = KWH for electric furnace
KWH × 0.039 = Gal. for oil-fired furnace
KWH × 0.054 = Therms for gas-fired furnace

Electric Furnace to Fuel-Fired Furnace
KWH × 0.032 = Gal. for oil-fired furnace
KWH × 0.045 = Therms for gas-fired furnace

Ceiling Cable to Warm Air Furnace
KWH × 1.06 = KWH for electric furnace
KWH × 0.034 = Gal. for oil-fired furnace
KWH × 0.048 = Therms for gas-fired furnace

Heat Pump to Warm Air Furnace
KWH × 1.88 = KWH for electric furnace
KWH × 0.061 = Gal. for oil-fired furnace
KWH × 0.085 = Therms for gas-fired furnace

Warm Air Systems to Hydronic Baseboard System
Gal. oil for W.A. × 0.857 = Gal. for hydronics
Therms gas for W.A. × 0.857 = Therms for hydronics
Gal. oil for W.A. × 1.2 = Therms for hydronics
Therms gas for W.A. × 0.612 = Gal. for hydronics

Electric Baseboard to Hydronic Baseboard
KWH × 1.19 = KWH for electric boiler
KWH × 0.033 = Gal. for oil-fired boiler
KWH × 0.046 = Therms for gas-fired boiler

SPECIFIC HEATS OF SUBSTANCES

Substance	Specific Heat (Btu/lb./degree)
Air	.24
Alcohol	.615
Aluminum	.22
Ammonia	1.098
Brass	.092
Brick	.220
Bronze	.104
Carbon	.126
Carbon dioxide	.215
Cement, Portland	.271
Coal	.201
Concrete	.156
Copper	.092
Fuel oil	.5
Gasoline	.5
Glass	.18
Gold	.032
Hydrogen	3.41
Ice	.504
Iron, cast	.113
Kerosene	.5
Lead	.03
Limestone	.217
Marble	.206
Mercury	.033
Nickel	.109
Nitrogen	.224
Oxygen	.224
Rubber, hard	.339
Sand	.195
Silicon	.175
Silver	.056
Steam	.48
Steel	.118
Tin	.054
Water	1.0
Wood	.327
Zinc	.093

R-VALUES	
Material	R-Value per Inch or as Specified
Concrete	.11
Mortar	.20
Brick	.20
Concrete block	1.11 for 8 in.
Soft wood	1.25
Hard wood	.09
Plywood	1.25
Hardboard	.75
Glass (single pane)	.88 to 1.1
Glass (double pane)	1.72
Air space in wall (vertical)	1.35 for .75-in. space
Inside air film (vertical)	.68
Outside air film at 7.5 mph (vertical)	.17
Carpet	2.08
Vinyl flooring	.05
Mineral wool insulation (loose fill)	3.70 per in.
Extruded polystyrene	5.0 per in.
Urethane	7.2 to 8.0 per in.
Urethane with foil .75 in. thick	10.0 per in.
Fiberglass batt	3.17 per in.

SIZING R-VALUES FOR INSTALLATION

Insulation	R/in.	Approximate Inches Needed						
		R11	R19	R22	R34	R38	R49	
Loose Fill Machine-Blown								
Fiberglass	R2.25	5	8.5	10	15.5	17	22	
Mineral wool	R3.125	3.5	6	7	11	12.5	16	
Cellulose	R3.7	3	5.5	6	9.5	10.5	13.5	
Loose Fill Hand-Poured								
Cellulose	R3.7	3	5.5	6	9.5	10.5	13.5	
Mineral wool	R3.125	3.5	6	7	11	12.5	16	
Fiberglass	R2.25	5	8.5	10	15.5	17	22	
Vermiculite	R2.	5.5	9	10.5	16.5	18	23.5	
Batts or Blankets								
Fiberglass	R3.14	3.5	6	7	11	12.5	16	
Mineral wool	R3.14	3.5	6	7	11	12.5	16	
Rigid Board								
Polystyrene beadboard	R3.6	3	5.5	6.5	9.5	10.5	14	
Extruded polystyrene (Styrofoam)	R4–5.41	3–2	5–3.5	5.5–4	8.5–6.5	9.5–7	12.5	
Urethane	R6.2	2	3	3.5	5.5	6.5	8	
Fiberglass	R4.0	3	5	5.5	8.5	9.5	12.5	

EFFECT OF INSULATION COMPRESSION ON R-VALUES

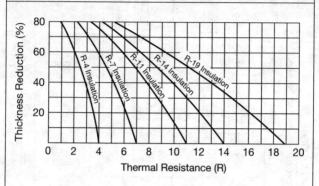

ARI AND ASHRAE STANDARDS

ASHRAE Standard 34

Toxicity and Flammability of refrigerants:

- Class A refrigerants are those that show no evidence of toxicity at concentrations below 400 ppm. (Most refrigerants are Class A.)
- Class B refrigerants are those that show evidence of toxicity at concentrations below 400 ppm.

Flammability Falls into Three Groups or Classes:

- Class 1 refrigerants show no flame propagation when tested in air at 65°F @ 14.7 psia.
- Class 2 refrigerants are flammable and require caution.
- Class 3 refrigerants are considered highly flammable and require even greater caution.

ASHRAE Standard 15

Equipment Room Safety and the Use of Sensors and Alarms to Detect Refrigerants Leaks

- This standard applies to all refrigerants in all classifications.
- Each machine room must have an alarm that will activate before the refrigerant concentrations exceed toxicity limits.
- Every system installed in an equipment room must be equipped with a safety pressure relief valve vented to the outdoors.
- Machine rooms must be provided with ventilation meeting ASHRAE Standard 15R.
- At least one approved self-contained breathing apparatus (SCBA) should be located near an equipment room.
- The lower limit of acceptable oxygen levels in a room is 19.5% (20.9% normal).

ARI Standard 700

Standard for the Level of Purity of Refrigerants

- Refrigerant must meet this standard or it cannot be sold to a customer.
- It is legal for a technician to recover, filter, and recharge refrigerant into equipment owned by the same customer without meeting the ARI 700 standard.

ARI Standard 740

Certification of Refrigerant Recovery Equipment by an Independent Agency

- All recovery equipment must be listed as certified to ARI Standard 740.

TYPES OF COPPER TUBING

Nominal Size (in.)	Actual OD (in.)	Type K Wall (in.)	Type L Wall (in.)	Type M Wall (in.)
¼	.375	.035	.03	.025
⅜	.5	.049	.035	.025
½	.625	.049	.04	.028
⅝	.75	.049	.042	.030
¾	.875	.065	.045	.032
1	1.125	.065	.050	.035
1¼	1.375	.065	.055	.042
1½	1.625	.072	.060	.049
2	2.125	.083	.070	.058

- AC/R tubing is not the same as standard plumbing copper. It is charged with dry nitrogen and the ends sealed.
- AC/R tubing is specified by its outside diameter while plumbing tubing is specified by its inside diameter.
- AC/R copper tubing comes in hard and soft types. Soft tubing comes in 50 ft. lengths and hard comes in 20 ft. lengths and cannot be formed unless annealed.
- Hard tubing is used with fittings.

MAXIMUM SPACING BETWEEN SUPPORTS FOR COPPER TUBING

Copper Tubing OD (in.)	Maximum Spacing Between Supports
5/8 OD and under	5 feet
7/8	6 feet
1⅛	7 feet
1⅝	9 feet
2⅛	10 feet
3⅛	12 feet
3⅝	13 feet
4⅛	14 feet

EQUIVALENT LENGTH IN FEET OF STRAIGHT PIPE FOR COPPER PIPE FITTINGS

Tubing Size (in.)	90° Ell	45° Ell	Line Tee	Branch Tee
½	.9	.4	.6	2.0
⅝	1.0	.5	.8	2.5
⅞	1.5	.7	1.0	3.5
1⅛	1.8	.9	1.5	4.5
1⅜	2.4	1.2	1.8	6.0
1⅝	2.8	1.4	2.0	7.0
2⅛	3.9	1.8	3.0	10.0
2⅝	4.6	2.2	3.5	12.0

Note:
- The equivalent length in feet for the fitting used is to be added to the actual length of the piping run to get the total length.

RECOMMENDED LIQUID LINE SIZES FOR R-22

Capacity	50 Foot Run (in.)	100 Foot Run (in.)
36,000 (3-tons)	½	½
48,000 (4-tons)	½	⅝
60,000 (5-tons)	½	⅝
75,000 (6.25-tons)	½	⅝
100,000 (8.33-tons)	⅝	⅞
150,000 (12.5-tons)	⅞	⅞
200,000 (16.6-tons)	⅞	⅞
300,000 (25-tons)	1⅛	1⅛

The pipe sizes listed are approximate values only. Always follow the manufactures' recommendations for the system you are installing.

RECOMMENDED SUCTION LINE SIZES FOR R-22 AT 40 DEGREES EVAPORATING

Capacity	Fifty Foot Run (in.)		One Hundred Foot Run (in.)	
	Horizontal	Vertical	Horizontal	Vertical
24,000 (2-ton)	⅞	⅞	⅞	⅞
36,000 (3-ton)	⅞	⅞	1⅛	⅞
48,000 (4-ton)	1⅛	1⅛	1⅛	1⅛
60,000 (5-ton)	1⅛	1⅛	1⅛	1⅛
75,000 (6.25 ton)	1⅛	1⅛	1⅜	1⅛
100K (8.33 ton)	1⅜	1⅜	1⅜	1⅜
150K (12.5 ton)	1⅜	1⅜	1⅝	1⅝
200K (16.6 ton)	1⅝	1⅝	2⅛	1⅝
300K (25 ton)	2⅛	2⅛	2⅛	2⅛

Undersized suction lines create increased pressure drop and decreased capacity. Oversized suction lines cause the refrigerant velocity to decrease which then fails to properly return oil to the compressor crankcase.

CALCULATING THERMAL EXPANSION OF COPPER PIPE

The following equation determines the change in length of copper tubing when it is heated or cooled.

Change In Length = Coefficient of Linear Expansion × Total Length × ΔT

The Coefficient of Linear Expansion for copper tubing is .0000094 inches/ft./degree F.

A temperature change may require the installation of one or more piping offsets to absorb the change in length.

LINEAR EXPANSION OF COPPER PIPE

Temperature Change in Degrees F	Expansion in Inches per 100 ft.
50	.57
60	.64
70	.74
80	.85
90	.95
100	1.06
110	1.17
120	1.28
130	1.38
140	1.49
150	1.60
160	1.70
170	1.81
180	1.91
190	2.02
200	2.13
250	2.66
300	3.19

USEFUL HVAC VALUES AND MULTIPLIERS

- 3.414 Btu per Watt
- One Kilowatt equals 3,415 Btu per hour
- 746 Watts per horsepower
- One Ton of cooling equals
  - 288,000 Btu per 24 hours
  - 12,000 Btu per hour
  - 200 Btu per minute
- 1 Therm = 100,000 Btu
- 7,000 grains = 1 pound
- One horsepower per ton @ 40 degree evaporator saturation temperature and 100 degree condensing temperature
- Atmospheric pressure @ sea level is 14.70 psia
- Gauge pressure equals psia minus 14.70
- Psia equals psig plus 14.70
- .24 specific heat of air
- .075 pounds per cubic foot (Density of standard air)
- 13.33 cubic feet per pound (Specific volume of standard air)
- 25,400 microns per inch
- 29.92 inches Hg (Standard sea level air pressure)
- 8760 hours in a year
- R-value of an inside air film is .68

Standard air is air at 70 degrees fahrenheit dry bulb, 0% relative humidity, at sea level.

USEFUL HVAC CONVERSION MEASURES

Length

1 Mile = 1,760 yds. = 5,280 ft. = 63,360 in. = 1.609 km
1 ft. = 0.3048 m = 30.48 cm = 304.8 mm
1 in. = 2.54 cm = 25.4 mm
1 cm = 0.3937 in.
1 m = 39.37 in. = 3.2808 ft. = 1.094 yds.
1 km = 3281 ft. = 0.6214 miles = 1094 yds.
1 Fathom = 6 feet = 1.828804 meters
1 Furlong = 660 feet

Weight

1 gal. H_2O = 8.33 lbs. H_2O
1 lb. = 16 oz. = 7,000 Grains = 0.4536 kg
1 ton = 2,000 lbs. = 907 kg
1 kg = 2.205 lbs.
1 lb. Steam = 1 lb. H_2O

Area

1 sq. ft. = 144 sq. in.
1 Acre = 43,560 sq. ft. = 4840 sq. yds. = 0.4047 Hectares
1 sq. mile = 640 acres
1 sq. yd. = 9 sq. ft. = 1296 sq. in.
1 Hectare = 2.417 Acres
1 sq. m = 1,550 sq. in. = 0.0929 sq. ft. = 1.1968 sq. yds.

USEFUL HVAC CONVERSION MEASURES (cont.)

Volume

1 cu. yd. = 27 cu. ft. = 46,656 cu. in. = 1616 Pints = 807.9 Quarts = 764.6 Liters

1 cu. ft. = 1,728 cu. in.

1 Liter = 0.2642 Gallons = 1.057 Quarts = 2.113 Pints

1 Gallon = 4 Quarts = 8 Pints = 3.785 Liters

1 cu. Meter = 61,023 cu. in. = 0.02832 cu. ft. = 1.3093 cu. yds.

1 Barrel Oil = 42 Gallons Oil

1 Barrel Beer = 31.5 Gallons Beer

1 Barrel Wine = 31.0 Gallons Wine

1 Bushel = 1.2445 cu. ft. = 32 Quarts (Dry) = 64 Pints (Dry) = 4 Pecks

1 Hogshead = 63 Gallon = 8.42184 cu. ft.

Velocity

1 MPH = 5280 ft./hr. = 88 ft./min. = 1.467 ft./sec. = 0.8684 Knots

1 Knot = 1.1515 Mph = 1.8532 km/hr. = 1.0 Nautical Miles/hr.

1 League = 3.0 Miles (Approx.)

Pressure

14.7 psi = 33.95 ft. H_2O = 29.92 in. Hg = 407.2 in. W.G. = 2,116.8 lbs./sq. ft.

1 psi = 2.307 ft. H_2O = 2.036 in. Hg = 16 ounces = 27.7 in. WC

1 ft. H_2O = 0.4335 psi = 62.43 lbs./sq. ft.

1 ounce = 1.73 in. WC

Air Density
Standard Air @ 60°F., 14.7 psi:

13.329 cu. ft./lb. = 0.0750 lbs./cu. ft.
1 lb./cu. ft. = 177.72 cu. ft./lbs.
1 cu. ft./lb. = 0.00563 lbs./cu. ft.
1 kg/cu. m = 16.017 lbs./cu. ft.
1 cu. m/kg = 0.0624 cu. ft./lb.

Energy

1 HP = 0.746 KW = 746 Watts = 2,545 Btu/h ≈ 1.0 kVA
1 KW = 1,000 Watts = 3,413 Btu/h = 1.341 hp
1 Watt = 3.413 Btu/h
1 Ton Ac = 12,000 Btu/h Cooling = 15,000 Btu/h Heat
  Rejection
1 Btu/h = 1 Btu per hour
1 BHP = 34,500 Btu/h (33,472 Btu/h) = 34.5 lb.
  Steam per hour = 34.5 lb. H_2O/hr. = 0.069 GPM = 4.14
  GPH = 140 EDR (sq. ft. of Equivalent Radiation)
1 Therm = 100,000 Btu/h
1 MBH = 1,000 btu/h
1 lb. Stm./hr. = 0.002 GPM
1 GPM = 500 lbs. Steam per hour
EDR = Equivalent Direct Radiation
1 EDR = 0.000496 GPM = 0.25 lbs. Steam
  Condensate per hour
1000 EDR = 0.496 GPM
1 EDR Hot Water = 150 Btu/h
1 EDR Steam = 240 Btu/h
1 EDR = 240 Btu/h (Up to 1,000 ft. Above Sea Level)
1 EDR = 230 Btu/h (1,000 ft. 3,000 ft. Above
  Sea Level)

USEFUL HVAC CONVERSION MEASURES *(cont.)*

1 EDR = 223 Btu/h (3,000 ft.–5,000 ft. Above Sea Level)
1 EDR = 216 Btu/h (5,000 ft.–7,000 ft. Above Sea Level)
1 EDR = 209 Btu/h (7,000 ft.–10,000 ft. Above Sea Level)

Flow

1 MGD (million gal. per day) = 1.547 cu. ft./sec.
= 694.4 GPM
1 cu. ft/min. = 62.43 lbs. H_2O/min. = 448.8 gal. per hour

Metric Conversions

KJ/Hr.	=	Btu/h × 1.055
CMM	=	CFM × 0.02832
LPM	=	GPM × 3.785
KJ/Lb.	=	Btu/lb. × 2.326
Meters	=	Feet × 0.3048
sq. Meters	=	sq. ft. × 0.0929
cu. Meters	=	cu. ft. × 0.02832
kg	=	Pounds × 0.4536
kg/cu. Meter	=	Pounds/cu. ft. × 16.017 (Density)
cu. Meters/kg	=	cu. ft./Pound × 0.0624 (Specific Volume)
kg H_2O/kg DA	=	Gr H_2O/Lb. DA/7,000 = lb.H_2O/Lb DA

METRIC LIQUID VOLUME EQUIVALENTS

Metric	U.S.
3.7854 L	1 gallon
0.946 L	1 quart
0.473 L	1 pint
1 L	0.264 gallons
1 L	33.814 ounces
29.576 ml	1 fluid ounce
236.584 ml	1 cup

METRIC LENGTH EQUIVALENTS

Metric	U.S.
1 m	39.37 inches
1 m	3.28 feet
1 m	1.094 yards
1 m	.0016 mile
1 km	0.625 miles
1.609 km	1 mile
25.4 mm	1 inch
2.54 cm	1 inch
304.8 mm	1 foot
1 mm	0.03937 inch
1 cm	0.3937 inch
1 dm	3.937 inches

METRIC PRESSURE CONVERSIONS

Measurement	Equivalent
1 pound per square inch (psi)	6.8947 kPa
1 m column of water	9.794 kPa
10.2 cm of water	1 kPa
1 cm column of mercury	1.3332 kPa
1 inch of mercury (in. Hg)	3.3864 kPa
6 cm of mercury	8 kPa

UNDERSTANDING PRESSURE IN KILOPASCALS

The kilopascal (kPa) is a unit of measurement for fluid pressure. When working with kilopascals, note that atmospheric pressure is measured at 101.3 kPa metric and 14.7 psi English. The following facts will help you to understand pressures measured in kilopascals.

- Atmospheric pressure of 101.3 kPa will support a column of mercury 76 cm high.
- To find head pressure in decimeters when pressure is given in kPa, divide pressure by 0.9794.
- To find head pressure in kPa of a column of water given in decimeters, multiply the number of decimeters by 0.9794.
- 1 kilopascal (kPa) of air pressure elevates water approximately 10.2 cm under atmospheric conditions of 101 kPa.
- 10.2 cm of water equals 1 kPa.
- 51 cm of water equals 5 kPa.
- 1 meter of water equals 9.8 kPa.
- 10,000 square centimeters equals 1 square meter.
- 1 cubic meter equals 1,000,000 cubic centimeters (cm^3) or 1000 cubic decimeters (dm^3).
- 1 liter of cold water weighs 1 kilogram.

AIR PRESSURE

- 1 cubic inch of mercury weighs 0.49 lbs.
- Generally speaking, 2 inches of mercury is equivalent to 1 pound of pressure (psi).
- 1 cubic foot of air weighs 0.075 lbs.
- 1 cubic meter of air weighs 1.214 kilograms.

ABSOLUTE ZERO

- Absolute zero is −459.69°F.
- Absolute zero is −273.16°C.

ALTITUDE–PRESSURE RELATIONSHIP

Altitude (ft.)	Mercury (in.)
0 sea level	29.92
1,000	28.86
2,000	27.72
3,000	26.81
4,000	25.84
5,000	24.89
6,000	23.98
7,000	23.09
8,000	22.22
9,000	21.38
10,000	20.58

WATER BASICS

1 ft.³ of water contains 7.48 gal., 1,728 in.³, and weighs 62.48 lb.

1 ft.³ of ice weighs 57.2 lb.

1 gal. of water weighs 8.33 lb. and contains 231 in.³ or 0.1337 ft.³

1 lb. of water equals 27.72 in.³

One BTU is the heat needed to raise one pound of water one degree F.

1 ft. of water equals 0.434 psi

2.31 ft. of water equals 1.0 psi

The height of a column of water, equal to a pressure of 1.0 psi, is 2.31 ft.

To find the pressure in psi of a column of water, multiply the height of the column in feet by 0.434.

The average pressure of the atmosphere is estimated at 14.7 psi so that with a perfect vacuum it will sustain a column of water 34 ft. high.

Water expands $\frac{1}{23}$ of its volume when heated from 40° to 212°F.

Water is at its greatest density at 39.2°F

The friction of water in pipes varies as the square of the velocity.

To evaporate 1 ft.³ of water requires the consumption of 7½ lb. of ordinary coal or about 1 lb. of coal to 1 gal. of water.

1 in.³ of water evaporated at atmospheric pressure is converted into approximately 1 ft.³ of steam.

BOILING POINT OF WATER AT VARIOUS PRESSURES

Vacuum (in. Hg)	Boiling Point (°F)
0	212.00
1	210.25
2	208.50
3	206.70
4	201.85
5	202.25
6	200.96
7	198.87
8	196.73
9	194.50
10	192.19
11	189.75
12	187.21
13	184.61
14	181.82
15	178.91
16	175.80
17	172.51
18	169.00
19	165.24
20	161.19
21	156.75
22	151.87
23	146.45
24	140.31
25	133.22
26	124.77
27	114.22
28	99.93
29	76.62

BOILING TEMPERATURE OF WATER AT VARIOUS ALTITUDES

Altitude (ft.)	Boiling Temperature of Water (°F)
Sea Level	212.0
1,000	210.0
2,000	208.4
3,000	206.5
4,000	204.7
5,000	202.9
6,000	201.0
7,000	199.2
8,000	197.4
9,000	195.6
10,000	193.8
11,000	192.0
12,000	190.2
13,000	188.3
14,000	186.5
15,000	184.7
16,000	182.9
17,000	181.1
18,000	179.3
19,000	177.4
20,000	175.6
30,000	160.0
40,000	138.0
50,000	119.0
60,000	101.0
80,000	72.0
105,000	41.5

The boiling point decreases approximately 1°F for each 550' increase in altitude

CHAPTER 2
Heating

CHIMNEY CONNECTOR AND VENT CONNECTOR CLEARANCE FROM COMBUSTIBLE MATERIALS	
Description of Appliance	**Minimum Clearance (in.)**
Residential Appliances	
Single-Wall, Metal Pipe Connectors	
Electric, gas, and oil incinerators	18
Oil and solid-fuel appliances	18
Oil appliances listed as suitable for use with type L venting systems, but only when connected to chimneys	9
Type L Venting System Piping Connectors	
Electric, gas, and oil incinerators	9
Oil and solid-fuel appliances	9
Oil appliances listed as suitable for use with type L venting systems	As listed
Commercial and Industrial Appliances	
Low-Heat Appliances **Single-Wall, Metal Pipe Connectors**	
Gas, oil, and solid-fuel boilers, furnaces, and water heaters	18
Ranges, restaurant type	18
Oil unit heaters	18
Other low-heat industrial appliances	18
Medium-Heat Appliances **Single-Wall, Metal Pipe Connectors**	
All gas, oil, and solid-fuel appliances	36

CLEARANCE TOLERANCE PER ROOF SLOPE FOR GAS VENT TERMINATIONS—VENT CAPS

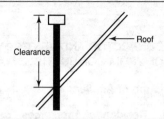

Clearance (In.)	Roof Slope
12	Flat to 6/12
15	6/12 to 7/12
18	7/12 to 8/12
24	8/12 to 9/12
30	9/12 to 10/12
39	10/12 to 11/12
48	11/12 to 12/12

ROOF TERMINATIONS FOR CHIMNEYS AND SINGLE WALL VENTS—NO CAPS

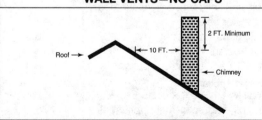

2-2

CONVERSION FOR FLUE/VENT LINERS

Liner Size (in.)	Inside Dimensions (in.)	Round Equiv. (in.)	Area sq. (in.)
4 × 8	2½ × 6½	4	12.2
		5	19.6
		6	28.3
		7	38.3
8 × 8	6¾ × 6¾	7.4	42.7
		8	50.3
8 × 12	6½ × 10½	9	63.6
		10	78.5
12 × 12	9¾ × 9¾	10.4	83.3
		11	95
12 × 16	9½ × 13½	11.8	107.5
		12	113.0
		14	153.9
16 × 16	13¼ × 13¼	14.5	162.9
		15	176.7
16 × 20	13 × 17	16.2	206.1
		18	254.4
20 × 20	16¾ × 16¾	18.2	260.2
		20	314.1
20 × 24	16½ × 20½	20.1	314.2
		22	380.1
24 × 24	20¼ × 20¼	22.1	380.1
		24	452.3

THICKNESS FOR SINGLE-WALL METAL PIPE CONNECTORS

Diameter of Connector (in)	Galvanized Sheet Metal Gauge	Minimum Thickness (in.)
Less than 6	26	0.019
6 to 10	24	0.024
more than 10	22	0.029

STANDARD CLEARANCE FOR HEAT-PRODUCING APPLIANCES IN RESIDENTIAL INSTALLATIONS

Residential Type Appliances for Installation in Rooms that Are Large	Clearance in Inches					
	Above Top of Casing or Appliance	From Top and Sides of Warm-Air Bonnet or Plenum	From Front	From Back	From Sides
Boilers and Water Heaters Steam boilers—15 psi; water boilers—250°F; water heaters—200°F; all water walled or jacketed	Automatic oil or comb. gas-oil	6	—	24	6	6
Furnaces—Central Gravity, upflow, downflow, horizontal and duct. warm air—250°F max.	Automatic oil or comb. gas-oil	6 or as listed	6 or as listed	24	6	6
Furnaces—Floor For mounting in combustible floors	Automatic oil or comb. gas-oil	36	—	12	12	12

STANDARD CLEARANCE FOR HEAT-PRODUCING APPLIANCES IN COMMERCIAL AND INDUSTRIAL INSTALLATIONS

Commercial-Industrial Type Low Heat Appliances (Any and All Physical Sizes Except as Noted)		Clearance in Inches				
		Above Top of Casing or Appliance	From Top and Sides of Warm-Air Bonnet or Plenum	From Front	From Back	From Sides
Boiler and Water Heaters 100 cu. ft. or less, any psi,	All fuels	18	—	48	18	18
steam 50 psi or less, any size	All fuels	18	—	48	18	18
Unit Heaters Floor mounted or suspended—any size	Steam or hot water	1	—	—	1	1
Suspended 100 cu. ft. or less	Oil or comb. gas-oil	6	—	24	18	18
Suspended More than 100 cu. ft.	All fuels	18	—	48	18	18
Floor Mounted Any size	All fuels	18	—	48	18	18

2-5

APPLIANCE CLEARANCE WITH AND WITHOUT PROTECTION

Required Clearance with No Protection

Type of Protection	36" Above	36" Sides & Rear	36" Chimney or Vent Connector	18" Above	18" Sides & Rear	18" Chimney or Vent Connector	12" Above	12" Sides & Rear	12" Chimney or Vent Connector	9" Above	9" Sides & Rear	9" Chimney or Vent Connector	6" Above	6" Sides & Rear	6" Chimney or Vent Connector
1/4" asbestos millboard spaced out 1"	30"	18"	30"	15"	9"	12"	9"	6"	6"	9"	6"	6"	3"	2"	3"
28-gauge sheet metal on 1/4" asbestos millboard	24"	18"	24"	12"	9"	12"	9"	6"	4"	9"	6"	4"	3"	2"	2"
28-gauge sheet metal spaced out 1"	18"	12"	18"	9"	6"	9"	6"	4"	4"	6"	4"	4"	2"	2"	2"
28-gauge sheet metal on 1/4" in asbestos millboard spaced out 1"	18"	12"	18"	9"	6"	9"	6"	4"	4"	6"	4"	4"	2"	2"	2"

APPLIANCE CLEARANCE WITH AND WITHOUT PROTECTION (cont.)

Required Clearance with No Protection

Type of Protection	36"			18"			12"			9"			6"		
	Above	Sides & Rear	Chimney or Vent Connector	Above	Sides & Rear	Chimney or Vent Connector	Above	Sides & Rear	Chimney or Vent Connector	Above	Sides & Rear	Chimney or Vent Connector	Above	Sides & Rear	Chimney or Vent Connector
1¼" asbestos cement covering on heating appliance	18"	12"	36"	9"	6"	18"	6"	4"	6"	2"	4"	9"	2"	1"	6"
¼" asbestos millboard on 1" mineral fiber bats reinforced with wire mesh or equivalent	18"	12"	18"	6"	6"	6"	4"	4"	6"	2"	4"	4"	2"	2"	2"
22-gauge sheet metal on 1" mineral fiber bats reinforced with wire or equivalent	18"	12"	12"	4"	3"	3"	2"	2"	3"	2"	2"	2"	2"	2"	2"
¼" asbestos millboard	36"	36"	36"	18"	18"	18"	12"	12"	18"	4"	12"	9"	4"	4"	4"
¼" cellular asbestos	36"	36"	36"	18"	18"	18"	12"	12"	18"	3"	12"	9"	3"	3"	3"

ATMOSPHERIC PRESSURE AND BAROMETER READINGS FOR VARIOUS ALTITUDES ABOVE SEA LEVEL

Altitude (ft.)	Pressure (psi)	Barometer (in. of Hg)
Sea Level	14.69	29.92
500	14.42	29.38
1000	14.16	28.86
1500	13.91	28.33
2000	13.66	27.82
2500	13.41	27.31
3000	13.16	26.81
3500	12.92	26.32
4000	12.68	25.84
4500	12.45	25.36
5000	12.22	24.89
5500	11.99	24.43
6000	11.77	23.98
6500	11.55	23.53
7000	11.33	23.09
7500	11.12	22.65
8000	10.91	22.22
8500	10.70	21.80
9000	10.50	21.38
9500	10.30	20.98
10,000	10.10	20.58
11,000	9.71	19.75
12,000	9.34	19.03
13,000	8.97	18.29
14,000	8.62	17.57
15,000	8.28	16.88

WINTER DESIGN TEMPERATURES FOR MAJOR U.S. CITIES

State	City	Outside Design Temperature Commonly Used (°F)	State	City	Outside Design Temperature Commonly Used (°F)
Alabama	Birmingham	10	New Hampshire	Concord	−15
Arizona	Tucson	25	New Jersey	Atlantic City	5
Arkansas	Little Rock	5		Trenton	0
California	San Francisco	35	New Mexico	Albuquerque	0
	Los Angeles	35	New York	Albany	−10
Colorado	Denver	−10		Buffalo	−5
Connecticut	New Haven	0		New York City	0
Dist. of Columbia	Washington	0	North Carolina	Asheville	0
Florida	Jacksonville	25		Charlotte	10
	Key West	45	North Dakota	Bismarck	−30
Georgia	Atlanta	10	Ohio	Akron	0
	Savannah	20		Cincinnati	0

WINTER DESIGN TEMPERATURES FOR MAJOR U.S. CITIES (cont.)

State	City	Outside Design Temperature Commonly Used (°F)	State	City	Outside Design Temperature Commonly Used (°F)
Idaho	Boise	−10		Columbus	−10
Illinois	Cairo	0	Oklahoma	Oklahoma City	0
	Chicago	−10	Oregon	Portland	10
Indiana	Indianapolis	−10	Pennsylvania	Erie	−5
Iowa	Des Moines	−15		Harrisburg	0
	Sioux City	−20		Philadelphia	0
Kansas	Topeka	−10	Rhode Island	Providence	0
Kentucky	Louisville	0	South Carolina	Charleston	15
Louisiana	New Orleans	20	South Dakota	Huron	−20
Maine	Portland	−5	Tennessee	Knoxville	0
Maryland	Baltimore	0	Texas	Abilene	15
Massachusetts	Boston	0		Austin	20
Michigan	Detroit	−10		Brownsville	30

2-10

State	City	Temp
Michigan	Escanaba	-15
	Sault Ste. Marie	-20
	Duluth	-20
Minnesota	Minneapolis	-20
Mississippi	Vicksburg	10
Missouri	Kansas City	-10
	St. Louis	0
Montana	Billings	-25
	Helena	-20
	Miles City	-35
Nebraska	Lincoln	-10
	Valentine	-25
Nevada	Reno	-5

State	City	Temp
Texas	Corpus Christi	20
	Dallas	0
	Houston	20
Utah	Salt Lake City	-10
Vermont	Burlington	-10
Virginia	Lynchburg	5
	Norfolk	15
Washington	Seattle	15
	Spokane	-15
West Virginia	Parkersburg	-10
Wisconsin	Green Bay	-20
	Madison	-15
Wyoming	Cheyenne	-15

GROUND TEMPERATURES (U.S.) BELOW THE FROST LINE

State	City	Ground Temperature (°F)	State	City	Ground Temperature (°F)
Alabama	Birmingham	66	Nevada	Reno	52
Arizona	Tucson	60	New Hampshire	Concord	47
Arkansas	Little Rock	65	New Jersey	Atlantic City	57
California	San Francisco	62	New Mexico	Albuquerque	57
	Los Angeles	67	New York	Albany	48
Colorado	Denver	48		New York City	52
Connecticut	New Haven	52	North Carolina	Greensboro	62
Dist. of Columbia	Washington	57	North Dakota	Bismarck	42
Florida	Jacksonville	70	Ohio	Cleveland	52
	Key West	78		Cincinnati	57
Georgia	Atlanta	65	Oklahoma	Oklahoma City	62
Idaho	Boise	52	Oregon	Portland	52
Illinois	Cairo	60	Pennsylvania	Pittsburgh	52
	Chicago	52		Philadelphia	52
	Peoria	55	Rhode Island	Providence	52

2-12

State	City	Value		State	City	Value
Indiana	Indianapolis	55		South Carolina	Greenville	67
Iowa	Des Moines	52		South Dakota	Huron	47
Kentucky	Louisville	57		Tennessee	Knoxville	61
Louisiana	New Orleans	72		Texas	Abilene	62
Maine	Portland	45			Dallas	67
Maryland	Baltimore	57			Corpus Christi	72
Massachusetts	Boston	48		Utah	Salt Lake City	52
Michigan	Detroit	48		Vermont	Burlington	46
Minnesota	Duluth	41		Virginia	Richmond	57
	Minneapolis	44		Washington	Seattle	52
Mississippi	Vicksburg	67		West Virginia	Parkersburg	52
Missouri	Kansas City	57		Wisconsin	Green Bay	44
Montana	Billings	42			Madison	47
Nebraska	Lincoln	52		Wyoming	Cheyenne	42

2-13

REDUCTION IN DEGREE DAY TEMPERATURE BASE WHEN CALCULATED HEAT LOSS IS LESS THAN 1000

Calculated Heat loss (Btu per 1°F Temp. Diff.)	Revised Degree day Temp. base in °F
200	55
300	60
400	61
500	62
600	63
700	64
800	64
900	64
1,000	65

INFILTRATION RATES THROUGH VARIOUS TYPES OF WALL CONSTRUCTION (CU. FT. PER FOOT OF CRACK PER HR.)

Wall Type	Wind Velocity (mph)					
	5	10	15	20	25	30
Brick Wall (8½" plain)	2	4	8	12	19	23
with plaster or wallboard	.02	.04	.07	.11	.16	.24
Brick wall (13" plain)	1	4	7	12	16	21
with plaster or wallboard.)	.01	.01	.03	.04	.07	.10
Frame wall with lath and plaster or wallboard	.03	.07	.13	.18	.23	.26

INFILTRATION RATES THROUGH VARIOUS TYPES OF WINDOWS
(CU. FT. PER FOOT OF CRACK PER HR.)

Window		Wind Velocity (mph)						
		5	10	15	20	25	30
Double-hung wood sash windows (unlocked)	Around frame in masonry wall—not caulked	3	8	14	20	27	35
	Around frame in masonry wall—caulked	1	2	3	4	5	6
	Around frame in wood-frame construction	2	6	11	17	23	30
Double-hung metal windows	Non-weather-stripped, locked	20	45	70	96	125	154
	Non-weather-stripped, unlocked	20	47	74	104	137	170
	Weather-stripped, unlocked	6	19	32	46	60	76
Rolled section steel sash windows	Industrial pivoted, 1⁄16" crack	52	108	176	244	304	372
	Architectural projected, 1⁄32" crack	15	36	62	86	112	139
	Architectural projected, 3⁄64" crack	20	52	88	116	152	182
	Residential casement, 1⁄64" crack	6	18	33	47	60	74
	Residential casement, 1⁄32" crack	14	32	52	76	100	128
	Heavy casement section, projected, 1⁄64" crack	3	10	18	26	36	48
	Heavy casement section, projected, 1⁄32" crack	8	24	38	54	72	92
Hollow metal, vertically pivoted window		30	88	145	186	221	242

HEAT GAIN FACTORS (WALLS, FLOOR, AND CEILING)
BTU PER SQ. FT. (24 HR.)

Insulation	1	Temp. Difference (Ambient Temp. Minus Storage Temp. °F)							
		40	45	50	55	60	65	70	75
Cork or equivalent (in., k = 0.3)									
3	2.4	96	108	120	132	144	156	168	180
4	1.8	72	81	90	99	108	117	126	135
5	1.44	58	65	72	79	87	94	101	108
6	1.2	48	54	60	66	72	78	84	90
7	1.03	41	46	52	57	62	67	72	77
8	0.90	36	41	45	50	54	59	63	68
9	0.80	32	36	40	44	48	52	56	60
10	0.72	29	32	36	40	43	47	50	54
11	0.66	26	30	33	36	40	43	46	50
12	0.60	24	27	30	33	36	39	42	45
13	0.55	22	25	28	30	33	36	39	41
14	0.51	20	23	26	28	31	33	36	38
Single glass	27.0	1,080	1,220	1,350	1,490	1,620	1,760	1,890	2,030
Double glass	11.0	440	500	550	610	660	715	770	825
Triple glass	7.0	280	320	350	390	420	454	490	525

Cork or equivalent (in., k = 0.3)	80	85	90	95	100	105	110	115	120
3	192	204	216	228	240	252	264	276	288
4	144	153	162	171	180	189	198	207	216
5	115	122	130	137	144	151	159	166	172
6	96	102	108	114	120	126	132	138	144
7	82	88	93	98	103	108	113	118	124
8	72	77	81	86	90	95	99	104	108
9	64	68	72	76	80	84	88	92	94
10	58	61	63	68	72	76	79	83	86
11	53	53	60	63	66	69	73	76	79
12	48	51	54	57	60	63	66	69	72
13	44	47	50	52	55	58	61	63	66
14	41	43	46	49	51	54	56	59	61
Single glass	2,160	2,290	2,440	2,560	2,700	2,840	2,970	3,100	3,240
Double glass	880	936	990	1,050	1,100	1,160	1,210	1,270	1,320
Triple glass	560	595	630	665	700	740	770	810	840

HEATING FUELS—COST COMPARISON FORMULAS

Note that the formulas below are averages and are for example purposes only

Natural Gas

$$\frac{\text{price per cubic ft.} \times 100{,}000 \text{ Btu}}{1028 \text{ (Btu per cubic foot)}} = \text{cost per } 100{,}000 \text{ Btu}$$

Propane LP Gas

$$\frac{\text{price per gallon} \times 100{,}000 \text{ Btu}}{91{,}333 \text{ (Btu per gallon)}} = \text{cost per } 100{,}000 \text{ Btu}$$

Fuel Oil

$$\frac{\text{price per gallon} \times 100{,}000 \text{ Btu}}{140{,}000 \text{ (Btu per gallon)}} = \text{cost per } 100{,}000 \text{ Btu}$$

Electric (kWH)

$$\frac{\text{price per kWH} \times 100{,}000 \text{ Btu}}{3412 \text{ (Btu per kWH)}} = \text{cost per } 100{,}000 \text{ Btu}$$

AVERAGE BTU CONTENT OF VARIOUS HEATING FUELS

Fuel Type	Btu per unit of Measurement
Fuel oil	140,000/gallon
Natural gas	1,025,000/thousand cubic feet
Propane	91,330/gallon
Electricity	3,412/kilowatt hour
Coal	28,000,000/ton
Wood (air dried)	20,000,000/cord or 8000/pound
Kerosene	135,000/gallon
Pellets	16,500,000/ton

AVERAGE FUEL CONVERSION EFFICIENCY FOR HEATING

Fuel Type	Heating Source	Fuel Efficiency (%)
Fuel oil	High-efficiency central heating unit	89
	Typical central heating system unit	80
	Water heater (50 gal.)	60
Gas	High-efficiency central heating furnace	97
	Typical central heating boiler	85
	Minimum efficiency central heating furnace	78
	Room heater (unvented)	99
	Room heater (vented)	65
	Water heater (50 gal.)	62
Electricity	Baseboard (resistance)	99
	Forced-air central heating unit	97
	Heat pump central heating system	200+
	Ground source heat pump	300+
	Water heater (50 gal.)	97
Coal	Hand-fired central heating unit	45
	Stoker-fired central heating unit	60
Wood and pellets	Franklin stove	30 to 40
	Stoves with circulating fans	40 to 70
	Catalytic stoves	65 to 75
	Pellet stoves	85 to 90

TYPICAL GAS DEMAND—APPLIANCES

Appliance Type	Input Btu per hour
Boiler or furnace	100,000 to 250,000
Small broiler	30,000
Large broiler	60,000
Combination boiler and roaster	66,000
Oven	25,000
Stove top burners	40,000
Range	65,000
Clothes dryer	35,000
Water heater, 30 to 40 gal. cap.	45,000
Water heater, 50 gal. cap.	55,000
Log lighter	25,000
Barbeque/grill	50,000
Gas engine, per horsepower	10,000
Steam boiler, per horsepower	50,000

CHARACTERISTICS OF FUEL OIL GRADES

Characteristics	No. 2	No. 4	No. 5	No. 6
Type	Light distillate	Light distillate or blend	Light residual	Residual
Color	Amber	Black	Black	Black
Specific gravity	.8654	.9279	.9529	.9861
BTU/gal.	141,000	146,000	148,000	150,000
BTU/lb.	19,500	19,100	18,950	18,750

2-20

BTU CONTENT OF FUEL OILS AND GASES		
Grade or Type	Unit	Btu
No. 1 oil	Gal.	137,400
No. 2 oil	Gal.	139,600
No. 3 oil	Gal.	141,800
No. 4 oil	Gal.	145,100
No. 5 oil	Gal.	148,800
No. 6 oil	Gal.	152,400
Natural gas	Cu. ft.	950–1150
Propane	Cu. ft.	2,550
Butane	Cu. ft.	3,200
PROPERTIES OF LP GAS		
Property	Butane	Propane
Btu per cu. ft. 60°F	3,280	2,516
Btu per lb.	21,221	21,591
Btu per gal.	102,032	91,547
Cu. ft. per lb.	6.506	8.58
Cu. ft. per gal.	31.26	36.69
Lb. per gal.	4.81	4.24

HEATING VALUES AND CHEMICAL COMPOSITION OF STANDARD GRADES OF COAL

Coal Grade by Rank	Heating Value (Btu/lb.)	Chemical Composition (%)						
		Oxygen	Hydrogen	Carbon	Nitrogen	Sulfur	Ash
Anthracite	12,910	5.0	2.9	80.0	0.9	0.7	10.5
Semi-anthracite	13,770	5.0	3.9	80.4	1.1	1.1	8.5
Low-volatile bituminous	14,340	5.0	4.7	81.7	1.4	1.2	6.0
Medium-volatile bituminous	13,840	5.0	5.0	79.0	1.4	1.5	8.1
High-volatile bituminous A	13,090	9.2	5.3	73.2	1.5	2.0	8.8
High-volatile bituminous B	12,130	13.8	5.5	68.0	1.4	2.1	9.2
High-volatile bituminous C	10,750	21.0	5.8	60.6	1.1	2.1	9.4
Subbituminous B	9,150	29.5	6.2	52.5	1.0	1.0	9.8
Subbituminous C	8,940	35.8	6.5	46.7	0.8	0.6	9.6
Lignite	6,900	44.0	6.9	40.1	0.7	1.0	7.3

HEAT VALUE PER CORD (MILLION BTU) OF WOOD WITH 12% MOISTURE CONTENT

Name of Wood	Heat Value	Equivalent Coal Heat Value
Ash, white	28.3	1.09
Beech	31.1	1.20
Birch, yellow	30.4	1.17
Chestnut	20.7	0.80
Cottonwood	19.4	0.75
Elm, white	24.2	0.93
Hickory	35.3	1.36
Maple, sugar	30.4	1.17
Maple, red	26.3	1.01
Oak, red	30.4	1.17
Oak, white	32.5	1.25
Pine, yellow	26.0	1.00
Pine, white	18.1	0.70

HEAT VALUE PER CORD (MILLION BTU) OF GREEN WOODS

Name of Wood	Heat Value	Equivalent Coal Heat Value
Ash, white	26.0	1.00
Beech	27.1	1.04
Birch, yellow	27.2	1.05
Chestnut	19.2	0.75
Cottonwood	18.0	0.69
Elm, white	22.2	0.85
Hickory	29.0	1.12
Maple, sugar	27.4	1.05
Maple, red	23.7	0.91
Oak, red	27.5	1.06
Oak, white	28.7	1.10
Pine, yellow	23.7	0.91
Pine, white	17.3	0.67

PERFORMANCE OF STORAGE WATER HEATERS

Input Heat Units	Eff. (%)	Usable BTU/h	GPH 100°F Rise	Tank Size Gal.	Avail. Hot-Water Storage plus Recovery 100°F Rise				Continuous Draw, Gph
					15 Min.	30 Min.	45 Min.	60 Min.	
Electricity, kWh									
1.5	92.5	4,750	5.7	20	21.4	22.8	24.3	25.7	5.7
2.5	92.5	7,900	9.5	20	32.4	34.8	37.1	39.5	9.5
4.5	92.5	14,200	17.1	50	54.3	58.6	62.9	67.1	17.1
6.0	92.5	19,000	22.8	66	71.6	77.2	82.8	88.8	22.8
7.0	92.5	22,100	26.5	80	86.6	93.2	99.8	106.5	26.5
Gas, BTU/h									
34,000	75	25,500	30.6	30	37.7	45.3	53.0	55.6	25.6
42,000	75	31,600	38.0	30	39.5	49.0	58.8	61.7	31.7
50,000	75	37,400	45.0	40	51.3	62.6	73.9	77.6	37.6
60,000	75	45,000	54.0	50	63.5	77.0	90.5	95.0	45.0
Oil, Gph									
0.05	75	52,500	63.0	30	45.8	61.6	77.4	82.5	52.5
0.75	75	78,700	94.6	30	53.6	77.2	100.8	109.0	79.0
0.85	75	89,100	107.0	30	57.7	83.4	110.1	119.1	89.0
1.00	75	105,000	126.0	50	81.5	113.0	144.5	155.0	105.0
1.20	75	126,000	151.5	50	87.9	125.8	163.7	176.0	126.0
1.35	75	145,000	174.0	50	93.5	137.0	180.5	195.0	145.0
1.50	75	157,000	188.5	85	132.1	179.2	226.3	242.0	157.0
1.65	75	174,000	204.5	85	136.1	187.2	238.4	259.0	174.0

HOT-WATER STORAGE TANK DIMENSIONS WITH CAPACITIES (GAL.)

Tank Length (feet)	Tank Diameter (inches)									
	20"	22"	24"	30"	36"	42"	48"	54"	60"
1'	16	20	24	37	53	72	94	120	145
2'	32	40	48	74	106	144	188	240	290
3'	48	60	72	110	159	216	282	360	435
4'	66	80	96	147	212	288	376	480	580
5'	82	100	120	184	265	360	470	600	725
6'	98	120	144	220	317	432	564	720	870
7'	114	140	163	257	370	504	658	840	1015
8'	131	160	192	294	423	576	752	960	1160
9'	147	180	216	330	476	648	846	1080	1305
10'	163	200	240	367	529	720	940	1200	1450

FORMULA FOR SIZING GAS PIPE

Size the gas supply pipe size according to volume of gas used and to prevent undue pressure drop. The diameter must be equal to the manual shutoff valve in the supply riser.

To obtain the size of piping required, determine the number of cubic feet of gas per hour consumed. Note that the heating value per cubic foot of gas is 1000 Btu.

$$\text{cu. ft. of gas per hr.} = \frac{\text{total Btu per hr. required}}{1000\ (\text{Btu per cu. ft.})}$$

THREADING IRON GAS PIPE

Pipe Size (in.)	Length of Threaded Portion (in.)	Number of Threads to Be Cut
½	¾	10
¾	¾	10
1	⅞	10
1¼	1	11
1½	1	11
2	1	11
2½	1½	12
3	1½	12
4	1¾	13

PIPING SUPPORT FOR IRON GAS PIPE

Pipe Size (in.)	Support Spacing (ft.)
up to ¾	10
1 or greater	12

NATURAL GAS IRON PIPE CAPACITIES (CU. FT.)

Pipe Size (in.)	Pipe Length in Feet						
	10	20	30	40	50	60	70
½	170	118	95	80	71	64	60
¾	360	245	198	169	150	135	123
1	670	430	370	318	282	255	235
1¼	1320	930	740	640	565	510	470
1½	1990	1370	1100	950	830	760	700
2	3880	2680	2150	1840	1610	1480	1350
2½	6200	4120	3420	2950	2600	2360	2180
3	10900	7500	6000	5150	4600	4150	3820
4	22500	15500	12400	10600	9300	8500	7900

Pipe Size (in.)	Pipe Length in Feet						
	80	90	100	125	150	200	250
½	55	52	49	44	40	34	30
¾	115	108	102	92	83	71	63
1	220	205	192	172	158	132	118
1¼	440	410	390	345	315	270	238
1½	650	610	570	510	460	400	350
2	1250	1180	1100	1000	910	780	690
2½	2000	1900	1800	1600	1450	1230	1100
3	3550	3300	3120	2810	2550	2180	1930
4	7300	6800	6400	5700	5200	4400	3950

Pipe Size (in.)	Pipe Length in Foot						
	300	350	400	450	500	550	600
½	27	25	23	22	21	20	19
¾	57	52	48	45	43	41	39
1	108	100	92	86	81	77	74
1¼	215	200	185	172	162	155	150
1½	320	295	275	255	240	230	220
2	625	570	535	500	470	450	430
2½	1000	920	850	800	760	720	690
3	1750	1600	1500	1400	1320	1250	1200
4	3600	3250	3050	2850	2700	2570	2450

Note: For gas pressure of 0.5 psi or less; specific gravity 0.6

ORIFICE CAPACITIES FOR NATURAL GAS

1000 Btu/ft.³, Manifold Pressure 3½" Water Columns

Wire Gauge Drill Size	Rate (ft.³/hr.)	Rate (Btu/hr.)
70	1.34	1,340
68	1.65	1,650
66	1.80	1,800
64	2.22	2,220
62	2.45	2,450
60	2.75	2,750
58	3.50	3,500
56	3.69	3,695
54	5.13	5,130
52	6.92	6,925
50	8.35	8,350
48	9.87	9,875
46	11.25	11,250
44	12.62	12,625
42	15.00	15,000
40	16.55	16,550
38	17.70	17,700
36	19.50	19,500
34	21.05	21,050
32	23.70	23,700
30	28.50	28,500
28	34.12	34,125
26	37.25	37,250
24	38.75	38,750
22	42.50	42,500
20	44.75	44,750

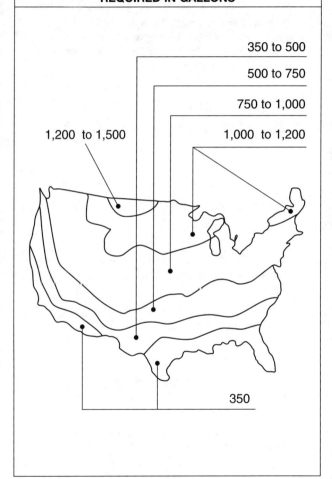

WEATHER ZONES—SIZE OF LP STORAGE TANKS REQUIRED IN GALLONS

350 to 500

500 to 750

750 to 1,000

1,000 to 1,200

1,200 to 1,500

350

2-29

OBTAINING PROPER OIL BURNER NOZZLE FLOW RATES

The proper size nozzle for a given burner unit is normally stamped on the nameplate. However, if it is missing use the following formulas for determining the proper flow rates.

For a unit rating given in Btu per hour input:

GPH = Btu input / 140,000

For a unit rating given in Btu per hour output:

GPH = Btu output / (efficiency %) × 140,000

For a steam heating system in which the total square feet of steam radiation including piping is known:

GPH = total square feet of steam × 240 / (efficiency %) × 140,000

For a hydronic (hot-water) heating system operating at 180^0 in which the total square feet of radiation including piping is known:

GPH = total square feet of hot water × 165 / (efficiency %) × 140,000

CALCULATING OIL PUMP CAPACITY

The capacity of an oil burner fuel pump is calculated by obtaining the total vacuum in the system. The vacuum is expressed in inches and can be determined by the following:

- 1 inch of vacuum for each foot of lift
- 1 inch of vacuum for each 90° elbow in either the suction and/or return lines
- 1 inch of vacuum for each 10 feet of horizontal run (⅜-inch OD line) or
- 1 inch of vacuum for each 20 feet of horizontal run (½-inch OD line)

OIL PUMPS FOR DIFFERENT VACUUMS

Total Vacuum (in.)	Pump Type
Up to 3	Single-stage pump
4 to 13	Two-stage pump
14+	Single-stage pump for burner plus separate lift pump with reservoir

ADJUSTING OIL PUMP PRESSURE

The oil pressure regulator on the fuel pump is normally pre-set to give nozzle oil pressures of 100 psig. The firing rate is indicated on the nameplate. The firing rate can be obtained with standard nozzles by inserting a gauge in the pump gauge port and turning the adjusting screw clockwise to increase pressure or counterclockwise to decrease pressure. Never exceed maximum recommended pressure.

RECOMMENDED LIMIT CONTROL SETTINGS

Heating System		Setting
Hot-water system	(gravity)	180°F
Hot-water system	(forced)	160°F
Warm air system	(forced)	200°F
Warm air system	(gravity)	300°F
Steam system		"off" 3 lb. — "on" 1 lb.
Vapor system		"off" 4 oz. — "on" 2 oz.

COMMON TERMINAL IDENTIFICATIONS

Letter	Wire Color	Terminal Function
R	Red	Power supply; transformer
W	White	Heating control; heating relay or valve coil
Y	Yellow	Cooling control; cooling contactor coil
G	Green	Fan relay coil
O	Orange	Cooling damper
B	Brown	Heating damper
X	—	Malfunction light
P	—	Heat pump contactor coil
Z	—	Low-voltage fan switch

BASIC HVAC THERMOSTAT CIRCUIT

Heating element

Magnetic switch

Thermostat

M

M

Low voltage circuit

Transformer

115 or 230 V supply lines

2-33

TWIN-TYPE THERMOSTAT

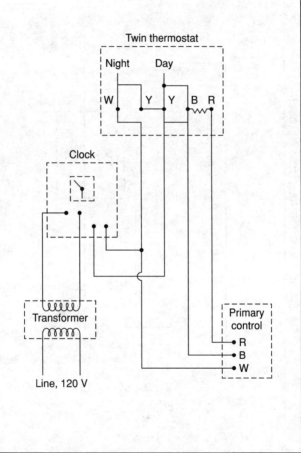

MILLIVOLT CONTROL

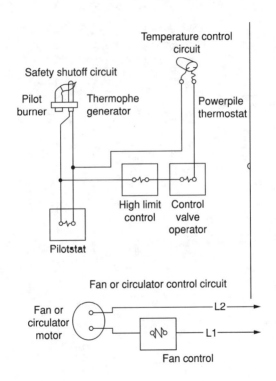

GAS BURNER CONTROLS

Limit control

Automatic electric pilot valve (if used)

Thermostat

Gas valve

Transformer

Line

2-36

RELAY POSITION AND EFFECT ON GAS BURNER

Relay Position	Description	Effect
Standby	Load and flame relays both "out."	Motor and gas valve both deenergized.
Starting	Load relay "in."; flame relay "out."	Motor energized, gas valve deenergized, trial for ignition period, safety lockout occurs after 45 sec.
Running	Load and flame relays both "in."	Motor and gas valve both energized
Abnormal conditions due to flame simulating failure	Load relay "out"; flame relay "in."	Motor and gas valve both deenergized; safety lockout occurs after 45 sec.

2-37

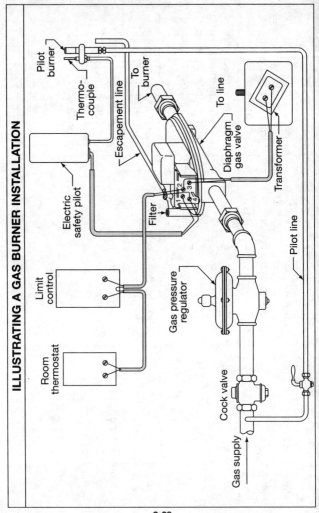

ILLUSTRATING A GAS BURNER INSTALLATION

Pilot burner

Thermo-couple

Escapement line

To burner

Electric safety pilot

Filter

Diaphragm gas valve

To line

Transformer

Limit control

Room thermostat

Gas pressure regulator

Pilot line

Cock valve

Gas supply

2-38

CONNECTIONS FOR A STOKER INSTALLATION

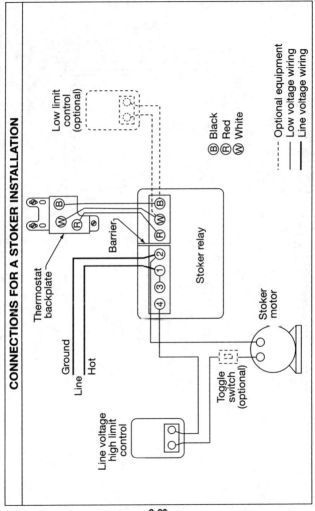

Low limit control (optional)

Thermostat backplate

Barrier

Stoker relay

Ground
Line
Hot

Stoker motor

Toggle switch (optional)

Line voltage high limit control

Ⓑ Black
Ⓡ Red
Ⓦ White

----- Optional equipment
— Low voltage wiring
— Line voltage wiring

2-39

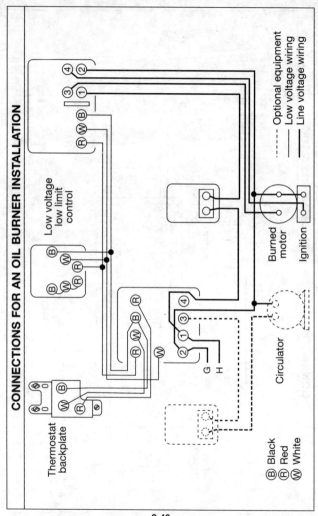

CONNECTIONS FOR AN OIL BURNER INSTALLATION

Low voltage low limit control

Thermostat backplate

Burned motor

Ignition

Circulator

G H

- - - - Optional equipment
——— Low voltage wiring
——— Line voltage wiring

Ⓑ Black
Ⓡ Red
Ⓦ White

2-40

DAMPER-CONTROL LOW-VOLTAGE INSTALLATION

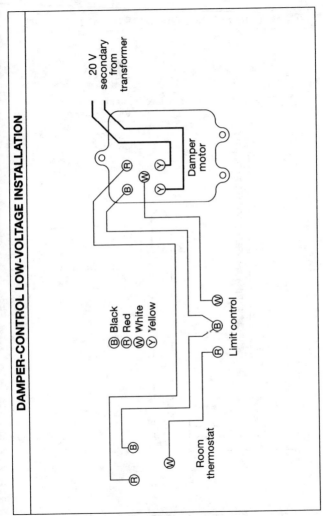

20 V secondary from transformer

Damper motor

(B) Black
(R) Red
(W) White
(Y) Yellow

Limit control

Room thermostat

COMPOUND METAL THERMOSTAT AND CIRCUIT

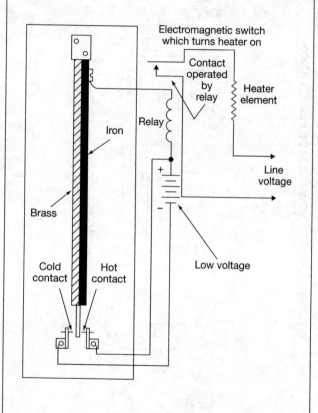

LIMIT SWITCH INCLUDED IN A BASEBOARD HEATER

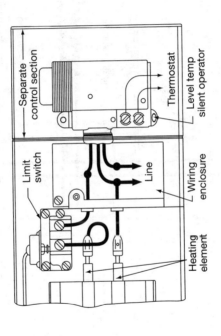

Separate control section

Limit switch

Line

Wiring enclosure

Thermostat

Level temp silent operator

Heating element

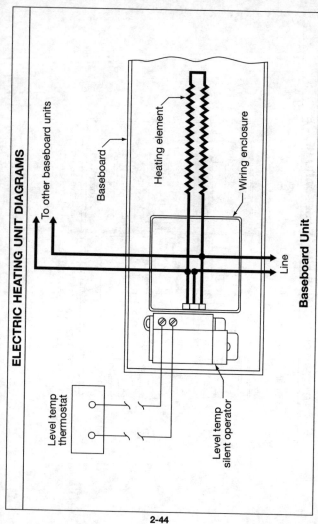

ELECTRIC HEATING UNIT DIAGRAMS

To other baseboard units

Baseboard

Heating element

Wiring enclosure

Line

Baseboard Unit

Level temp thermostat

Level temp silent operator

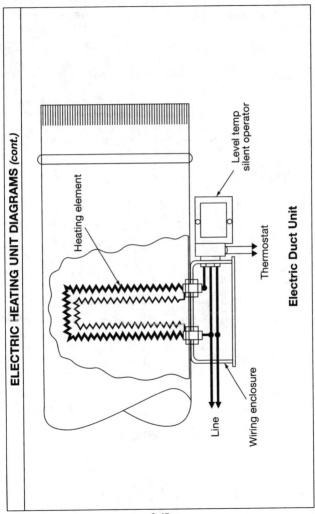

Heating element

Level temp silent operator

Thermostat

Line

Wiring enclosure

Electric Duct Unit

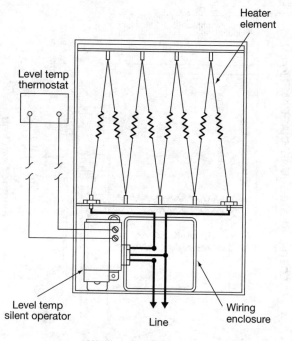

Wall or Ceiling Unit

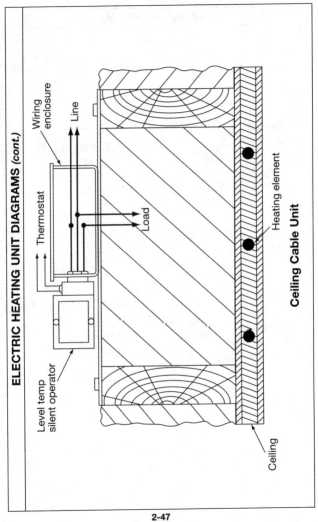

ELECTRIC HEATING UNIT DIAGRAMS *(cont.)*

Wiring enclosure

Line

Thermostat

Load

Level temp silent operator

Heating element

Ceiling

Ceiling Cable Unit

HEATING CONTROL

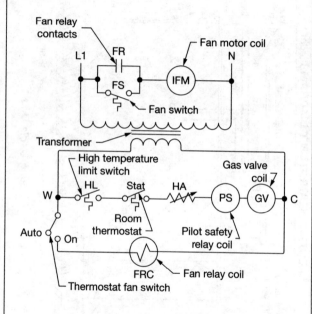

LIMIT THERMOSTATS

Pressure switch

Fan energizes electrical/ pneumatic switch

Normally open steam valve

Direct acting

Limit thermostat

Outside air

Return air

Room thermostat

M B

M

Electrical/ pneumatic switch

Actuator

2-49

PACKAGED UNIT CONTROL

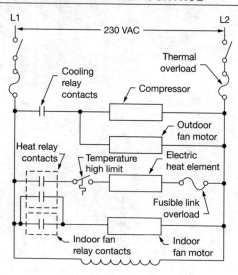

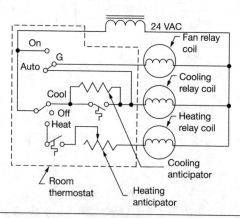

2-50

BASIC HEAT PUMP CONTROL

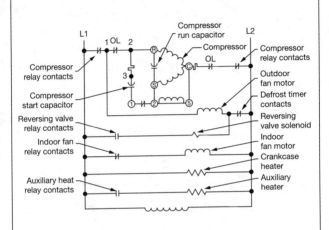

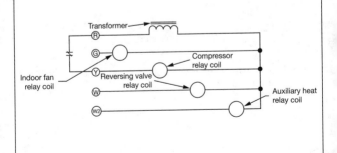

2-51

HOT WATER CONTROL

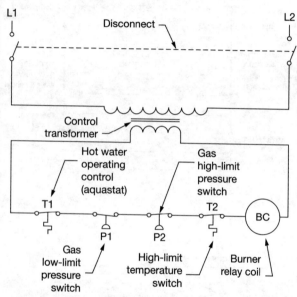

L1

L2

Disconnect

Control transformer

Hot water operating control (aquastat)

Gas high-limit pressure switch

T1

T2

BC

Gas low-limit pressure switch

P1

P2

High-limit temperature switch

Burner relay coil

Note: Other safety controls may be needed to comply with all necessary codes.

BOILER CLASSIFICATIONS

Name	Description
Automatic boiler	Equipped with controls and limit devices
Boiler	Closed vessel used for heating water or other liquid, or for generating steam or vapor by direct application of heat.
Boiler plant	One or more boilers, connecting piping, and vessels within the same premises.
Hot water supply boiler	Low-pressure hot water heating boiler having a volume exceeding 120 gallons or heat input exceeding 200,000 Btu per hour, or an operating temperature exceeding 200°F that provides hot water to be used externally to itself.
Low-pressure hot water heating boiler	Boiler in which water is heated for the purpose of supplying heat at pressures not exceeding 160 psi or temperatures not exceeding 250°F
Low-pressure steam heating boiler	Boiler operated at pressures not exceeding 15 psi for steam.
Power hot water boiler	Boiler used for heating water or liquid to pressure exceeding 160 psi or a temperature exceeding 250°F.
Power steam boiler	Boiler in which steam or vapor is generated at pressures exceeding 15 psi.
Small power boiler	Boiler with pressures exceeding 15 psi, but not exceeding 100 psi and having less than 440,000 Btu per hour input.

SIZING HYDRONIC (HOT-WATER) BOILERS

To size a hydronic (hot-water) boiler, first make a heat-loss calculation for the structure. Then refer to the figure in the I=B=R Net Rating column that most closely matches the calculated heat loss. If the figure is in between, always choose the higher figure.

SIZES OF HOT-WATER MAINS

Radiation (cu. ft.)	Pipe size (in.)
75 to 125	1¼
125 to 175	1½
175 to 300	2
300 to 475	2½
475 to 700	3
700 to 950	3½
950 to 1200	4
1200 to 1575	4½
1575 to 1975	5
1975 to 2375	5½
2375 to 2850	6

SIZING CIRCULATORS
(CENTRIFUGAL WATER PUMPS)

A method for sizing a circulator for a hydronic heating system requires the following:

- The estimated total heating load for the structure expressed in Btu per hour.
- The design temperature drop

$$\text{GPM} = \frac{\text{total heating load (in Btu/h)}}{T \times 60 \times 8.33}$$

GPM = gallons per minute
T = design temperature drop
60 = minutes per hour
8.33 = weight (in lbs.) of a gallon of water

FEET HEAD OF WATER TO PSI @ 62°F

Feet Head	psi	Feet Head	psi	Feet Head	psi
1	.43	40	17.32	200	86.62
2	.87	50	21.65	250	108.27
3	1.30	60	25.99	300	129.93
4	1.73	70	30.32	350	151.58
5	2.17	80	34.65	400	173.24
6	2.60	90	38.98	500	216.55
7	3.03	100	43.31	600	259.85
8	3.46	120	51.97	700	303.16
9	3.90	140	60.63	800	346.47
10	4.33	160	69.29	900	389.78
20	8.66	180	77.96	1000	433.00
30	12.99				

To find psi for feet head not given, multiply feet head by 0.433.

WEIGHT OF HOT WATER PER CUBIC FOOT (ft³) AT VARIOUS TEMPS.

Temp. (°F)	lb./ft³	Temp. (°F)	lb./ft³	Temp. (°F)	lb./ft³
110	61.86	310	57.08	510	48.31
120	61.71	320	56.75	520	47.62
130	61.55	330	56.40	530	46.95
140	61.37	340	56.02	540	46.30
150	61.10	350	55.65	550	45.66
160	60.99	360	55.25	560	44.84
170	60.79	370	54.85	570	44.05
180	60.57	380	54.47	580	43.29
190	60.35	390	54.05	590	42.37
200	60.11	400	53.62	600	41.49
210	59.87	410	53.19	610	40.49
220	59.67	420	52.74	620	39.37
230	59.42	430	52.33	630	38.31
240	59.17	440	51.87	640	37.17
250	58.89	450	51.28	650	35.97
260	58.62	460	51.02	660	34.48
270	58.34	470	50.51	670	32.89
280	58.04	480	50.00	680	31.06
290	57.74	490	49.50	690	28.82
300	57.41	500	48.78	700	25.38

HYDRONIC (HOT-WATER) SYSTEM DESIGN TEMPERATURES AND PRESSURES

Water Temperature (°F)	Vapor Pressure (psig)	System Operating Pressure Antiflash Margin			
		10°F	20°F	30°F	40°F
200	–3.2	–0.6	2.5	6	10
210	–0.6	2.5	6	10	15
212	0.0	3	7	11	16
215	0.9	4	8	13	18
220	2.5	6	10	15	21
225	4.2	8	13	18	24
230	6.1	10	15	21	27
240	10.3	15	21	27	34
250	15.1	21	27	34	43
260	20.7	27	34	43	52
270	27.2	34	43	52	63
275	30.7	39	47	58	69
280	34.5	43	52	63	75
290	42.8	52	63	75	88
300	52.3	63	75	88	103
310	62.9	75	88	103	120
320	74.9	88	103	120	138

Notes:

A. High-temperature hydronic systems, when operated at higher system temperatures and higher system pressures, will result in a lower chance of water hammer and the damaging effects of pipe leaks. These high-temperature systems are also safer than lower temperature heating water systems because system leaks subcool to temperatures below scalding due to the sudden decrease in pressure and the production of water vapor.

B. The antiflash margin of 40°F minimum is recommended for nitrogen or mechanically pressurized systems.

HYDRONIC (HOT-WATER) SYSTEM DESIGN TEMPERATURES AND PRESSURES (cont.)

Water Temperature (°F)	Vapor Pressure (psig)	System Operating Pressure Antiflash Margin		
		50°F	60°F	70°F
200	−3.2	15	21	27
210	−0.6	21	27	35
212	0.0	22	29	36
215	0.9	24	31	39
220	2.5	27	35	43
225	4.2	30	39	48
230	6.1	35	43	52
240	10.3	43	52	63
250	15.1	52	63	75
260	20.7	63	75	88
270	27.2	75	88	103
275	30.7	81	96	111
280	34.5	88	103	120
290	42.8	103	120	138
300	52.3	120	138	159
310	62.9	138	159	181
320	74.9	159	181	206

Notes:

A. High-temperature hydronic systems, when operated at higher system temperatures and higher system pressures, will result in a lower chance of water hammer and the damaging effects of pipe leaks. These high-temperature systems are also safer than lower temperature heating water systems because system leaks subcool to temperatures below scalding due to the sudden decrease in pressure and the production of water vapor.

B. The antiflash margin of 40°F minimum is recommended for nitrogen or mechanically pressurized systems.

HYDRONIC (HOT-WATER) SYSTEM DESIGN TEMPERATURES AND PRESSURES (cont.)

Water Temperature (°F)	Vapor Pressure (psig)	System Operating Pressure Antiflash Margin			
		10°F	20°F	30°F	40°F
325	81.4	96	111	129	148
330	88.3	103	120	138	159
340	103.2	120	138	159	181
350	119.8	138	159	181	206
360	138.2	159	181	206	232
370	158.5	181	206	232	262
375	169.5	193	219	247	277
380	180.9	206	232	262	294
390	205.5	232	262	294	329
400	232.4	262	294	329	367
410	261.8	294	329	367	407
420	293.8	329	367	407	452
425	310.9	347	387	429	475
430	328.6	367	407	452	500
440	366.5	407	452	500	551
450	407.4	452	500	551	606
455	429.1	475	525	578	635

Notes:

A. High-temperature hydronic systems, when operated at higher system temperatures and higher system pressures, will result in a lower chance of water hammer and the damaging effects of pipe leaks. These high-temperature systems are also safer than lower temperature heating water systems because system leaks subcool to temperatures below scalding due to the sudden decrease in pressure and the production of water vapor.

B. The antiflash margin of 40°F minimum is recommended for nitrogen or mechanically pressurized systems.

HYDRONIC (HOT-WATER) SYSTEM DESIGN TEMPERATURES AND PRESSURES (cont.)

Water Temperature (°F)	Vapor Pressure (psig)	System Operating Pressure Antiflash Margin		
		50°F	60°F	70°F
325	81.4	170	193	219
330	88.3	181	206	232
340	103.2	206	232	262
350	119.8	232	262	294
360	138.2	262	294	329
370	158.5	294	329	367
375	169.5	311	347	387
380	180.9	329	367	407
390	205.5	367	407	452
400	232.4	407	452	500
410	261.8	452	500	551
420	293.8	500	551	606
425	310.9	524	578	635
430	328.6	551	606	665
440	366.5	606	665	729
450	407.4	665	729	797
455	429.1	697	762	832

Notes:

A. High-temperature hydronic systems, when operated at higher system temperatures and higher system pressures, will result in a lower chance of water hammer and the damaging effects of pipe leaks. These high-temperature systems are also safer than lower temperature heating water systems because system leaks subcool to temperatures below scalding due to the sudden decrease in pressure and the production of water vapor.

B. The antiflash margin of 40°F minimum is recommended for nitrogen or mechanically pressurized systems.

EXPANSION TANK SIZING FOR LOW-TEMPERATURE SYSTEMS—EXPRESSED AS PERCENTAGE OF SYSTEM VOLUME

Maximum System Temperature (F°)	Expansion Tank Type	
	Closed Tank	Open Tank
100	2.21	1.37
110	3.08	1.87
120	3.71	2.24
130	4.81	2.87
140	5.67	3.37
150	6.77	3.99
160	7.87	4.61
170	9.20	5.36
180	10.53	6.11
190	11.87	6.86
200	13.20	7.61
210	14.77	—
220	16.34	—
230	17.90	—
240	19.71	—
250	21.51	—

Notes: Based on an initial temperature of 50°F, an initial pressure of 10 psig, and a maximum operating pressure of 30 psig. For initial and maximum pressures different from those listed, multiply the tank size only by the correction factors in this section.

EXPANSION TANK SIZING FOR LOW-TEMPERATURE SYSTEMS—EXPRESSED AS PERCENTAGE OF SYSTEM VOLUME *(cont.)*

Maximum System Temperature (F°)	Diaphragm Type Tank	
	Tank Volume	Acceptance Volume
100	1.32	0.59
110	1.83	0.82
120	2.21	0.99
130	2.86	1.28
140	3.37	1.51
150	4.03	1.80
160	4.68	2.10
170	5.48	2.45
180	6.27	2.81
190	7.06	3.16
200	7.06	3.52
210	8.79	3.93
220	9.72	4.35
230	10.66	4.77
240	11.73	5.25
250	12.80	5.73

Notes: Based on an initial temperature of 50°F, an initial pressure of 10 psig, and a maximum operating pressure of 30 psig. For initial and maximum pressures different from those listed, multiply the tank size only by the correction factors in this section.

CLOSED EXPANSION TANK SIZING CORRECTION FACTORS FOR LOW-TEMPERATURE SYSTEMS

Initial Pressure (psig)	Pressure Increase (psig) Initial Pressure + Pressure Increase = Maximum Operating Pressure				
	5	10	15	20	25
5	1.76	1.06	0.83	0.71	0.64
10	2.66	1.55	1.18	**1.00**	0.89
15	3.73	2.14	1.60	1.34	1.18
20	4.99	2.81	2.08	1.72	1.50
25	6.43	3.57	2.62	2.15	1.86
30	8.05	4.43	3.22	2.62	2.26
35	9.85	5.37	3.88	3.14	2.69
40	11.83	6.41	4.60	3.70	3.16
45	13.99	7.54	5.39	4.31	3.66
50	16.34	8.75	6.23	4.96	4.21
55	18.86	10.06	7.13	5.66	4.78
60	21.57	11.46	8.09	6.41	5.40
65	24.46	12.95	9.11	7.20	6.05
70	27.53	14.53	10.20	8.03	6.73
75	30.77	16.20	11.34	8.91	7.45
80	34.21	17.96	12.55	9.84	8.21
85	37.82	19.81	13.81	10.81	9.01
90	41.61	21.75	15.13	11.83	9.84
95	45.59	23.79	16.52	12.89	10.71
100	49.74	25.91	17.97	13.99	11.61

Notes: Based on an initial temperature of 50°F, an initial pressure of 10 psig, and a maximum operating pressure of 30 psig.

CLOSED EXPANSION TANK SIZING CORRECTION FACTORS FOR LOW-TEMPERATURE SYSTEMS *(cont.)*

Initial Pressure (psig)	Pressure Increase (psig) Initial Pressure + Pressure Increase = Maximum Operating Pressure				
	30	35	40	45	50
5	0.59	0.56	0.53	0.51	0.50
10	0.82	0.76	0.72	0.69	0.67
15	1.07	0.99	0.94	0.89	0.86
20	1.36	1.25	1.17	1.11	1.06
25	1.67	1.53	1.43	1.35	1.29
30	2.02	1.84	1.71	1.61	1.53
35	2.39	2.18	2.02	1.89	1.80
40	2.80	2.54	2.35	2.20	2.07
45	3.23	2.93	2.70	2.52	2.37
50	3.70	3.34	3.07	2.86	2.69
55	4.20	3.78	3.46	3.22	3.02
60	4.72	4.24	3.88	3.60	3.37
65	5.28	4.73	4.32	4.00	3.75
70	5.87	5.25	4.78	4.42	4.13
75	6.48	5.79	5.27	4.86	4.54
80	7.13	6.36	5.78	5.33	4.96
85	7.81	6.95	6.31	5.81	5.41
90	8.52	7.57	6.86	6.31	5.87
95	9.25	8.22	7.44	6.83	6.35
100	10.02	8.89	8.04	7.37	6.84

Notes: Based on an initial temperature of 50°F, an initial pressure of 10 psig, and a maximum operating pressure of 30 psig.

CLOSED EXPANSION TANK SIZING CORRECTION FACTORS FOR LOW-TEMPERATURE SYSTEMS *(cont.)*

Initial Pressure (psig)	Pressure Increase (psig) Initial Pressure + Pressure Increase = Maximum Operating Pressure				
	55	60	65	70	75
5	0.48	0.47	0.47	0.46	0.45
10	0.65	0.63	0.62	0.61	0.59
15	0.83	0.80	0.78	0.77	0.75
20	1.03	0.99	0.96	0.94	0.92
25	1.24	1.19	1.16	1.13	1.10
30	1.47	1.41	1.37	1.33	1.29
35	1.71	1.65	1.59	1.54	1.50
40	1.98	1.89	1.82	1.77	1.71
45	2.26	2.16	2.07	2.00	1.94
50	2.55	2.44	2.34	2.26	2.18
55	2.86	2.73	2.62	2.52	2.44
60	3.19	3.04	2.91	2.80	2.70
65	3.54	3.36	3.21	3.09	2.98
70	3.90	3.70	3.53	3.39	3.27
75	4.27	4.05	3.87	3.71	3.57
80	4.67	4.42	4.21	4.04	3.88
85	5.08	4.81	4.58	4.38	4.21
90	5.51	5.21	4.95	4.73	4.54
95	5.95	5.62	5.34	5.10	4.89
100	6.41	6.05	5.74	5.48	5.26

Notes: Based on an initial temperature of 50°F, an initial pressure of 10 psig, and a maximum operating pressure of 30 psig.

CLOSED EXPANSION TANK SIZING
CORRECTION FACTORS FOR
LOW-TEMPERATURE SYSTEMS *(cont.)*

Initial Pressure (psig)	Pressure Increase (psig) Initial Pressure + Pressure Increase = Maximum Operating Pressure				
	80	85	90	95	100
5	0.44	0.44	0.43	0.43	0.43
10	0.59	0.58	0.57	0.56	0.56
15	0.74	0.73	0.72	0.71	0.70
20	0.90	0.89	0.87	0.86	0.85
25	1.08	1.06	1.04	1.02	1.00
30	1.26	1.24	1.21	1.19	1.17
35	1.46	1.43	1.40	1.37	1.35
40	1.67	1.63	1.59	1.56	1.53
45	1.89	1.84	1.80	1.76	1.73
50	2.12	2.06	2.01	1.97	1.93
55	2.36	2.30	2.24	2.19	2.14
60	2.02	2.54	2.48	2.42	2.36
65	2.88	2.80	2.72	2.65	2.59
70	3.16	3.06	2.98	2.90	2.83
75	3.45	3.34	3.24	3.16	3.08
80	3.75	3.63	3.52	3.43	3.34
85	4.06	3.92	3.81	3.70	3.61
90	4.38	4.23	4.10	3.99	3.88
95	4.71	4.55	4.41	4.28	4.17
100	5.06	4.88	4.73	4.59	4.46

Notes: Based on an initial temperature of 50°F, an initial pressure of 10 psig, and a maximum operating pressure of 30 psig.

DIAPHRAGM EXPANSION TANK SIZING CORRECTION FACTORS FOR LOW-TEMPERATURE SYSTEMS

Initial Pressure (psig)	Pressure Increase (psig) Initial Pressure + Pressure Increase = Maximum Operating Pressure				
	5	10	15	20	25
5	2.21	1.33	1.04	0.89	0.80
10	2.66	1.55	1.18	1.00	0.89
15	3.11	1.78	1.33	1.11	0.98
20	3.55	2.00	1.48	1.22	1.07
25	4.00	2.22	1.63	1.34	1.16
30	4.45	2.45	1.78	1.45	1.25
35	4.89	2.67	1.93	1.56	1.34
40	5.34	2.89	2.08	1.67	1.43
45	5.79	3.12	2.23	1.78	1.52
50	6.24	3.34	2.38	1.89	1.61
55	6.68	3.57	2.53	2.01	1.69
60	7.13	3.79	2.68	2.12	1.78
65	7.58	4.01	2.82	2.23	1.87
70	8.03	4.24	2.97	2.34	1.96
75	8.47	4.46	3.12	2.45	2.05
80	8.92	4.68	3.27	2.57	2.14
85	9.37	4.91	3.42	2.68	2.23
90	9.82	5.13	3.57	2.79	2.32
95	10.26	5.36	3.72	2.90	2.41
100	10.71	5.58	3.87	3.01	2.50

Notes: Based on an initial temperature of 50°F, an initial pressure of 10 psig, and a maximum operating pressure of 30 psig.

DIAPHRAGM EXPANSION TANK SIZING CORRECTION FACTORS FOR LOW-TEMPERATURE SYSTEMS (cont.)

Initial Pressure (psig)	Pressure Increase (psig) Initial Pressure + Pressure Increase = Maximum Operating Pressure				
	30	35	40	45	50
5	0.74	0.70	0.67	0.64	0.62
10	0.82	0.76	0.72	0.69	0.67
15	0.89	0.83	0.78	0.74	0.71
20	0.96	0.89	0.84	0.79	0.76
25	1.04	0.95	0.89	0.84	0.80
30	1.11	1.02	0.95	0.89	0.85
35	1.19	1.08	1.00	0.94	0.89
40	1.26	1.15	1.06	0.99	0.94
45	1.34	1.21	1.12	1.04	0.98
50	1.41	1.27	1.17	1.09	1.03
55	1.49	1.34	1.23	1.14	1.07
60	1.56	1.40	1.28	1.19	1.12
65	1.64	1.47	1.34	1.24	1.16
70	1.71	1.53	1.39	1.29	1.21
75	1.79	1.59	1.45	1.34	1.25
80	1.86	1.66	1.51	1.39	1.29
85	1.93	1.72	1.56	1.44	1.34
90	2.01	1.79	1.62	1.49	1.38
95	2.08	1.85	1.67	1.54	1.43
100	2.16	1.91	1.73	1.59	1.47

Notes: Based on an initial temperature of 50°F, an initial pressure of 10 psig, and a maximum operating pressure of 30 psig.

DIAPHRAGM EXPANSION TANK SIZING CORRECTION FACTORS FOR LOW-TEMPERATURE SYSTEMS *(cont.)*

Initial Pressure (psig)	Pressure Increase (psig) Initial Pressure + Pressure Increase = Maximum Operating Pressure				
	55	60	65	70	75
5	0.61	0.59	0.58	0.57	0.56
10	0.65	0.63	0.62	0.61	0.59
15	0.69	0.67	0.65	0.64	0.62
20	0.73	0.71	0.69	0.67	0.65
25	0.77	0.74	0.72	0.70	0.68
30	0.81	0.78	0.76	0.73	0.71
35	0.85	0.82	0.79	0.77	0.74
40	0.89	0.86	0.82	0.80	0.77
45	0.93	0.89	0.86	0.83	0.80
50	0.97	0.93	0.89	0.86	0.83
55	1.01	0.97	0.93	0.89	0.86
60	1.06	1.00	0.96	0.92	0.89
65	1.10	1.04	1.00	0.96	0.92
70	1.14	1.08	1.03	0.99	0.95
75	1.18	1.12	1.06	1.02	0.98
80	1.22	1.15	1.10	1.05	1.01
85	1.26	1.19	1.13	1.08	1.04
90	1.30	1.23	1.17	1.12	1.07
95	1.34	1.27	1.20	1.15	1.10
100	1.38	1.30	1.24	1.18	1.13

Notes: Based on an initial temperature of 50°F, an initial pressure of 10 psig, and a maximum operating pressure of 30 psig.

DIAPHRAGM EXPANSION TANK SIZING CORRECTION FACTORS FOR LOW-TEMPERATURE SYSTEMS *(cont.)*

Initial Pressure (psig)	Pressure Increase (psig) Initial Pressure + Pressure Increase = Maximum Operating Pressure				
	80	85	90	95	100
5	0.56	0.55	0.55	0.54	0.54
10	0.59	0.58	0.57	0.56	0.56
15	0.61	0.60	0.60	0.59	0.58
20	0.64	0.63	0.62	0.61	0.60
25	0.67	0.66	0.64	0.63	0.63
30	0.70	0.68	0.67	0.66	0.65
35	0.73	0.71	0.69	0.68	0.67
40	0.75	0.74	0.72	0.71	0.69
45	0.78	0.76	0.74	0.73	0.71
50	0.81	0.79	0.77	0.75	0.74
55	0.84	0.81	0.79	0.78	0.76
60	0.07	0.84	0.82	0.80	0.78
65	0.89	0.87	0.84	0.82	0.80
70	0.92	0.89	0.87	0.85	0.83
75	0.95	0.92	0.89	0.87	0.85
80	0.98	0.95	0.92	0.89	0.87
85	1.01	0.97	0.94	0.92	0.89
90	1.03	1.00	0.97	0.94	0.92
95	1.06	1.02	0.99	0.96	0.94
100	1.09	1.05	1.02	0.99	0.96

Notes: Based on an initial temperature of 50°F, an initial pressure of 10 psig, and a maximum operating pressure of 30 psig.

EXPANSION TANK SIZING FOR MEDIUM-TEMPERATURE SYSTEMS—EXPRESSED AS PERCENTAGE OF SYSTEM VOLUME

Maximum System Temperature (F°)	Expansion Tank Type	
	Closed Tank	Open Tank
250	263.25	–
260	285.30	–
270	310.23	–
280	335.16	–
290	360.08	–
300	387.88	–
310	415.67	–
320	443.47	–
330	474.13	–
340	504.80	–
350	538.33	–

Notes: Based on an initial temperature of 50°F, an initial pressure of 200 psig, and a maximum operating pressure of 300 psig. For initial and maximum pressures different from those listed, multiply the tank size only by the correction factors in this section.

EXPANSION TANK SIZING FOR MEDIUM-TEMPERATURE SYSTEMS—EXPRESSED AS PERCENTAGE OF SYSTEM VOLUME *(cont.)*

Maximum System Temperature (F°)	Diaphragm Type Tank	
	Tank Volume	Acceptance Volume
250	18.02	5.73
260	19.53	6.21
270	21.24	6.75
280	22.95	7.29
290	24.65	7.83
300	26.56	8.44
310	28.46	9.04
320	30.36	9.65
330	32.46	10.32
340	34.56	10.98
350	36.86	11.71

Notes: Based on an initial temperature of 50°F, an initial pressure of 200 psig, and a maximum operating pressure of 300 psig. For initial and maximum pressures different from those listed, multiply the tank size only by the correction factors in this section.

CLOSED EXPANSION TANK SIZING CORRECTION FACTORS FOR MEDIUM-TEMPERATURE SYSTEMS

Initial Pressure (psig)	Pressure Increase (psig) Initial Pressure + Pressure Increase = Maximum Operating Pressure				
	10	20	30	40	50
30	0.36	0.21	0.16	0.14	0.13
40	0.52	0.30	0.23	0.19	0.17
50	0.72	0.41	0.30	0.25	0.22
60	0.94	0.52	0.39	0.32	0.28
70	1.19	0.66	0.48	0.39	0.34
80	1.47	0.80	0.58	0.47	0.41
90	1.78	0.97	0.70	0.56	0.48
100	2.12	1.14	0.82	0.66	0.56
110	2.49	1.34	0.95	0.76	0.64
120	2.88	1.54	1.09	0.87	0.74
130	3.31	1.76	1.25	0.99	0.83
140	3.77	2.00	1.41	1.11	0.94
150	4.26	2.25	1.58	1.25	1.05
160	4.78	2.52	1.76	1.39	1.16
170	5.32	2.80	1.96	1.54	1.28
180	5.90	3.09	2.16	1.69	1.41
190	6.50	3.40	2.37	1.85	1.54
200	7.14	3.73	2.59	2.02	1.68
210	7.81	4.07	2.82	2.20	1.83
220	8.50	4.42	3.06	2.39	1.98
230	9.22	4.79	3.32	2.58	2.13
240	9.98	5.18	3.58	2.78	2.30
250	10.76	5.58	3.85	2.98	2.47
260	11.57	5.99	4.13	3.20	2.64

Notes: Based on an initial temperature of 50°F, an initial pressure of 200 psig, and a maximum operating pressure of 300 psig.

CLOSED EXPANSION TANK SIZING CORRECTION FACTORS FOR MEDIUM-TEMPERATURE SYSTEMS *(cont.)*

Initial Pressure (psig)	Pressure Increase (psig) Initial Pressure + Pressure Increase = Maximum Operating Pressure				
	60	70	80	90	100
30	0.12	0.11	0.10	0.10	0.10
40	0.15	0.14	0.14	0.13	0.13
50	0.20	0.18	0.17	0.16	0.16
60	0.25	0.23	0.21	0.20	0.19
70	0.30	0.28	0.26	0.24	0.23
80	0.36	0.33	0.31	0.29	0.27
90	0.43	0.39	0.36	0.34	0.32
100	0.49	0.45	0.41	0.39	0.36
110	0.57	0.51	0.47	0.44	0.41
120	0.65	0.58	0.54	0.50	0.47
130	0.73	0.66	0.60	0.56	0.52
140	0.82	0.73	0.67	0.62	0.58
150	0.91	0.82	0.75	0.69	0.65
160	1.01	0.90	0.82	0.76	0.71
170	1.11	0.99	0.90	0.83	0.78
180	1.22	1.09	0.99	0.91	0.85
190	1.34	1.19	1.08	0.99	.92
200	1.45	1.29	1.17	1.08	**1.00**
210	1.58	1.40	1.27	1.16	1.08
220	1.71	1.51	1.37	1.25	1.16
230	1.84	1.63	1.47	1.35	1.25
240	1.98	1.75	1.58	1.44	1.34
250	2.12	1.87	1.69	1.54	1.43
260	2.27	2.00	1.80	1.65	1.52

Notes: Based on an initial temperature of 50°F, an initial pressure of 200 psig, and a maximum operating pressure of 300 psig.

CLOSED EXPANSION TANK SIZING CORRECTION FACTORS FOR MEDIUM-TEMPERATURE SYSTEMS *(cont.)*

Initial Pressure (psig)	Pressure Increase (psig) Initial Pressure + Pressure Increase = Maximum Operating Pressure				
	110	120	130	140	150
30	0.09	0.09	0.09	0.09	0.09
40	0.12	0.12	0.12	0.11	0.11
50	0.15	0.15	0.14	0.14	0.14
60	0.19	0.18	0.17	0.17	0.17
70	0.22	0.21	0.21	0.20	0.20
80	0.26	0.25	0.24	0.23	0.23
90	0.30	0.29	0.28	0.27	0.26
100	0.35	0.33	0.32	0.31	0.30
110	0.39	0.38	0.36	0.35	0.34
120	0.44	0.42	0.41	0.39	0.38
130	0.50	0.47	0.45	0.44	0.42
140	0.55	0.52	0.50	0.48	0.47
150	0.61	0.58	0.55	0.53	0.51
160	0.67	0.63	0.61	0.58	0.56
170	0.73	0.69	0.66	0.63	0.61
180	0.80	0.76	0.72	0.69	0.66
190	0.87	0.82	0.78	0.75	0.72
200	0.94	0.89	0.84	0.81	0.77
210	1.01	0.96	0.91	0.87	0.83
220	1.09	1.03	0.97	0.93	0.89
230	1.17	1.10	1.04	1.00	0.95
240	1.25	1.18	1.12	1.06	1.02
250	1.33	1.26	1.19	1.13	1.08
260	1.42	1.34	1.27	1.20	1.15

Notes: Based on an initial temperature of 50°F, an initial pressure of 200 psig, and a maximum operating pressure of 300 psig.

CLOSED EXPANSION TANK SIZING
CORRECTION FACTORS FOR
MEDIUM-TEMPERATURE SYSTEMS *(cont.)*

Initial Pressure (psig)	Pressure Increase (psig) Initial Pressure + Pressure Increase = Maximum Operating Pressure				
	160	170	180	190	200
30	0.08	0.08	0.08	0.08	0.08
40	0.11	0.11	0.11	0.10	0.10
50	0.13	0.13	0.13	0.13	0.13
60	0.16	0.16	0.16	0.15	0.15
70	0.19	0.19	0.18	0.18	0.18
80	0.22	0.22	0.21	0.21	0.21
90	0.26	0.25	0.25	0.24	0.24
100	0.29	0.28	0.28	0.27	0.27
110	0.33	0.32	0.31	0.31	0.30
120	0.37	0.36	0.35	0.34	0.33
130	0.41	0.40	0.39	0.38	0.37
140	0.45	0.44	0.43	0.42	0.41
150	0.49	0.48	0.47	0.46	0.44
160	0.54	0.52	0.51	0.50	0.48
170	0.59	0.57	0.55	0.54	0.53
180	0.64	0.62	0.60	0.58	0.57
190	0.69	0.67	0.65	0.63	0.61
200	0.74	0.72	0.70	0.68	0.66
210	0.80	0.77	0.75	0.73	0.71
220	0.86	0.83	0.80	0.78	0.75
230	0.92	0.88	0.85	0.83	0.81
240	0.98	0.94	0.91	0.88	0.86
250	1.04	1.00	0.97	0.94	0.91
260	1.10	1.06	1.03	0.99	0.96

Notes: Based on an initial temperature of 50°F, an initial pressure of 200 psig, and a maximum operating pressure of 300 psig.

DIAPHRAGM EXPANSION TANK SIZING CORRECTION FACTORS FOR MEDIUM-TEMPERATURE SYSTEMS

Initial Pressure (psig)	Pressure Increase (psig) Initial Pressure + Pressure Increase = Maximum Operating Pressure				
	10	20	30	40	50
30	1.74	1.03	0.79	0.67	0.60
40	2.06	1.19	0.90	0.75	0.67
50	2.37	1.35	1.00	0.83	0.73
60	2.69	1.50	1.11	0.91	0.79
70	3.01	1.66	1.21	0.99	0.86
80	3.33	1.82	1.32	1.07	0.92
90	3.64	1.98	1.43	1.15	0.98
100	3.96	2.14	1.53	1.23	1.05
110	4.28	2.30	1.64	1.31	1.11
120	4.60	2.46	1.74	1.39	1.17
130	4.92	2.62	1.85	1.47	1.24
140	5.23	2.78	1.96	1.55	1.30
150	5.55	2.93	2.06	1.63	1.36
160	5.87	3.09	2.17	1.71	1.43
170	6.19	3.25	2.27	1.79	1.49
180	6.50	3.41	2.38	1.86	1.56
190	6.82	3.57	2.49	1.94	1.62
200	7.14	3.73	2.59	2.02	1.68
210	7.46	3.89	2.70	2.10	1.75
220	7.78	4.05	2.80	2.18	1.81
230	8.09	4.21	2.91	2.26	1.87
240	8.41	4.36	3.02	2.34	1.94
250	8.73	4.52	3.12	2.42	2.00
260	9.05	4.68	3.23	2.50	2.06

Notes: Based on an initial temperature of 50°F, an initial pressure of 200 psig, and a maximum operating pressure of 300 psig.

DIAPHRAGM EXPANSION TANK SIZING CORRECTION FACTORS FOR MEDIUM-TEMPERATURE SYSTEMS *(cont.)*

Initial Pressure (psig)	Pressure Increase (psig) Initial Pressure + Pressure Increase = Maximum Operating Pressure				
	60	70	80	90	100
30	0.55	0.52	0.50	0.48	0.46
40	0.61	0.57	0.54	0.51	0.49
50	0.66	0.61	0.57	0.55	0.52
60	0.71	0.66	0.61	0.58	0.56
70	0.77	0.70	0.65	0.62	0.59
80	0.82	0.75	0.69	0.65	0.62
90	0.87	0.79	0.73	0.69	0.65
100	0.93	0.84	0.77	0.72	0.68
110	0.98	0.88	0.81	0.76	0.71
120	1.03	0.93	0.85	0.79	0.75
130	1.08	0.97	0.89	0.83	0.78
140	1.14	1.02	0.93	0.86	0.81
150	1.19	1.07	0.97	0.90	0.84
160	1.24	1.11	1.01	0.93	0.87
170	1.30	1.16	1.05	0.97	0.90
180	1.35	1.20	1.09	1.01	0.94
190	1.40	1.25	1.13	1.04	0.97
200	1.45	1.29	1.17	1.08	**1.00**
210	1.51	1.34	1.21	1.11	1.03
220	1.56	1.38	1.25	1.15	1.06
230	1.61	1.43	1.29	1.18	1.10
240	1.67	1.47	1.33	1.22	1.13
250	1.72	1.52	1.37	1.25	1.16
260	1.77	1.56	1.41	1.29	1.19

Notes: Based on an initial temperature of 50°F, an initial pressure of 200 psig, and a maximum operating pressure of 300 psig.

DIAPHRAGM EXPANSION TANK SIZING CORRECTION FACTORS FOR MEDIUM-TEMPERATURE SYSTEMS *(cont.)*

Initial Pressure (psig)	Pressure Increase (psig) Initial Pressure + Pressure Increase = Maximum Operating Pressure				
	110	120	130	140	150
30	0.45	0.44	0.43	0.42	0.41
40	0.48	0.46	0.45	0.44	0.43
50	0.50	0.49	0.48	0.46	0.45
60	0.53	0.52	0.50	0.49	0.48
70	0.56	0.54	0.52	0.51	0.50
80	0.59	0.57	0.55	0.53	0.52
90	0.62	0.60	0.57	0.56	0.54
100	0.65	0.62	0.60	0.58	0.56
110	0.68	0.65	0.62	0.60	0.58
120	0.71	0.67	0.65	0.62	0.60
130	0.74	0.70	0.67	0.65	0.62
140	0.76	0.73	0.70	0.67	0.65
150	0.79	0.75	0.72	0.69	0.67
160	0.82	0.78	0.74	0.71	0.69
170	0.85	0.81	0.77	0.74	0.71
180	0.88	0.83	0.79	0.76	0.73
190	0.91	0.86	0.82	0.78	0.75
200	0.94	0.89	0.84	0.81	0.77
210	0.97	0.91	0.87	0.83	0.79
220	1.00	0.94	0.89	0.85	0.81
230	1.02	0.97	0.92	0.87	0.84
240	1.05	0.99	0.94	0.90	0.86
250	1.08	1.02	0.96	0.92	0.88
260	1.11	1.05	0.99	0.94	0.90

Notes: Based on an initial temperature of 50°F, an initial pressure of 200 psig, and a maximum operating pressure of 300 psig.

DIAPHRAGM EXPANSION TANK SIZING CORRECTION FACTORS FOR MEDIUM-TEMPERATURE SYSTEMS (cont.)

Initial Pressure (psig)	Pressure Increase (psig) Initial Pressure + Pressure Increase = Maximum Operating Pressure				
	160	170	180	190	200
30	0.41	0.40	0.40	0.39	0.39
40	0.43	0.42	0.41	0.41	0.40
50	0.45	0.44	0.43	0.43	0.42
60	0.47	0.46	0.45	0.44	0.44
70	0.49	0.48	0.47	0.46	0.45
80	0.51	0.49	0.48	0.48	0.47
90	0.53	0.51	0.50	0.49	0.48
100	0.55	0.53	0.52	0.51	0.50
110	0.57	0.55	0.54	0.53	0.52
120	0.59	0.57	0.56	0.54	0.53
130	0.61	0.59	0.57	0.56	0.55
140	0.63	0.61	0.59	0.58	0.56
150	0.64	0.63	0.61	0.59	0.58
160	0.66	0.64	0.63	0.61	0.60
170	0.68	0.66	0.64	0.63	0.61
180	0.70	0.68	0.66	0.64	0.63
190	0.72	0.70	0.68	0.66	0.64
200	0.74	0.72	0.70	0.68	0.66
210	0.76	0.74	0.71	0.69	0.67
220	0.78	0.76	0.73	0.71	0.69
230	0.80	0.78	0.75	0.73	0.71
240	0.82	0.79	0.77	0.74	0.72
250	0.84	0.81	0.79	0.76	0.74
260	0.86	0.83	0.80	0.78	0.75

Notes: Based on an initial temperature of 50°F, an initial pressure of 200 psig, and a maximum operating pressure of 300 psig.

EXPANSION TANK SIZING FOR HIGH-TEMPERATURE SYSTEMS—EXPRESSED AS PERCENTAGE OF SYSTEM VOLUME

Maximum System Temperature (°F)	Expansion Tank Type	
	Closed Tank	Open Tank
350	1,995.03	—
360	2,119.30	—
370	2,243.58	—
380	2,378.48	—
390	2,524.02	—
400	2,669.56	—
410	2,815.10	—
420	2,981.90	—
430	3,138.07	—
440	3,315.51	—
450	3,492.95	—

Notes: Based on an initial temperature of 50°F, an initial pressure of 600 psig, and a maximum operating pressure of 800 psig. For initial and maximum pressures different from those listed, multiply the tank size only by correction factors on the following pages.

EXPANSION TANK SIZING FOR HIGH-TEMPERATURE SYSTEMS—EXPRESSED AS PERCENTAGE OF SYSTEM VOLUME (cont.)

Maximum System Temperature (°F)	Diaphragm Type Tank	
	Tank Volume	Acceptance Volume
350	47.71	11.71
360	50.68	12.44
370	53.65	13.17
380	56.88	13.96
390	60.36	14.82
400	63.84	15.67
410	67.32	16.53
420	71.31	17.51
430	75.04	18.42
440	79.29	19.46
450	83.53	20.51

Notes: Based on an initial temperature of 50°F, an initial pressure of 600 psig, and a maximum operating pressure of 800 psig. For initial and maximum pressures different from those listed, multiply the tank size only by correction factors on the following pages.

CLOSED EXPANSION TANK
SIZING CORRECTION FACTORS FOR
HIGH-TEMPERATURE SYSTEMS

Initial Pressure (psig)	Pressure Increase (psig) Initial Pressure + Pressure Increase = Maximum Operating Pressure				
	20	40	60	80	100
160	0.68	0.37	0.27	0.22	0.19
180	0.83	0.46	0.33	0.27	0.23
200	1.01	0.55	0.39	0.32	0.27
220	1.19	0.64	0.46	0.37	0.31
240	1.40	0.75	0.53	0.43	0.36
260	1.62	0.86	0.61	0.49	0.41
280	1.85	0.98	0.70	0.55	0.46
300	2.10	1.11	0.78	0.62	0.52
320	2.37	1.25	0.88	0.69	0.58
340	2.65	1.40	0.98	0.77	0.64
360	2.95	1.55	1.08	0.85	0.71
380	3.27	1.71	1.19	0.94	0.78
400	3.60	1.88	1.31	1.02	0.85
420	3.95	2.06	1.43	1.12	0.93
440	4.31	2.25	1.56	1.21	1.01
460	4.69	2.44	1.69	1.31	1.09
480	5.08	2.64	1.83	1.42	1.17
500	5.50	2.85	1.97	1.53	1.26
520	5.92	3.07	2.12	1.64	1.36
540	6.37	3.29	2.27	1.76	1.45
560	6.82	3.53	2.43	1.88	1.55
580	7.30	3.77	2.59	2.00	1.65
600	7.79	4.02	2.76	2.13	1.75
620	8.30	4.28	2.93	2.26	1.86
640	8.82	4.54	3.11	2.40	1.97
660	9.36	4.81	3.30	2.54	2.09
680	9.91	5.10	3.49	2.69	2.20
700	10.49	5.39	3.69	2.84	2.33

Notes: Based on an initial temperature of 50°F, an initial pressure of 600 psig, and a maximum operating pressure of 800 psig.

CLOSED EXPANSION TANK
SIZING CORRECTION FACTORS FOR
HIGH-TEMPERATURE SYSTEMS *(cont.)*

Initial Pressure (psig)	Pressure Increase (psig) Initial Pressure + Pressure Increase = Maximum Operating Pressure				
	120	140	160	180	200
160	0.17	0.16	0.15	0.14	0.13
180	0.20	0.19	0.17	0.16	0.15
200	0.24	0.22	0.20	0.19	0.18
220	0.28	0.25	0.23	0.22	0.20
240	0.32	0.29	0.26	0.25	0.23
260	0.36	0.32	0.30	0.28	0.26
280	0.41	0.37	0.33	0.31	0.29
300	0.46	0.41	0.37	0.35	0.32
320	0.51	0.45	0.41	0.38	0.36
340	0.56	0.50	0.46	0.42	0.39
360	0.62	0.55	0.50	0.46	0.43
380	0.68	0.60	0.55	0.50	0.47
400	0.74	0.66	0.59	0.55	0.51
420	0.80	0.71	0.65	0.59	0.55
440	0.87	0.77	0.70	0.64	0.59
460	0.94	0.03	0.76	0.69	0.64
480	1.01	0.90	0.81	0.74	0.69
500	1.09	0.96	0.87	0.79	0.73
520	1.17	1.03	0.93	0.85	0.78
540	1.25	1.10	0.99	0.90	0.84
560	1.33	1.17	1.05	0.96	0.89
580	1.41	1.25	1.12	1.02	0.94
600	1.50	1.32	1.19	1.08	**1.00**
620	1.59	1.40	1.26	1.15	1.06
640	1.69	1.48	1.33	1.21	1.12
660	1.78	1.57	1.41	1.28	1.18
680	1.88	1.65	1.48	1.35	1.24
700	1.99	1.74	1.56	1.42	1.31

Notes: Based on an initial temperature of 50°F, an initial pressure of 600 psig, and a maximum operating pressure of 800 psig.

CLOSED EXPANSION TANK SIZING CORRECTION FACTORS FOR HIGH-TEMPERATURE SYSTEMS *(cont.)*

Initial Pressure (psig)	Pressure Increase (psig) Initial Pressure + Pressure Increase = Maximum Operating Pressure				
	220	240	260	280	300
160	0.13	0.12	0.12	0.11	0.11
180	0.15	0.14	0.14	0.13	0.13
200	0.17	0.16	0.16	0.15	0.15
220	0.19	0.19	0.18	0.17	0.17
240	0.22	0.21	0.20	0.19	0.19
260	0.25	0.24	0.23	0.22	0.21
280	0.28	0.26	0.25	0.24	0.23
300	0.31	0.29	0.28	0.27	0.26
320	0.34	0.32	0.31	0.29	0.28
340	0.37	0.35	0.33	0.32	0.31
360	0.40	0.38	0.37	0.35	0.34
380	0.44	0.42	0.40	0.38	0.37
400	0.48	0.45	0.43	0.41	0.39
420	0.52	0.49	0.46	0.44	0.43
440	0.56	0.53	0.50	0.48	0.46
460	0.60	0.56	0.54	0.51	0.49
480	0.64	0.60	0.57	0.55	0.52
500	0.69	0.65	0.61	0.58	0.56
520	0.73	0.69	0.65	0.62	0.59
540	0.78	0.73	0.69	0.66	0.63
560	0.83	0.78	0.74	0.70	0.67
580	0.88	0.83	0.78	0.74	0.71
600	0.93	0.87	0.83	0.78	0.75
620	0.98	0.92	0.87	0.83	0.79
640	1.04	0.97	0.92	0.87	0.83
660	1.10	1.03	0.97	0.92	0.88
680	1.15	1.08	1.02	0.97	0.92
700	1.21	1.14	1.07	1.01	0.87

Notes: Based on an initial temperature of 50°F, an initial pressure of 600 psig, and a maximum operating pressure of 800 psig.

CLOSED EXPANSION TANK
SIZING CORRECTION FACTORS FOR
HIGH-TEMPERATURE SYSTEMS *(cont.)*

Initial Pressure (psig)	Pressure Increase (psig) Initial Pressure + Pressure Increase = Maximum Operating Pressure				
	320	340	360	380	400
160	0.11	0.11	0.10	0.10	0.10
180	0.13	0.12	0.12	0.12	0.12
200	0.14	0.14	0.14	0.13	0.13
220	0.16	0.16	0.15	0.15	0.15
240	0.18	0.18	0.17	0.17	0.17
260	0.20	0.20	0.19	0.19	0.19
280	0.23	0.22	0.21	0.21	0.20
300	0.25	0.24	0.24	0.23	0.22
320	0.27	0.27	0.26	0.25	0.25
340	0.30	0.29	0.28	0.27	0.27
360	0.32	0.31	0.31	0.30	0.29
380	0.35	0.34	0.33	0.32	0.31
400	0.38	0.37	0.36	0.35	0.34
420	0.41	0.40	0.38	0.37	0.36
440	0.44	0.42	0.41	0.40	0.39
460	0.47	0.45	0.44	0.43	0.41
480	0.50	0.49	0.47	0.45	0.44
500	0.54	0.52	0.50	0.48	0.47
520	0.57	0.55	0.53	0.51	0.50
540	0.61	0.58	0.56	0.54	0.53
560	0.64	0.62	0.60	0.58	0.56
580	0.68	0.65	0.63	0.61	0.59
600	0.72	0.69	0.66	0.64	0.62
620	0.76	0.73	0.70	0.68	0.66
640	0.80	0.76	0.74	0.71	0.69
660	0.84	0.80	0.77	0.75	0.72
680	0.88	0.84	0.81	0.78	0.76
700	0.92	0.89	0.85	0.82	0.80

Notes: Based on an initial temperature of 50°F, an initial pressure of 600 psig, and a maximum operating pressure of 800 psig.

DIAPHRAGM EXPANSION TANK SIZING CORRECTION FACTORS FOR HIGH-TEMPERATURE SYSTEMS

Initial Pressure (psig)	Pressure Increase (psig) Initial Pressure + Pressure Increase = Maximum Operating Pressure				
	20	40	60	80	100
160	2.39	1.32	0.96	0.78	0.67
180	2.64	1.44	1.04	0.84	0.72
200	2.88	1.56	1.12	0.90	0.77
220	3.13	1.69	1.21	0.97	0.82
240	3.37	1.81	1.29	1.03	0.87
260	3.62	1.93	1.37	1.09	0.92
280	3.86	2.05	1.45	1.15	0.97
300	4.11	2.18	1.53	1.21	1.02
320	4.35	2.30	1.61	1.27	1.07
340	4.60	2.42	1.70	1.33	1.12
360	4.84	2.55	1.78	1.40	1.17
380	5.09	2.67	1.86	1.46	1.21
400	5.34	2.79	1.94	1.52	1.26
420	5.58	2.91	2.02	1.58	1.31
440	5.83	3.04	2.11	1.64	1.36
460	6.07	3.16	2.19	1.70	1.41
480	6.32	3.28	2.27	1.76	1.46
500	6.56	3.40	2.35	1.82	1.51
520	6.81	3.53	2.43	1.89	1.56
540	7.05	3.65	2.52	1.95	1.61
560	7.30	3.77	2.60	2.01	1.66
580	7.55	3.90	2.68	2.07	1.71
600	7.79	4.02	2.76	2.13	1.75
620	8.04	4.14	2.84	2.19	1.80
640	8.28	4.26	2.92	2.25	1.85
660	8.53	4.39	3.01	2.32	1.90
680	8.77	4.51	3.09	2.38	1.95
700	9.02	4.63	3.17	2.44	2.00

Notes: Based on an initial temperature of 50°F, an initial pressure of 600 psig, and a maximum operating pressure of 800 psig.

DIAPHRAGM EXPANSION TANK SIZING CORRECTION FACTORS FOR HIGH-TEMPERATURE SYSTEMS *(cont.)*

Initial Pressure (psig)	Pressure Increase (psig) Initial Pressure + Pressure Increase = Maximum Operating Pressure				
	120	140	160	180	200
160	0.60	0.55	0.51	0.48	0.46
180	0.64	0.59	0.54	0.51	0.48
200	0.68	0.62	0.57	0.54	0.51
220	0.73	0.66	0.61	0.57	0.53
240	0.77	0.69	0.64	0.59	0.56
260	0.81	0.73	0.67	0.62	0.58
280	0.85	0.76	0.70	0.65	0.61
300	0.89	0.80	0.73	0.67	0.63
320	0.93	0.83	0.76	0.70	0.66
340	0.97	0.87	0.79	0.73	0.68
360	1.01	0.90	0.82	0.76	0.71
380	1.05	0.94	0.85	0.78	0.73
400	1.09	0.97	0.88	0.81	0.75
420	1.13	1.01	0.91	0.84	0.78
440	1.18	1.04	0.94	0.87	0.80
460	1.22	1.08	0.97	0.89	0.83
480	1.26	1.11	1.00	0.92	0.85
500	1.30	1.15	1.04	0.95	0.88
520	1.34	1.18	1.07	0.97	0.90
540	1.38	1.22	1.10	1.00	0.93
560	1.42	1.25	1.13	1.03	0.95
580	1.46	1.29	1.16	1.06	0.98
600	1.50	1.32	1.19	1.08	**1.00**
620	1.54	1.36	1.22	1.11	1.02
640	1.58	1.39	1.25	1.14	1.05
660	1.63	1.43	1.28	1.17	1.07
680	1.67	1.46	1.31	1.19	1.10
700	1.71	1.50	1.34	1.22	1.12

Notes: Based on an initial temperature of 50°F, an initial pressure of 600 psig, and a maximum operating pressure of 800 psig.

DIAPHRAGM EXPANSION TANK SIZING CORRECTION FACTORS FOR HIGH-TEMPERATURE SYSTEMS (cont.)

Initial Pressure (psig)	Pressure Increase (psig) Initial Pressure + Pressure Increase = Maximum Operating Pressure				
	220	240	260	280	300
160	0.44	0.42	0.41	0.40	0.39
180	0.46	0.44	0.43	0.42	0.40
200	0.49	0.47	0.45	0.43	0.42
220	0.51	0.49	0.47	0.45	0.44
240	0.53	0.51	0.49	0.47	0.45
260	0.55	0.53	0.50	0.49	0.47
280	0.57	0.55	0.52	0.50	0.49
300	0.60	0.57	0.54	0.52	0.50
320	0.62	0.59	0.56	0.54	0.52
340	0.64	0.61	0.58	0.56	0.54
360	0.66	0.63	0.60	0.57	0.55
380	0.69	0.65	0.62	0.59	0.57
400	0.71	0.67	0.64	0.61	0.58
420	0.73	0.69	0.66	0.63	0.60
440	0.75	0.71	0.67	0.64	0.62
460	0.78	0.73	0.69	0.66	0.63
480	0.80	0.75	0.71	0.68	0.65
500	0.82	0.77	0.73	0.70	0.67
520	0.84	0.79	0.75	0.71	0.68
540	0.86	0.81	0.77	0.73	0.70
560	0.89	0.83	0.79	0.75	0.72
580	0.91	0.85	0.81	0.77	0.73
600	0.93	0.87	0.73	0.78	0.75
620	0.95	0.89	0.84	0.80	0.76
640	0.98	0.92	0.86	0.82	0.78
660	1.00	0.94	0.88	0.84	0.80
680	1.02	0.96	0.90	0.85	0.81
700	1.04	0.98	0.92	0.87	0.83

Notes: Based on an initial temperature of 50°F, an initial pressure of 600 psig, and a maximum operating pressure of 800 psig.

DIAPHRAGM EXPANSION TANK SIZING CORRECTION FACTORS FOR HIGH-TEMPERATURE SYSTEMS *(cont.)*

Initial Pressure (psig)	Pressure Increase (psig) Initial Pressure + Pressure Increase = Maximum Operating Pressure				
	320	340	360	380	400
160	0.38	0.37	0.36	0.36	0.35
180	0.39	0.39	0.38	0.37	0.36
200	0.41	0.40	0.39	0.38	0.38
220	0.43	0.41	0.41	0.40	0.39
240	0.44	0.43	0.42	0.41	0.40
260	0.46	0.44	0.43	0.42	0.41
280	0.47	0.46	0.45	0.44	0.43
300	0.49	0.47	0.46	0.45	0.44
320	0.50	0.49	0.47	0.46	0.45
340	0.52	0.50	0.49	0.47	0.46
360	0.53	0.52	0.50	0.49	0.48
380	0.55	0.53	0.51	0.50	0.49
400	0.56	0.54	0.53	0.51	0.50
420	0.58	0.56	0.54	0.53	0.51
440	0.59	0.57	0.56	0.54	0.52
460	0.61	0.59	0.57	0.55	0.54
480	0.63	0.60	0.58	0.57	0.55
600	0.64	0.62	0.60	0.58	0.56
520	0.66	0.63	0.61	0.59	0.57
540	0.67	0.65	0.62	0.60	0.59
560	0.69	0.66	0.64	0.62	0.60
580	0.70	0.67	0.65	0.63	0.61
600	0.72	0.69	0.66	0.64	0.62
620	0.73	0.70	0.68	0.66	0.64
640	0.75	0.72	0.69	0.67	0.65
660	0.76	0.73	0.71	0.68	0.66
680	0.78	0.75	0.72	0.69	0.67
700	0.79	0.76	0.73	0.71	0.68

Notes: Based on an initial temperature of 50°F, an initial pressure of 600 psig, and a maximum operating pressure of 800 psig.

BIOCIDES

Biocide	Effectiveness Against			Comments
	Bacteria	Fungi	Algae	
Oxidizing Biocides				
Chlorine (Cl_2)	E	G	G	Usable pH range 5 to 8 Reacts with $-NH_2$ groups
Chlorine dioxide (ClO_2)	E	G	G	Insensitive to pH levels Insensitive to presence of $-NH_2$ groups
Bromine	E	G	P	Usable pH range 5 to 10 Substitute for chlorine
Ozone	E	G	G	pH range 7 to 9
Non-oxidizing Biocides				
Carbamate	E	E	G	pH range of 5 to 9 Good in high suspended solids systems Incompatible with chromate treatment programs
Organo-bromide (DBNPA)	E	P	P	pH range 6 to 8.5
Methylenebis-thiocyanate (MBT)	E	P	P	Decomposes above a pH of 8

				Notes
Isothiazoline	E	G	G	Insensitive to pH levels Deactivated by HS and $-NH_2$ groups
Quaternary ammonium salts	E	G	G	Tendency to foam Surface active Ineffective in organic-fouled systems
Organo-tin/quaternary ammonia salts	E	G	E	Tendency to foam Functions best in alkaline pH
Glutaraldehyde	E	E	G	Effective over broad pH range Deactivated by $-NH_2$ groups
Dodecylguanidine (DGH)	E	E	G	pH range of 6 to 9
Triazine	N	N	E	pH range of 6 to 9 Specific for algae control Must be used with other biocides

Notes:
E = Excellent
G = Good
P = Poor
N = No control

2-91

BYPASS AND WARMING VALVES

Main Valve Nominal Pipe Size in Inches	Nominal Pipe Size in Inches	
	Series A Warming Valves	Series B Bypass Valves
4	½	1
5	¾	1¼
6	¾	1¼
8	¾	1½
10	1	1½
12	1	2
14	1	2
16	1	3
18	1	3
20	1	3
24	1	4
30	1	4
36	1	6
42	1	6
48	1	8
54	1	8
60	1	10
72	1	10
84	1	12
96	1	12

SIZING STEAM BOILERS

Size steam boilers by calculating the square feet of steam produced by each radiator in the structure and adding them together. Then refer to the square feet of steam column for boilers and find the figure closest to the total. If the closest figure is less, always choose the next larger size.

SIZE OF STEAM MAINS

Radiation (sq. ft.)	One-Pipe Work (in.)	Two-Pipe Work (in.)
125	1½	1¼ × 1
250	2½	1½ × 1¼
400	3	2 × 1½
650	3½	2½ × 2
900	4	3 × 2½
1250	4½	3½ × 3
1600	5	4 × 3½
2050	5½	4½ × 4
2500	6	5 × 4½
3600	7	6 × 5
5000	8	7 × 6
6500	9	8 × 6

EFFECTS OF AIR ON TEMPERATURE (°F) OF STEAM

psig	Pure Steam	5% Air	10% Air	15% Air
2	219°	216°	213°	210°
5	227°	225°	222°	219°
10	239°	237°	233°	230°
20	259°	256°	252°	249°

EFFECT OF BACK PRESSURE ON STEAM TRAP CAPACITY (%)

Back Pressure %	Inlet Pressure psig			
	5	25	100	200
25	6	3	0	0
50	20	12	6	5
75	38	30	25	23

SIZING TRAPS FOR STEAM MAINS

Condensation Load in lbs. per hour

Steam pressure (psig)	Main size													0°F Correction Factor*
	2"	2½"	3"	4"	5"	6"	8"	10"	12"	14"	16"	18"	
10	6	7	9	11	13	16	20	24	29	32	36	39	1.58
30	8	9	11	14	17	20	26	32	38	42	48	51	1.50
60	10	12	14	18	24	27	33	41	49	54	62	67	1.45
100	12	15	18	22	28	33	41	51	61	67	77	83	1.41
125	13	16	20	24	30	36	45	56	66	73	84	90	1.39
175	16	19	23	26	33	38	53	66	78	86	98	107	1.38
250	18	22	27	34	42	50	62	77	92	101	116	126	1.36
300	20	25	30	37	46	54	68	85	101	111	126	138	1.35
400	23	28	34	43	53	63	80	99	118	130	148	162	1.33
500	27	33	39	49	61	73	91	114	135	148	170	185	1.32

Note: Based on 1000 ft. of 80% efficient steam main with AMB. Temp @ 70°F
*For outdoor temperature of 0°F, multiply load value in table for each main size by correction factor.

2-94

STEAM TRAP COMPARISON

Characteristic	Steam Trap Type		
	Inverted Bucket	Float & Thermostatic	Liquid Expansion Thermostatic
Method of Operation	Intermittent: condensate drainage is continuous, discharge is intermittent	Continuous	Intermittent
No load	Small dribble	No action	No action
Light load	Intermittent	Usually continuous but may cycle at high pressures	Continuous, usually dribble action
Normal load	Intermittent	Usually continuous but may cycle at high pressures	May blast at high pressures
Full or overload	Continuous	Continuous	Continuous
Energy conservation	Excellent	Good	Fair
Resistance to wear	Excellent	Good	Fair
Corrosion resistance	Excellent	Good	Good
Resistance to hydraulic shock	Excellent	Poor	Poor

2-95

STEAM TRAP COMPARISON (cont.)

Characteristic	Steam Trap Type		
	Inverted Bucket	Float & Thermostatic	Liquid Expansion Thermostatic
Vents air and CO_2 at steam temperature	Yes	No	No
Ability to vent air at very low pressure (¼ psig)	Poor	Excellent	Good
Ability to handle start-up air loads	Fair	Excellent	Excellent
Operation against back pressure	Excellent	Excellent	Excellent
Resistance to damage from freezing, cast iron trap not recommended	Good	Poor	Good
Ability to purge system	Excellent	Fair	Good
Performance on very light loads	Excellent	Excellent	Excellent

2-96

	Immediate	Immediate	Delayed
Responsiveness to slugs of condensate			
Ability to handle dirt	Excellent	Poor	Fair
Comparative physical size	Large	Large	Small
Ability to handle flash steam	Fair	Poor	Poor
Usual mechanical failure mode	Open	Closed with air vent open	Open or closed
Subcooling	No	No	Yes
Venting	Fair	Excellent	Excellent
Seat pressure rating	Yes	Yes	—
Advantages	Rugged Tolerates water hammer without damage —	Continuous condensate discharge Handles rapid pressure changes High non-condensible capacity	Utilizes sensible heat of condensate Allows discharge of non-condensibles at startup to the set point temperature Not affected by superheated steam, water hammer, or vibration

STEAM TRAP COMPARISON *(cont.)*

Characteristic	Steam Trap Type			
	Inverted Bucket	Float & Thermostatic	Liquid Expansion Thermostatic	
Advantages	–	–	Resists freezing	
Disadvantages	Discharges non-condensibles slowly (additional air vent required)	Float can be damaged by water hammer	Element subject to corrosion damage	
	Level of condensate can freeze, damaging the trap body	Level of condensate in chamber can freeze, damaging float and body	Condensate backs up into the drain line and/or process	
	Must have water seal to operate—subject to loosing prime	Some thermostatic air vent designs are susceptible to corrosion	–	
	Pressure fluctuations and superheated steam can cause loss of water seal	–	–	

Recommended services	Continuous operation where non-condensible venting is not critical and rugged construction is important	Heat exchangers with high and variable heat transfer rates	Ideal for tracing used for freeze protection
	—	When condensate pump is required	Freeze protection—water and condensate lines and traps
	—	Batch processes that require frequent startup of an air-filled system	Non-critical temperature control of heated tanks

STEAM TRAP COMPARISON (cont.)

Characteristic	Balanced Pressure Thermostatic	Bimetal Thermostatic	Thermodynamic
		Steam Trap Type	
Method of Operation	Intermittent	Intermittent	Intermittent
No load	No action	No action	No action
Light load	Continuous, usually dribble action	Continuous, usually dribble action	Intermittent
Normal load	May blast at high pressures	May blast at high pressures	Intermittent
Full or overload	Continuous	Continuous	Continuous
Energy conservation	Fair	Fair	Poor
Resistance to wear	Fair	Fair	Poor
Corrosion resistance	Good	Good	Excellent
Resistance to hydraulic shock	Good	Good	Excellent
Vents air and CO_2 at steam temperature	No	No	No
Ability to vent air at very low pressure (¼ psig)	Good	Good	Not for low-pressure applications

2-100

Ability to handle start-up air loads	Excellent	Excellent	Poor
Operation against back pressure	Excellent	Excellent	Poor
Resistance to damage from freezing, cast iron trap not recommended	Good	Good	Good
Ability to purge system	Good	Good	Excellent
Performance on very light loads	Excellent	Excellent	Poor
Responsiveness to slugs of condensate	Delayed	Delayed	Delayed
Ability to handle dirt	Fair	Fair	Poor
Comparative physical size	Small	Small	Small
Ability to handle flash steam	Poor	Poor	Poor
Usual mechanical failure mode	Open or closed	Open or closed	Open, dirt can cause to fail closed

STEAM TRAP COMPARISON (cont.)

Characteristic	Steam Trap Type		
	Balanced Pressure Thermostatic	Bimetal Thermostatic	Thermodynamic
Subcooling	Yes	Yes	No
Venting	Excellent	Excellent	Fair
Seat pressure rating	—	—	—
Advantages	Small and lightweight	Small and lightweight	Withstands corrosion, water hammer, high pressure, and superheated steam
	Maximum discharge of non-condensibles at startup	Maximum discharge of non-condensibles at startup	Wide pressure range
	Unlikely to freeze	Unlikely to freeze and unlikely to be damaged if it does freeze	Compact and simple
	—	Withstands corrosion, water hammer, high pressure, and superheated steam	Audible operations warns when repair is needed

Disadvantages	Some types of damage by water hammer, corrosion, and superheated steam	Responds slowly to load and pressure changes	Poor operation with very low-pressure steam or high back pressure
	Condensate backs up into the drain line and/or process	More condensate backup than balance pressure thermostatic trap	Requires slow pressure buildup to remove air at startup to prevent air binding
	—	Back pressure changes operating characteristics	Noisy operation
Recommended services	Batch processing requiring rapid discharge of non-condensibles at startup	Drip legs on constant-pressure steam mains	Steam main drips, tracers
	Drip legs on steam mains and tracing	Installations subject to ambient conditions below freezing	Constant-pressure, constant-load applications
	Installations subject to ambient conditions below freezing	—	Installations subject to ambient conditions below freezing

STEAM TRAP INSPECTION VERSUS FAILURE RATE

Trap Failure Rate	Steam Trap Inspection Frequency
More than 10%	Every 2 months
5 to 10%	Every 3 months
Less than 5%	Every 6 months
System Pressure	**Steam Trap Inspection Frequency**
0 to 30 psig	Annually
30 to 100 psig	Semi-annually
100 to 250 psig	Quarterly or monthly
More than 250 psig	Monthly or weekly

CHAPTER 3
Ventilation

MAXIMUM DUCT AND PIPE SIZES THROUGH STEEL JOISTS (K TYPE)			
Joist Depth (in.)	Round Duct or Pipe Size (in.)	Square Duct Size (in.)	Rectangular Duct Size (in.)
8	5	4 × 4	3 × 8
10	6	5 × 5	3 × 8
12	7	6 × 6	4 × 9
14	8	6 × 6	5 × 9
16	9	7 × 7	6 × 10
18	11	8 × 8	7 × 11
20	11	9 × 9	7 × 12
22	12	9 × 9	8 × 12
24	13	10 × 10	8 × 13
26	15	12 × 12	9 × 18
28	16	13 × 13	9 × 18
30	17	14 × 14	10 × 18

CONVERTING ROUND DUCT AREAS TO SQUARE FEET

Duct Diameter (in.)	Duct Diameter (mm)	Area (ft.²)	Area (m²)
8	203	0.3491	0.032
10	254	0.5454	0.051
12	305	0.7854	0.073
14	356	1.069	0.099
16	406	1.396	0.130
18	457	1.767	0.290
20	508	2.182	0.203
22	559	2.640	0.245
24	609	3.142	0.292
26	660	3.687	0.342
28	711	4.276	0.397
30	762	4.900	0.455
32	813	5.585	0.519
34	864	6.305	0.586
36	914	7.069	0.657
38	965	7.876	0.732
40	1,016	8.727	0.811
42	1,067	9.62	0.894
44	1,119	10.56	0.981
46	1,168	11.54	1.072
48	1,219	12.57	1.168
50	1,270	13.67	1.270
52	1,321	14.75	1.370
54	1,372	15.90	1.477
56	1,422	17.10	1.586
58	1,473	18.35	1.705
60	1,524	19.63	1.824

DUCT LONGITUDINAL SEAMS

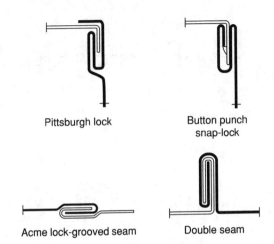

Pittsburgh lock

Button punch snap-lock

Acme lock-grooved seam

Double seam

Approximately 2" spacing between "buttons"

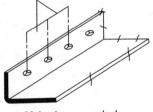

Male piece snap-lock

DUCT CROSS JOINTS

(A)
Drive slip

Air flow →
(B)
Plain "S" slip

Air flow →
(C)
Hemmed "S" slip

Air flow →
(E)
Bar slip

Air flow →
(F)
Alternate bar slip
(Standing "S" slip)

Air flow →
(G)
Reinforced
Bar slip (cleat)

3-4

DUCT CROSS JOINTS (cont.)

Air flow →

(H)
Angle slip

Air flow →

(I)
Standing seam

Air flow →

(J)
Angle reinforced standing seam

Air flow →

(K)
Pocket lock

Air flow →

(L)
Angle reinforced pocket lock

Air flow →

(M)
Companion angles (caulk or gasket)

3-5

AIR-DUCT SIZING

3-6

RECOMMENDED AIR VELOCITIES

Designation	Recommended Air Velocities (FPM)		
	Residences, Broadcasting Studios, Etc.	Schools, Theaters, Public Buildings	Industrial Applications
Initial air intake	750	800	1,000
Air washers	500	500	500
Extended surface heaters or coolers (face velocity)	450	500	500
Suction connections	750	800	1,000
Through fan outlet			
For 1.5" static pressure	—	2,200	2,400
For 1.25" static pressure	—	2,000	2,200
For 1" static pressure	1,700	1,800	2,000
For 0.75" static pressure	1,400	1,550	1,800
For 0.5" static pressure	1,200	1,300	1,600
Horizontal ducts	700	900	1,000–2,000
Branch ducts and risers	550	600	1,000–1,600
Supply grilles and openings	300	300 grille	400 opening
Exhaust grilles and openings	350	400 grille	500 opening
Duct outlets at high elevation	—	1,000	—

SUGGESTED DUCT VELOCITIES FOR VARIOUS LOW VELOCITY SYSTEMS

Application	Main Trunk and Risers (FPM)	Branch Duct and Small Risers (FPM)	Returns (FPM)
Residences	800	600	600
Concert halls	900	700	700
Apartments	1,000	800	800
Hotel, motel bedrooms	1,200	1,100	1,000
Theaters, Schools	1,300	1,100	1,000
Executive offices, libraries	1,500	1,200	1,200
Dining rooms	1,800	1,400	1,200
General offices	2,200	1,400	1,200
Stores	2,200	1,600	1,300
Cafeterias	2,300	1,800	1,300
Industrial buildings	2,600	1,800	1,500

CONVERSION DATA FOR PRESSURES		
Multiply	**By**	**To Obtain**
psi	16	ounces per inch squared
psi	2.31	feet of water
psi	27.73	inches of water
psi	0.0703	kilograms per centimeter squared
psi	2.036	inches of mercury
inches of water	0.07342	inches of mercury
inches of water	0.5770	ounces per inch squared
inches of water	0.03606	pounds per inch
inches of water	5.196	pounds per foot
feet of water	0.4328	pounds per inch
feet of water	62.32	pounds per foot

VENTILATION AIR CHANGES PER STRUCTURE		
Space to Be Ventilated	Air Changes per Hour	Minutes per Change
Auditoriums	6	10
Bakeries	20	3
Bowling alleys	12	5
Club rooms	12	5
Churches	6	10
Dining rooms (restaurants)	12	5
Factories	10	6
Foundries	20	3
Garages	12	5
Kitchens (restaurants)	30	2

VENTILATION AIR CHANGES PER STRUCTURE *(cont.)*		
Space to Be Ventilated	**Air Changes per Hour**	**Minutes per Change**
Laundries	20	3
Machine shops	10	6
Offices	10	6
Projection booths	60	1
Recreation rooms	10	6
Sheet-metal shops	10	6
Ship holds	6	10
Stores	10	6
Toilets	20	3
Tunnels	6	10

FRESH AIR REQUIREMENTS FOR STRUCTURES

Type of Building or Room	Minimum Air Changes per Hour	Cubic Ft. of Air per Minute per Occupant
Attic spaces (for cooling)	12 to 15	—
Boiler room	15 to 20	—
Churches, auditoriums	8	20 to 30
College classrooms		25 to 30
Dining rooms (hotel)	5	—
Engine rooms	4 to 6	—
Factory buildings (ordinary manufacturing)	2 to 4	—
Factory buildings (extreme fumes or moisture)	10 to 15	—
Foundries	15 to 20	—
Galvanizing plants	20 to 30	—
Garages (repair)	20 to 30	—
Garages (storage)	4 to 6	—
Homes (night cooling)	9 to 17	—

FRESH AIR REQUIREMENTS FOR STRUCTURES *(cont.)*

Type of Building or Room	Minimum Air Changes per Hour	Cubic Ft. of Air per Minute per Occupant
Hospitals (general)	—	40 to 50
Hospitals (children's)	—	35 to 40
Hospitals (contagious diseases)	—	80 to 90
Kitchens (hotel)	10 to 20	—
Kitchens (restaurant)	10 to 20	—
Libraries (public)	4	—
Laundries	10 to 15	—
Mills (paper)	15 to 20	—
Mills (textile—general buildings)	4	—
Mills (textile—dyehouses)	15 to 20	—
Offices (public)	3	—
Offices (private)	4	—
Pickling plants	10 to 15	—

FRESH AIR REQUIREMENTS FOR STRUCTURES (cont.)

Type of Building or Room	Minimum Air Changes per Hour	Cubic Ft. of Air per Minute per Occupant
Pump rooms	5	—
Schools (grade)	—	15 to 25
Schools (high)	—	30 to 35
Restaurants	8 to 12	—
Shops (machine)	5	—
Shops (paint)	15 to 20	—
Shops (railroad)	5	—
Shops (woodworking)	5	—
Substations (electric)	5 to 10	—
Theaters	—	10 to 15
Turbine rooms (electric)	5 to 10	—
Warehouses	2	—
Waiting rooms (public)	4	—

AVERAGE AIR CHANGES PER 24 HR. FOR STORAGE ROOMS DUE TO DOOR OPENINGS AND INFILTRATION (ABOVE 32°F)			
Volume (cu. ft.)	Air Changes per 24 hr.	Volume (cu. ft.)	Air Changes per 24 hr.
200	44.0	6,000	6.5
300	34.5	8,000	5.5
400	29.5	10,000	4.9
500	26.0	15,000	3.9
600	23.0	20,000	3.5
800	20.0	25,000	3.0
1,000	17.5	30,000	2.7
1,500	14.0	40,000	2.3
2,000	12.0	50,000	2.0
3,000	9.5	75,000	1.6
4,000	8.2	100,000	1.4
5,000	7.2	—	—

3-15

ATTIC VENT SIZING GUIDE

Attic Floor Area (sq. ft.)	Number of Turbine Ventilators Required	Size of Turbine (in.)	Minimum Inlet Louver Area Required (sq. ft.)	Minimum Number and Size EAVE Vents (in.)	Minimum Number and Size GABLE Vents (in.)
1,200	2 ea.	12	4.0	6 ea. 8 × 16	2 ea. 14 × 24
1,500	2 ea.	14	5.0	6 ea. 8 × 16	2 ea. 14 × 24
1,800	3 ea.	12	6.0	8 ea. 8 × 16	4 ea. 12 × 18
2,100	3 ea.	14	7.0	8 ea. 8 × 16	2 ea. 12 × 18 & 14 × 24
2,400	4 ea.	12	8.0	10 ea. 8 × 16	2 ea. 12 × 18 & 14 × 24

DUCT HEAT LOSS MULTIPLIERS

	Duct Loss Multipliers	
Duct Location and Insulation Value	Winter Design Below 15°F	Winter Design Above 15°F
Exposed to Outdoor Ambient Air: Attic, Garage, Exterior Wall, Etc.		
No insulation	1.30	1.25
R-2	1.20	1.15
R-4	1.15	1.10
R-6	1.10	1.05
Enclosed in Unheated Space: Vented or Unvented Crawl Space or Basement		
No insulation	1.20	1.15
R-2	1.15	1.10
R-4	1.10	1.05
R-6	1.05	1.00
Duct Buried in or under Concrete Slab: Edge Insulation		
No insulation	1.25	1.20
R-3 to 4	1.15	1.10
R-5 to 7	1.10	1.05
R-7 to 9	1.05	1.00

DUCT HEAT GAIN MULTIPLIERS

Duct Location and Insulation Value	Duct Gain Multiplier
Exposed to Outdoor Ambient Air: Attic, Garage, Exterior Wall, Etc.	
No insulation	1.30
R-2	1.20
R-4	1.15
R-6	1.10
Enclosed in Unconditioned Space: Vented or Unvented Crawl Space or Basement	
No insulation	1.15
R-2	1.10
R-4	1.05
R-6	1.00
Duct Buried in or under Concrete Slab: Edge Insulation	
No insulation	1.10
R-3 to 4	1.05
R-5 to 7	1.00
R-7 to 9	1.00

DUCT INSULATION AND OUTLET LOCATION

Below 2,000 HDD

Type of Construction	Insulation Range	Location of Supply Air Outlets
Slab	Slab edge R-0 to R-4	Ceiling
Open or vented crawl space	Below floor R-4 to R-14	Perimeter ceiling
Enclosed crawl space	Crawl space walls R-0 to R-8	Perimeter ceiling
Basement	Basement walls R-0 to R-8	Perimeter ceiling
HDD = Heating degree days		

DUCT INSULATION AND OUTLET LOCATION (cont.)

2,000 to 3,500 HDD

Type of Construction	Insulation Range	Location of Supply Air Outlets
Slab	Slab edge R-8 Slab edge R-4	Perimeter Ceiling
Open or vented crawl space	Below floor R-20 Below floor R-14	Perimeter ceiling
Enclosed crawl space	Crawl space walls R-8	Perimeter ceiling
Basement	Basement walls R-8	Perimeter ceiling
HDD = Heating degree days		

DUCT INSULATION AND OUTLET LOCATION *(cont.)*		
Above 3,500 HDD		
Type of Construction	**Insulation Range**	**Location of Supply Air Outlets**
Slab	Slab edge R-10 (HDD > 7,000) R-8 (HDD > 5,000) R-6 (HDD > 3,500)	Perimeter
Open or vented crawl space	Below floor R-20 (HDD > 6,000) R-14 (HDD > 3,500)	Perimeter
Enclosed crawl space	Crawl space walls R-16 (HDD > 6,000) R-10 (HDD > 3,500)	Perimeter (recommended) ceiling (acceptable)
Basement	Basement walls R-16 (HDD > 6,000) R-10 (HDD > 3,500)	Perimeter (recommended) ceiling (acceptable)
HDD = Heating degree days		

DUCT INSULATION THICKNESS			
	For Cooling		
Duct Location	**Annual Cooling Degree Days (Base 65°F)**	**Insulation R-Value**	**Insulation Thickness Minimum) (in.)**
Exterior of building	Below 500	3.3	0.75
	500 to 1150	5.0	1.5
	1151 to 2000	6.5	1.5
	Above 2000	8.0	2.0
Inside building or in unconditioned spaces			
$\Delta T < 15$	N/A	—	—
ΔT 15–40	N/A	3.3	0.75
$\Delta T > 40$	N/A	5.0	1.5

ΔT = Temperature difference

DUCT INSULATION THICKNESS *(cont.)*			
	For Heating		
Duct Location	**Annual Heating Degree Days (Base 65°F)**	**Insulation R-Value**	**Insulation Thickness (Minimum) (in.)**
Exterior of building	Below 1500	3.3	0.75
	1500 to 4500	5.0	1.5
	4501 to 7500	6.5	1.5
	Above 7500	8.0	2.0
Inside building or in unconditioned spaces			
$\Delta T < 15$	N/A	—	—
ΔT 15–40	N/A	3.3	0.75
$\Delta T > 40$	N/A	5.0	1.5
ΔT = Temperature difference			

RECOMMENDED APPLICATIONS FOR DUCT MATERIALS

Duct Location	Duct Board	Rigid Round Fiber Glass	Sheet Metal with Liner	Sheet Metal with Wrap	Sheet Metal Bare	Sheet Metal Rigid Exterior	Flexible Insulated
Attic	X	X	X	X			X
Basement—unconditioned	X	X	X	X			X
Basement—conditioned	X	X			X		X
Enclosed crawl space	X	X	X	X			X
Open crawl space			X			X	
Exterior wall cavity or chase	X	X	X				X
Interior wall cavity or chase	X	X			X		X
Soffit or ceiling plenum	X	X		X			X
In conditioned space	X	X			X		X
Roof or outdoor location						X	

3-24

SAMPLE DUCT SCHEDULE

Trunk Section	CFM	F/100	Diameter for Air Flow (in.)	Velocity (FPM)	Design Velocity (FPM)	Diameter for Velocity (in.)	Design Diameter (in.)
Fan to S-1	1000	0.125	13.5	1050	900	14.5	15
S-1 to S-2	900	0.125	13.0	1000	900	13.8	14
S-2 to S-3	500	0.125	10.3	890	900	10.2	11
Fan to R-1	1000	0.125	13.5	1050	700	16.5	17
R-1 to R-2	300	0.125	8.5	780	700	9.0	9

Runouts	CFM	F/100	Diameter for Air Flow (in.)	Velocity (FPM)	Design Velocity (FPM)	Diameter for Velocity (in.)	Design Diameter (in.)
S-1	100	0.125	5.5	600	900	4.5	6
S-2	400	0.125	9.5	840	900	9.1	10
S-3	500	0.125	10.3	880	900	10.2	11
R-1	700	0.125	11.7	960	600	14.8	15
R-2	300	0.125	8.5	780	600	9.6	10

3-25

RESIDENTIAL FORCED AIR SYSTEM

Air Distribution Sizes in Inches

Room Volume (cu. ft.)	Supply Duct		Outlet				Return Grille	Return Duct	
	Round (dia.)	Equiv.	Floor	Wall	Ceiling (dia.)			Round (dia.)	Equivalent
200	4	4½ × 3			4		6 × 10	6	8 × 4
300	4	4½ × 3			4		6 × 10	6	
400	4	4½ × 3			4		6 × 10	6	
500	4	4½ × 3			4		6 × 10	6	
600	5	10 × 2¼	2¼ × 10		4		6 × 10	6	
700	5	8 × 3¼	2¼ × 10	4 × 10	6		6 × 10	6	
800	5	5 × 4	2¼ × 10		6		6 × 10	6	
900	6	14 × 2¼	2¼ × 10		6		6 × 10	6	8 × 6
1,000	6	10 × 3¼	2¼ × 10		6		6 × 10	7	
1,100	6	8 × 4	2¼ × 12		6		6 × 10	7	
1,200	6	6 × 5	2¼ × 12	10 × 6	6		6 × 10	7	
1,300	6	6 × 5	2¼ × 12		6		6 × 10	7	8 × 7
1,400	7	14 × 3¼	2¼ × 14		6		6 × 14	8	
1,500	7	11 × 4	2¼ × 14	12 × 6	8		6 × 14	8	
1,600	7	8 × 5	4 × 10	14 × 6	8		6 × 14	8	

1,700	7	7 × 6	4 × 10	14 × 6	8	6 × 14	8	
1,800	7		4 × 12	14 × 6	8	6 × 14	8	
1,900	7		4 × 12	14 × 6	8	6 × 14	8	
2,000	7		4 × 12	14 × 6	8	6 × 14	8	
3,000	7½	13 × 4				8 × 14	10	8 × 11
4,000	9	8 × 8				6 × 24	11	8 × 13
5,000	10	8 × 11				6 × 30	12	8 × 16
6,000	11	8 × 13				8 × 30	13	8 × 18
7,000	11½	8 × 14				8 × 30	14	8 × 22
8,000	12	8 × 16				18 × 18	15	8 × 24
10,000	13	8 × 18				18 × 18	16	8 × 28
12,000	14	8 × 22				18 × 24	18	8 × 36
14,000	14½	8 × 24				24 × 24	20	8 × 46
16,000	15	8 × 26				24 × 24	20	
18,000	16	8 × 30				24 × 30	20	
20,000	17	8 × 34				24 × 30	22	8 × 60
25,000	18	8 × 39				24 × 30	22	

RECOMMENDED GAUGES FOR DUCT SYSTEMS

RECTANGULAR DUCTWORK, ½" wg static pressure positive or negative, up to 2,000 fpm, based on proper reinforcements spaced at 10' intervals.

Largest Dimension (in.)	Galvanized Steel Gauge	Aluminum, B&S Gauge	Copper, B&S Gauge
Through 26	26	24	24
27 to 30	24	22	20
31 to 36	22	20	18
37 to 48	20	18	18
49 to 60	18	16	14
73 to 84	16	14	12
73 to 84	16	But 8' reinforcement spacing required	
85 to 96	16	But 8' reinforcement spacing required	
Larger than 96	18	But 5' class H spacing	

RECOMMENDED GAUGES FOR DUCT SYSTEMS *(cont.)*

RECTANGULAR DUCTWORK, 1" wg static pressure positive or negative, up to 2,500 fpm, based on proper reinforcements spaced at 10' intervals.

Largest Dimension (in.)	Galvanized Steel Gauge	Aluminum, B&S Gauge	Copper, B&S Gauge
Through 14	26	24	24
15 to 24	24	22	20
25 to 30	22	20	18
31 to 36	20	18	18
37 to 42	18	16	14
43 to 54	16	14	12
55 to 60	18	But 8' reinforcement spacing required	
61 to 84	18	But 5' reinforcement spacing required	
85 to 96	16	But 5' reinforcement spacing required	
Larger than 96	18	But 2½' class H spacing	

RECOMMENDED GAUGES FOR DUCT SYSTEMS *(cont.)*

RECTANGULAR DUCTWORK, 2" wg static pressure positive or negative, up to 2,500 fpm.

Largest Dimension (in.)	Galvanized Steel Gauge	Reinforcement Spacing Intervals (ft.)
Through 18	22	10
19 to 26	20	10
27 to 30	18	10
31 to 36	16	10
37 to 48	16	8
49 to 60	18	5
61 to 72	16	5
73 to 84	18	4, class J
85 to 96	16	4, class K
Larger than 96	18	2½, class H

RECOMMENDED GAUGES FOR DUCT SYSTEMS *(cont.)*

RECTANGULAR DUCTWORK, 3" wg static pressure positive or negative, up to 4,000 fpm.

Largest Dimension (in.)	Galvanized Steel Gauge	Reinforcement Spacing Intervals (ft.)
Through 28	18	10
29 to 30	16	10
31 to 36	16	8
37 to 42	20	5
43 to 54	18	5
55 to 60	16	5, class H
61 to 72	16	4, class I
73 to 84	18	3, class J
85 to 96	16	3, class L
Larger than 96	18	2½, class H

RECOMMENDED GAUGES FOR DUCT SYSTEMS *(cont.)*

RECTANGULAR DUCTWORK, 4" wg static pressure positive, up to 4,000 fpm.

Largest Dimension (in.)	Galvanized Steel Gauge	Reinforcement Spacing Intervals (ft.)
Through 12	22	10
13 to 16	20	10
17 to 26	18	10
27 to 30	16	10
31 to 36	20	5
37 to 48	18	5
49 to 54	16	5, class H
55 to 60	16	5, class I
61 to 72	18	3, class I
73 to 84	16	3, class K
85 to 96	16	2½, class L
Larger than 96	18	2½, class H with tie rod

RECOMMENDED GAUGES FOR DUCT SYSTEMS *(cont.)*

RECTANGULAR DUCTWORK, 6" wg static pressure positive, velocities determined by designer.

Largest Dimension (in.)	Galvanized Steel Gauge	Reinforcement Spacing Intervals (ft.)
Through 14	20	10
15 to 18	18	10
19 to 22	16	10
23 to 24	18	8
25 to 28	16	8
29 to 36	18	5
37 to 42	16	5
43 to 48	18	4
49 to 54	16	4
55 to 60	18	3
61 to 72	16	3
73 to 84	16	2½
85 to 96	18	2, class L
Larger than 96	18	2, class H with tie rod

RECOMMENDED GAUGES FOR DUCT SYSTEMS *(cont.)*

RECTANGULAR DUCTWORK, 10" wg static pressure positive, velocities determined by designer.

Largest Dimension (in.)	Galvanized Steel Gauge	Reinforcement Spacing Intervals (ft.)
Through 14	18	8
15 to 20	16	8
21 to 28	18	5
29 to 36	16	5
37 to 42	16	4
43 to 48	18	3, class H
49 to 54	16	3, class I
55 to 60	16	3, class J
61 to 72	16	2½, class K
73 to 84	16	2
Larger than 85	16	2, class H with tie rod

RECOMMENDED GAUGES FOR DUCT SYSTEMS (cont.)

ROUND DUCTWORK; GALVANIZED STEEL, GAUGE SELECTION

Duct Diameter (in.)	Maximum 2" wg static positive		Maximum 10" wg static positive		Maximum 2" wg static negative	
	Spiral Seam Gauge	Longitudinal Seam Gauge	Spiral Seam Gauge	Longitudinal Seam Gauge	Spiral Seam Gauge	Longitudinal Seam Gauge
3 to 8	28	28	26	24	28	24
9 to 14	28	26	26	24	26	24
15 to 26	26	24	24	22	24	22
27 to 36	24	22	22	20	22	20
37 to 50	22	20	20	20	20	18
51 to 60	20	18	18	18	18	16
61 to 84	18	16	18	16	16	14

RECOMMENDED GAUGES FOR DUCT SYSTEMS (cont.)

ROUND DUCTWORK, ALUMINUM, GAUGE SELECTION

Duct Diameter (in.)	Maximum 2" wg static positive		Maximum 10" wg static positive	
	Spiral Seam Gauge (in.)	Longitudinal Seam Gauge (in.)	Spiral Seam Gauge (in.)	Longitudinal Seam Gauge (in.)
3 to 8	0.025	0.032	0.025	0.040
9 to 14	0.025	0.032	0.032	0.040
15 to 26	0.032	0.040	0.040	0.050
27 to 36	0.040	0.050	0.050	0.063
37 to 50	0.050	0.063	0.063	0.071
51 to 60	0.063	0.071	N/A	0.090
61 to 84	N/A	0.090	N/A	N/A

RECOMMENDED GAUGES FOR DUCT SYSTEMS *(cont.)*	
Largest Dimension (in.)	**Standard Gauge**
Industrial Ventilation	
Up to 8	24
9 to 15	22
16 to 22	20
23 to 30	18
Larger than 30	16
Boiler Breeching—Hot-Rolled Steel, Welded Seams	
Up to 12	18
13 to 24	16
25 to 36	14
37 to 60	12
Larger than 60	10

FIBROUS GLASS DUCT STRETCH-OUT AREAS

Material required for rectangular ducts fabricated from 1" thick fibrous glass board; includes allowance for overlap and 8" grooving.

Width plus Depth = Total (in.)	Sq. Ft. per Linear Foot	Width plus Depth = Total (in.)	Sq. Ft. per Linear Foot	Width plus Depth = Total (in.)	Sq. Ft. per Linear Foot
10	2.33	25	4.83	41	7.50
11	2.50	26	5.00	42	7.67
12	2.67	27	5.17	43	7.83
13	2.83	28	5.33	44	8.00
14	3.00	29	5.50	45	8.17
15	3.17	30	5.67	46	8.33
16	3.33	31	5.83	47	8.50
17	3.50	32	6.00	48	8.67
18	3.67	33	6.17	49	8.83
19	3.83	34	6.33	50	9.00
20	4.00	35	6.50	51	9.17
21	4.17	36	6.67	52	9.33
22	4.33	37	6.83	53	9.50
23	4.50	38	7.00	54	9.67
24	4.67	39	7.17	55	9.83
—	—	40	7.33	—	—

FIBROUS GLASS DUCT STRETCH-OUT AREAS (cont.)

Material required for rectangular ducts fabricated from 1" thick fibrous glass board; includes allowance for overlap and 8" grooving.

Width plus Depth = Total (in.)	Sq. Ft. per Linear Foot	Width plus Depth = Total (in.)	Sq. Ft. per Linear Foot	Width plus Depth = Total (in.)	Sq. Ft. per Linear Foot
56	10.00	71	12.50	86	15.00
57	10.17	72	12.67	87	15.17
58	10.33	73	12.83	88	15.33
59	10.55	74	13.00	89	15.55
60	10.67	75	13.17	90	15.67
61	10.83	76	13.33	91	15.83
62	11.00	77	13.50	92	16.00
63	11.17	78	13.67	93	16.17
64	11.33	79	13.83	94	16.33
65	11.55	80	14.00	95	16.55
66	11.67	81	14.17	96	16.67
67	11.83	82	14.33	97	16.83
68	12.0	83	14.50	98	17.00
69	12.17	84	14.67	99	17.17
70	12.33	85	14.83	100	17.33
—	—	—	—	—	—

WEIGHTS OF ALUMINUM, STEEL, AND COPPER DUCTWORK FOR STRAIGHT RUNS IN POUNDS PER LINEAR FEET

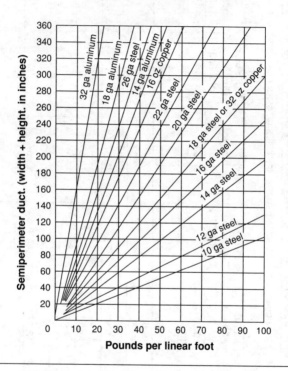

No allowances included. Add 20% allowance to each item for bracing and waste. For fittings add:

90° elbow	3'	Transition offset	6'
45° elbow	2.5'	Square-to-round transition	4'
Offset	4'	Reducing elbow	5'

WEIGHTS OF VARIOUS SHEET METAL DUCT MATERIALS

GALVANIZED STEEL

U.S. Gauge	Decimal Equiv. (in.)	Lb. per Sq. Ft.	Lb. per Sheet		
			36" × 96"	48" × 96"	48" × 120"
28	0.020	0.781	18.75	—	—
26	0.022	0.906	21.75	29.0	36.2
24	0.028	1.156	27.75	37.0	46.2
22	0.034	1.406	33.75	45.0	56.2
20	0.040	1.656	39.75	53.0	66.2
18	0.052	2.156	51.75	70.0	86.2
16	0.064	2.656	63.75	85.0	102.2
14	0.080	3.281	78.75	105.0	131.2
12	0.112	4.531	108.75	145.0	181.2
10	0.142	5.781	138.75	185.0	231.2

WEIGHTS OF VARIOUS SHEET METAL DUCT MATERIALS *(cont.)*

HOT-ROLLED-STEEL

U.S. Gauge	Decimal Equiv. (in.)	Lb. per Sq. Ft.	Lb. per Sheet		
			36" × 96"	48" × 96"	48" × 120"
26	0.018	0.750	18.0	24.0	30.0
24	0.024	1.000	24.0	32.0	40.0
22	0.030	1.250	30.0	40.0	50.0
20	0.036	1.500	36.0	48.0	60.0
18	0.048	2.000	48.0	64.0	80.0
16	0.057	2.500	60.0	80.0	100.0
14	0.075	3.125	75.0	100.0	125.0
12	0.108	4.250	102.0	138.0	170.0
10	0.135	5.625	137.0	180.0	225.0

WEIGHTS OF VARIOUS SHEET METAL DUCT MATERIALS (cont.)

STAINLESS STEEL

U.S. Gauge	Decimal Equiv. (in.)	Lb. per Sq. Ft.	Lb. per Sheet		
			36" × 96"	48" × 96"	48" × 120"
28	0.016	0.66	15.8	21.1	26.4
26	0.019	0.79	18.9	25.2	31.6
24	0.025	1.05	25.2	33.6	42.0
22	0.031	1.31	31.5	42.0	52.5
20	0.038	1.58	37.8	50.4	63.0
18	0.050	2.10	50.4	61.2	84.0
16	0.063	2.63	63.0	84.0	105.0
14	0.078	3.28	78.7	104.9	131.2
12	0.109	4.60	110.0	147.0	183.8

WEIGHTS OF NONFERROUS SHEET DUCT MATERIAL

ALUMINUM 3003

U.S. Gauge	Decimal Equiv. (in.)	Lb. per Sq. Ft.	Lb. per Sheet		
			36" × 96"	48" × 96"	48" × 120"
26	0.016	0.226	5.4	7.2	9.0
24	0.020	0.282	6.8	9.0	11.3
22	0.025	0.357	8.6	11.4	14.3
20	0.032	0.450	10.8	14.4	18.0
18	0.040	0.568	13.6	18.7	22.7
16	0.051	0.716	17.2	22.9	28.6
14	0.064	0.903	21.7	28.9	36.1
12	0.071	1.000	24.0	32.0	40.0
10	0.080	1.130	27.1	36.2	45.2

WEIGHTS OF NONFERROUS SHEET DUCT MATERIAL *(cont.)*

COLD-ROLLED COPPER

U.S. Gauge	Decimal Equiv. (in.)	Lb. per Sq. Ft.	36" × 96"	48" × 96"	48" × 120"
			Lb. per Sheet		
24	0.021	16	24	32	40
23	0.024	18	30	40	50
20	0.032	24	36	48	64
18	0.040	30	48	64	80
16	0.051	38	54	72	90
15	0.053	40	60	80	100

WEIGHTS OF GALVANIZED STEEL BANDS

Band Size (in.)	Lb. per Linear Ft.	Band Size (in.)	Lb. per Linear Ft.	Band Size (in.)	Lb. per Linear Ft.
1/8 × 1	0.425	3/16 × 1	0.670	1/4 × 1	0.90
1/8 × 1 1/4	0.531	3/16 × 1 1/4	0.837	1/4 × 1 1/2	1.35
1/8 × 1 1/2	0.638	3/16 × 1 1/2	1.00	1/4 × 2	1.78
1/8 × 2	0.850	3/16 × 2	1.34		

WEIGHTS OF METAL ANGLES, EQUAL LEGS IN LB. PER LINEAR FT.

Angle Size (in.)	Galvanized Steel	Hot-Rolled Steel	Aluminum
$\frac{1}{8} \times 1$	0.84	0.80	0.28
$\frac{1}{8} \times 1\frac{1}{4}$	1.06	1.01	0.36
$\frac{1}{8} \times 1\frac{1}{2}$	1.29	1.23	0.44
$\frac{1}{8} \times 1\frac{3}{4}$	1.51	1.44	0.51
$\frac{1}{8} \times 2$	1.73	1.65	0.59
$\frac{3}{16} \times 1$	1.22	1.16	0.41
$\frac{3}{16} \times 1\frac{1}{4}$	1.55	1.48	0.53
$\frac{3}{16} \times 1\frac{1}{2}$	1.89	1.80	0.64
$\frac{3}{16} \times 1\frac{3}{4}$	2.23	2.12	0.75
$\frac{3}{16} \times 2$	2.56	2.44	0.87
$\frac{1}{4} \times 1\frac{1}{2}$	2.46	2.34	0.83
$\frac{1}{4} \times 1\frac{3}{4}$	2.91	2.77	0.98
$\frac{1}{4} \times 2$	3.35	3.19	1.14
$\frac{1}{4} \times 2\frac{1}{2}$	4.26	4.10	1.45
$\frac{1}{4} \times 3$	5.15	4.90	1.70
$\frac{3}{8} \times 2$	4.93	4.70	1.65
$\frac{3}{8} \times 2\frac{1}{2}$	6.20	5.90	2.11
$\frac{3}{8} \times 3$	7.55	7.20	2.55

WEIGHTS OF FLOOR PLATES, SKID-RESISTANT, RAISED PATTERN

Thickness (in.)	1/8	3/16	1/4	5/16	3/8	7/16	1/2	5/8	3/4	7/8	1
Lb. per Sq. Ft.	6.15	8.70	11.25	13.80	16.35	18.90	21.45	26.55	31.65	36.75	41.85

WEIGHTS OF ROUND HANGER ROD, THREADED ENDS

Diameter (in.)	Steel	Aluminum	Brass
		Lb. per Linear Ft.	
1/4	C.167	0.06	0.181
5/16	0.261	0.09	0.283
3/8	0.376	0.13	0.407
7/16	0.511	0.18	0.554
1/2	0.668	0.24	0.723
5/8	1.043	0.37	1.130
3/4	1.502	0.54	1.163
7/8	2.044	0.73	2.220
1	2.670	0.96	2.900
1 1/4	4.172	1.50	4.520

3-47

PROPERTIES OF STEEL AND PLATE IRON

| Standard Gauge | Thickness | | Weight (lbs. per sq. ft.) |
	Fractions of an Inch	Decimal Equivalent	
0000000	1/2	.5	20.00
000000	15/32	.46875	18.75
00000	7/16	.4375	17.50
0000	13/32	.40625	16.25
000	3/8	.375	15.00
00	11/32	.34375	13.75
0	5/16	.3125	12.50
1	9/32	.28125	11.25
2	17/64	.265625	10.62
3	1/4	.25	10.00
4	15/64	.234375	9.37
5	7/32	.21875	8.75
6	13/64	.203125	8.12
7	3/16	.1875	7.50
8	11/64	.171875	6.87

PROPERTIES OF STEEL AND PLATE IRON *(cont.)*

| Standard Gauge | Thickness | | Weight (lbs. per sq. ft.) |
	Fractions of an Inch	Decimal Equivalent	
9	$5/32$	.15625	6.25
10	$9/64$	.140625	5.62
11	$1/8$	.125	5.00
12	$7/64$	.109375	4.375
13	$3/32$	.09375	3.750
14	$5/64$	.078125	3.125
15	$9/128$	.0703125	2.812
16	$1/16$	.0625	2.500
17	$9/16$	.05625	2.250
18	$1/20$	.05	2.00
19	$7/16$	.04375	1.75
20	$3/80$	.0375	1.50
21	$11/32$	.034375	1.37
22	$1/32$	.03125	1.25
23	$9/32$	.028125	1.12

PROPERTIES OF STEEL AND PLATE IRON *(cont.)*

| Standard Gauge | Thickness | | Weight (lbs. per sq. ft.) |
	Fractions of an Inch	Decimal Equivalent	
24	1/40	.025	1.00
25	7/32	.021875	0.87
26	3/16	.01875	0.75
27	11/64	.0171875	0.69
28	1/64	.015625	0.62
29	9/64	.0140625	0.56
30	1/8	.0125	0.50
31	7/64	.0109375	0.44
32	13/128	.01015625	0.41
33	3/32	.009375	0.37
34	11/128	.00859375	0.34
35	5/64	.0078125	0.31
36	9/128	.00703125	0.28
37	17/256	.0066406	0.27
38	1/16	.00625	0.25

CHAPTER 4
Air Conditioning and Refrigeration

REFRIGERANT COLOR CODES			
R-22	Light green	R-123	Light gray
R-134a	Light blue	R-401a	Dark brown
R-407c	Coral red		

QUANTITY LIMITS, TYPE A REFRIGERANTS	
Hospital kitchens	Up to 20 lbs.
Residential A/C systems	Up to 20 lbs.
Res. A/C systems, with precautions	Up to 50 lbs.
Indirect systems in places of public assembly	Up to 50 lbs.

SERIES NUMBERING FOR REFRIGERANT CLASSES			
–000	Methane-based	–100	Ethane-based
–200	Propane-based	–300	Cyclic organic
–400	Zeotropes	–500	Azeotropes
–600	Organic	–700	Inorganic
–1000	Unsaturated organic		

HVAC MATERIAL CHARACTERISTICS

Gas or Vapor	Molecular Weight	Specific Ratio Heat	Coefficient C	Specific Gravity
Acetylene	26.04	1.25	342	0.899
Air	28.97	1.40	356	1.000
Ammonia (R-717)	17.03	1.30	347	0.588
Argon	39.94	1.66	377	1.379
Benzene	78.11	1.12	329	2.696
N-butane	58.12	1.18	335	2.006
Iso-butane	58.12	1.19	336	2.006
Carbon dioxide	44.01	1.29	346	1.519
Carbon disulphide	76.13	1.21	338	2.628
Carbon monoxide	28.01	1.40	356	0.967
Chlorine	70.90	1.35	352	2.447
Cyclohexane	84.16	1.08	325	2.905
Ethane	30.07	1.19	336	1.038
Ethyl alcohol	46.07	1.13	330	1.590
Ethyl chloride	64.52	1.19	336	2.227
Ethylene	28.03	1.24	341	0.968
Helium	4.02	1.66	377	0.139
N-heptane	100.20	1.05	321	3.459
Hexane	86.17	1.06	322	2.974
Hydrochloric acid	36.47	1.41	357	1.259
Hydrogen	2.02	1.41	357	0.070
Hydrogen chloride	36.47	1.41	357	1.259
Hydrogen sulphide	34.08	1.32	349	1.176

HVAC MATERIAL CHARACTERISTICS *(cont.)*

Gas or Vapor	Molecular Weight	Specific Ratio Heat	Coefficient C	Specific Gravity
Methane	16.04	1.31	348	0.554
Methyl alcohol	32.04	1.20	337	1.106
Methyl butane	72.15	1.08	325	2.491
Methyl chloride	50.49	1.20	337	1.743
Natural gas	19.00	1.27	344	0.656
Nitric oxide	30.00	1.40	356	1.036
Nitrogen	28.02	1.40	356	0.967
Nitrous oxide	44.02	1.31	348	1.520
N-octane	114.22	1.05	321	3.943
Oxygen	32.00	1.40	356	1.105
N-pentane	72.15	1.08	325	2.491
Iso-pentane	72.15	1.08	325	2.491
Propane	44.09	1.13	330	1.522
R-11	137.37	1.14	331	4.742
R-12	120.92	1.14	331	4.174
R-22	86.48	1.18	335	2.985
R-114	170.93	1.09	326	5.900
R-123	152.93	1.10	327	5.279
R-134a	102.03	1.20	337	3.522
Sulfur dioxide	64.04	1.27	344	2.211
Toluene	92.13	1.09	326	3.180

REFRIGERANT PHYSICAL PROPERTIES

| Refrigerant | | Ashrae Std. 15 Group No. | Molecular Mass | Boiling Point at 14.7 psia (°F) | Freezing Point (°F) | Critical | | Volume (cu. ft./lb.) |
No.	Chemical Name					Temp. (°F)	Press. (psia)	
R-11	—	A1	137.38	74.87	−168.0	388.4	639.5	0.0289
R-12	—	A1	120.93	−21.62	−252.0	233.6	596.9	0.0287
R-13	—	A1	104.47	−114.60	−294.0	83.9	561.0	0.0277
R-13B1	—	A1	148.93	−71.95	−270.0	152.6	575.0	0.0215
R-14	—	A1	88.01	−198.30	−299.0	−50.2	543.0	0.0256
R-22	—	A1	86.48	−41.36	−256.0	204.8	721.9	0.0305
R-40	—	B2	50.49	−11.60	−144.0	289.6	968.7	0.0454
R-113	—	A1	187.39	117.63	−31.0	417.4	498.9	0.0278
R-114	—	A1	170.94	38.80	−137.0	294.3	473.0	0.0275
R-115	—	A1	154.48	−38.40	−159.0	175.9	457.6	0.0261
R-123	—	B1	152.93	82.17	−160.9	362.8	532.9	—
R-134a	—	A1	102.03	−15.08	−141.9	214.0	589.8	0.0290
R-142b	—	A2	100.50	14.40	−204.0	278.8	598.0	0.0368
R-152a	—	A2	66.05	−13.00	−178.6	236.3	652.0	0.0439

4-4

R-170	Ethane	A3	30.07	-127.85	-297.0	90.0	709.8	0.0830
R-290	Propane	A3	44.10	-43.73	-305.8	206.3	617.4	0.0728
R-C318	—	A1	200.04	21.50	-42.5	239.6	403.6	0.0258
R-500	—	A1	99.31	-28.30	-254.0	221.9	641.9	0.0323
R-502	—	A1	111.63	-49.80	—	179.9	591.0	0.0286
R-503	—	A1	87.50	-127.60	—	67.1	607.0	0.0326
R-600	Butane	A3	58.13	31.10	-217.3	305.6	550.7	0.0702
R-600a	Isobutane	A3	58.13	10.89	-255.5	275.0	529.1	0.0725
R-611	—	B2	60.05	89.20	-146.0	417.2	870.0	0.0459
R-717	Ammonia	B2	17.03	-28.00	-107.9	271.4	1657.0	0.0680
R-744	Carbon dioxide	A1	44.01	-109.20	-69.9	87.9	1070.0	0.0342
R-764	Sulfur dioxide	B1	64.07	14.00	-103.9	315.5	1143.0	0.0306
R-1150	Ethylene	A3	28.05	-154.7	-272.0	48.8	742.2	0.0700
R-1270	Propylene	A3	42.09	-53.86	-301.0	197.2	670.3	0.0720

PROPERTIES OF LIQUID AND SATURATED VAPOR OF R-12

Temp. (°F)	Pressure		Density of Liquid (lb./cu. ft.)	Volume of Vapor (cu. ft./lb.)	Heat Content (Enthalpy) (Btu/lb.)	
	psia	psig			Liquid	Vapor
−150	0.154	29.61*	104.36	178.65	−22.70	60.8
−125	0.516	28.67*	102.29	57.28	−17.59	63.5
−100	1.428	27.01*	100.15	22.16	−12.47	66.2
−75	3.388	23.02*	97.93	9.92	−7.31	69.0
−50	7.117	15.43*	95.62	4.97	−2.10	71.8
−25	13.556	2.32*	93.20	2.73	3.17	74.56
−15	17.141	2.45	92.20	2.19	5.30	75.65
−10	19.189	4.49	91.69	1.97	6.37	76.2
−5	21.422	6.73	91.18	1.78	7.44	76.73
0	23.849	9.15	90.66	1.61	8.52	77.27
5	26.483	11.79	90.14	1.46	9.60	77.80

4-6

10	29.335	14.64	89.61	1.32	10.68	78.335
25	39.310	24.61	87.98	1.00	13.96	79.9
50	61.394	46.70	85.14	0.66	19.51	82.43
75	91.682	76.99	82.09	0.44	25.20	84.82
86	108.04	93.34	80.67	0.38	27.77	85.82
100	131.86	117.16	78.79	0.31	31.10	87.03
125	183.76	169.06	75.15	0.22	37.28	88.97
150	249.31	234.61	71.04	0.16	43.85	90.53
175	330.64	315.94	66.20	0.11	51.03	91.48
200	430.09	415.39	60.03	0.08	59.20	91.28

*Inches of mercury below one atmosphere.

PROPERTIES OF LIQUID AND SATURATED VAPOR OF R-22

Temp (°F)	Pressure psia	Pressure psig	Density of Liquid (lb./cu. ft.)	Volume of Vapor (cu. ft./lb.)	Heat Content (Enthalpy) (Btu/lb.) Liquid	Heat Content (Enthalpy) (Btu/lb.) Vapor
−150	0.272	29.37*	98.24	141.23	−25.97	87.52
−125	0.886	28.12*	96.04	46.69	−20.33	90.43
−100	2.398	25.04*	93.77	18.43	−14.56	93.37
−75	5.610	18.50	91.43	8.36	−8.64	96.29
−50	11.674	6.15*	89.00	4.22	−2.51	99.14
−25	22.086	7.39	86.48	2.33	3.83	101.88
−15	27.865	13.17	85.43	1.87	6.44	102.94
−10	31.162	16.47	84.90	1.68	7.75	103.46
−5	34.754	20.06	84.37	1.52	9.08	103.96
0	38.657	23.96	83.83	1.37	10.41	104.47

5	42.888	28.19	83.28	1.24	11.75	104.96
10	47.464	32.77	82.72	1.13	13.10	105.44
25	63.450	48.75	81.02	0.86	17.22	106.84
50	98.727	84.03	78.03	0.56	24.28	108.95
75	146.91	132.22	74.80	0.37	31.61	110.74
86	172.87	158.17	73.28	0.32	34.93	111.40
100	210.60	195.91	71.24	0.26	39.27	112.11
125	292.62	277.92	67.20	0.18	47.37	112.88
150	396.19	381.50	62.40	0.12	56.14	112.73

*Inches of mercury below one atmosphere.

4-9

PROPERTIES OF LIQUID AND SATURATED VAPOR OF R-134A

Temp (°F)	Pressure psia	Pressure psig	Density of Liquid (lb./cu. ft.)	Volume of Vapor (cu. ft./lb.)	Heat Content (Enthalpy) (Btu./lb.) Liquid	Heat Content (Enthalpy) (Btu./lb.) Vapor
−150	0.07107	29.776*	102.344	457.0719	−32.781	80.212
−125	0.28333	29.344*	99.641	123.4418	−25.383	83.716
−100	0.89915	28.090*	96.891	41.5241	−17.939	87.245
−75	2.3866	25.062*	94.087	16.5646	−10.472	90.760
−50	5.4966	18.730*	91.220	7.5560	−2.995	94.248
−25	11.2964	6.9214*	88.278	3.8338	4.503	97.721
−15	14.6686	0.0555*	87.078	2.9960	7.518	99.109
−10	16.6293	1.9334	86.472	2.6610	9.030	99.804
−5	18.7906	4.0947	85.862	2.3705	10.546	100.499
0	21.1665	6.4706	85.248	2.1177	12.067	101.195

4-10

5	23.7710	9.0751	84.630	1.8969	13.593	101.891
10	26.619	11.923	84.007	1.7036	15.125	102.587
25	36.773	22.078	82.110	1.25178	19.763	104.677
50	60.032	45.335	78.836	0.78067	27.666	108.149
75	93.080	78.384	75.387	0.50743	35.851	111.553
86	111.321	96.626	73.799	0.42412	39.560	113.004
100	138.28	123.58	71.701	0.33993	44.393	114.782
125	198.27	183.57	67.679	0.23204	53.385	117.660
150	276.12	261.42	63.126	0.15914	62.989	119.879
175	375.69	36C.99	57.601	0.10705	73.581	120.788
200	502.54	487.85	49.439	0.06542	86.528	118.155

*Inches of mercury below one atmosphere.

4-11

PROPERTIES OF LIQUID AND SATURATED VAPOR OF R-500

Temp (°F)	Pressure		Density of Liquid (lb./cu. ft.)	Volume of Vapor (cu. ft./lb.)	Heat Content (Enthalpy) (Btu/lb.)	
	psia	psig			Liquid	Vapor
−40	10.95	7.62*	84.28	4.0	0.00	87.74
−30	14.10	1.22*	83.35	3.15	2.38	89.04
−20	17.92	3.23	82.40	2.52	4.79	90.31
−10	22.52	7.82	81.44	2.03	7.22	91.57
0	27.98	13.3	80.46	1.66	9.71	92.81
5	31.07	16.4	79.96	1.501	10.96	93.42
10	34.43	19.7	79.46	1.36	12.23	94.03
20	41.96	27.3	78.45	1.13	14.79	95.22
30	50.70	36.0	77.41	0.94	17.40	96.39
40	60.75	46.1	76.34	0.79	20.05	97.53

50	72.26	57.6	75.26	0.67	22.75	98.64
60	85.33	70.6	74.14	0.57	25.48	99.71
70	100.1	85.4	72.98	0.48	28.28	100.75
80	116.7	102.0	71.80	0.42	31.12	101.75
86	127.6	113.0	71.06	0.38	32.85	102.33
90	135.3	121.0	70.56	0.36	34.01	102.70
100	155.9	141.0	69.28	0.31	36.97	103.60
110	178.8	164.0	67.95	0.27	40.00	104.44
120	204.1	189.0	66.55	0.23	43.10	105.22
130	231.9	217.0	65.08	0.20	46.29	105.91
140	262.4	248.0	63.51	0.17	49.58	106.51

*Inches of mercury vacuum.

PROPERTIES OF LIQUID AND SATURATED VAPOR OF R-502

Temp (°F)	Pressure		Density of Liquid (lb./cu. ft.)	Volume of Vapor (cu. ft./lb.)	Heat Content (Enthalpy) (Btu/lb.)	
	psia	psig			Liquid	Vapor
-100	3.261	23.281*	97.857	10.461	-12.548	65.885
-75	7.281	15.097*	95.234	4.959	-7.597	68.919
-50	14.602	0.190*	92.513	2.596	-2.251	71.928
-25	26.817	12.121	89.673	1.465	3.496	74.866
-20	30.006	15.310	89.088	1.317	4.693	75.442
-15	33.480	18.784	88.496	1.187	5.905	76.012
-10	37.256	22.560	87.898	1.073	7.133	76.577
-5	41.349	26.653	87.293	0.973	8.376	77.137
0	45.775	31.079	86.681	0.881	9.633	77.690
5	50.553	35.857	86.062	0.801	10.906	78.237
10	55.697	41.001	85.434	0.731	12.193	78.777

4-14

15	61.225	46.529	84.797	0.666	13.494	79.310
20	67.155	52.459	84.152	0.612	14.809	79.836
25	73.503	58.807	83.497	0.557	16.138	80.353
50	112.12	97.42	80.058	0.367	22.977	82.800
75	163.81	149.11	76.269	0.248	30.122	84.958
86	191.28	176.59	74.453	0.210	33.359	85.789
100	230.89	216.19	71.967	0.171	37.563	86.711
125	316.04	301.35	66.838	0.118	45.361	87.834
150	423.06	408.55	60.092	0.079	53.850	87.757
160	473.38	458.69	56.429	0.066	57.732	87.013

*Inches of mercury below one atmosphere.

PROPERTIES OF LIQUID AND SATURATED VAPOR OF R-503						
Temp (°F)	Pressure		Density of Liquid (lb./cu. ft.)	Volume of Vapor (cu. ft./lb.)	Heat Content (Enthalpy) (Btu/lb.)	
	psia	psig			Liquid	Vapor
−140	9.234	11.1*	93.49	4.123	−26.45	52.88
−130	12.98	3.49*	92.47	2.998	−23.93	53.84
−120	17.83	3.13	91.39	2.227	−21.36	54.77
−110	23.98	9.28	90.25	1.685	−18.77	55.66
−100	31.64	16.9	89.05	1.296	−16.16	56.52
−90	41.05	26.3	87.78	1.012	−13.53	57.35
−80	52.42	37.7	86.44	0.8008	−10.87	58.13
−70	66.00	51.3	85.02	0.6409	−8.19	58.86
−60	82.05	67.4	83.52	0.5182	−5.49	59.54
−50	100.8	86.1	81.93	0.4227	−2.76	60.16
−40	122.6	108	80.25	0.3474	0.00	60.72

-30	147.6	133	78.46	0.2872	2.81	61.20
-20	176.2	161	76.56	0.2387	5.66	61.60
-10	208.6	194	74.52	0.1991	8.59	61.89
0	245.3	231.0	72.33	0.1664	11.60	62.05
5	265.9	251.5	70.99	0.15275	13.17	62.04
10	286.4	272.0	69.65	0.1391	14.74	62.04
20	332.6	318	67.35	0.1160	18.05	61.82
30	384.1	369	64.45	0.0962	21.60	61.31
40	440.6	426	61.12	0.0793	26.03	60.45
50	503.3	489	57.09	0.0640	26.69	58.95
60	574.8	560	51.40	0.0485	34.32	55.77

*Inches of mercury vacuum.

4-17

SATURATION TEMPERATURE–PRESSURE CURVES FOR GROUP A REFRIGERANTS

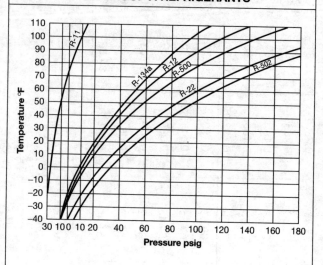

SATURATION TEMPERATURE–PRESSURE CURVES FOR GROUP B REFRIGERANTS

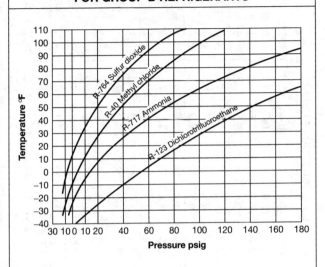

REFRIGERANT TEMPERATURE–PRESSURE CHART

°F	22	410A	12	134a	401a	409A	502	404a	507	408A	402A
-40	4.8	6.9	14.6	17.7	17.7	16.4	0.7	0.8	1.7	1.0	2.1
-44	2.0	9.1	12.9	16.2	16.0	14.8	2.3	2.5	3.5	1.1	3.9
-40	0.5	11.6	11.0	14.5	14.5	13.1	4.1	5.5	5.5	2.8	5.9
-36	2.2	14.2	8.9	12.8	12.5	11.2	6.0	7.5	7.6	4.6	8.0
-32	4.0	17.1	6.7	10.8	10.6	9.2	8.1	9.7	9.9	6.6	10.3
-28	5.9	20.1	4.3	8.6	8.3	6.9	10.3	12.0	12.4	8.7	12.8
-24	7.9	23.4	1.6	6.2	6.0	4.5	12.7	14.5	15.0	11.0	15.5
-20	10.1	26.9	0.6	3.6	3.5	1.9	15.3	17.1	17.8	13.5	18.4
-16	12.5	30.7	2.1	0.8	0.5	0.5	18.1	20.0	20.9	16.1	21.5
-12	15.1	34.7	3.7	1.1	1.4	2.0	21.0	23.0	24.1	18.9	24.8
-8	17.9	39.0	5.4	2.8	3.1	3.6	24.2	26.3	27.6	21.9	28.3
-4	20.8	43.7	7.2	4.5	4.8	5.3	27.5	29.8	31.3	25.2	32.1
0	24.0	48.6	9.2	6.5	6.7	7.2	31.1	33.5	35.2	28.7	36.1
2	25.6	51.1	10.2	7.5	8.0	8.2	32.9	34.8	37.3	30.5	38.1
4	27.3	53.8	11.2	8.5	8.8	9.2	34.9	37.4	39.4	32.3	40.4
6	29.1	56.7	12.3	9.6	9.9	10.2	36.9	39.4	41.6	34.3	42.6
8	30.9	59.4	13.5	10.8	11.0	11.3	38.9	41.6	43.8	36.3	44.9
10	32.8	62.3	14.6	12.0	12.2	12.5	41.0	43.7	46.2	38.3	47.3

12	34.7	65.4	15.8	13.1	13.4	13.6	43.2	46.0	48.5	40.4	49.7
14	36.7	68.6	17.1	14.4	14.6	14.8	45.4	48.3	51.0	42.6	52.2
16	38.7	71.9	18.4	15.7	15.9	16.1	47.7	50.7	53.5	44.9	54.8
18	40.9	75.2	19.7	17.0	17.2	17.4	50.0	53.1	56.1	47.2	57.5
20	43.0	78.3	21.0	18.4	18.6	18.7	52.5	55.6	58.8	49.6	60.2
22	45.3	82.3	22.4	19.9	20.0	20.1	54.9	58.2	61.5	52.0	63.0
24	47.6	85.9	23.9	21.4	21.5	21.5	57.5	60.9	64.3	54.5	65.9
26	49.9	89.7	25.4	22.9	23.0	22.9	60.1	63.6	67.2	57.1	68.9
28	52.4	93.5	26.9	24.5	24.6	24.4	62.8	66.5	70.2	59.8	72.0
30	54.9	96.8	28.5	26.1	26.2	26.0	65.6	69.4	73.3	62.5	75.1
32	57.5	101.6	30.1	27.8	27.9	27.6	68.4	72.3	76.4	65.3	78.3
34	60.1	105.7	31.7	29.5	29.6	29.2	71.3	75.4	79.6	68.2	81.6
36	62.8	110.0	33.4	31.3	31.3	30.9	74.3	78.5	82.9	71.2	85.0
38	65.6	114.4	35.2	33.2	33.2	32.7	77.4	81.8	86.3	74.2	88.5
40	68.5	118.0	36.9	35.1	35.0	34.5	80.5	85.1	89.8	77.4	92.1
42	71.5	123.6	38.8	37.0	37.0	36.3	83.8	88.5	93.4	80.6	95.7
44	74.5	128.3	40.7	39.1	39.0	38.2	87.0	91.9	97.0	83.9	99.5
46	77.6	133.2	42.7	41.1	41.0	40.2	90.4	95.5	100.8	87.3	103.4

Values in bold italic = Vacuum Other values = Vapor psig (calculating superheat)

4-21

REFRIGERANT TEMPERATURE–PRESSURE CHART (cont.)

°F	22	410A	12	134a	401a	409A	502	404a	507	408A	402A
48	80.7	138.2	44.7	43.3	43.1	42.2	93.9	99.2	104.6	90.7	107.3
50	84.0	142.2	46.7	45.5	45.3	44.3	97.4	102.9	108.6	94.3	111.4
52	87.3	148.5	48.8	47.7	60.0	63.6	101.0	109.0	112.6	97.9	120.0
56	94.3	159.3	53.2	52.3	65.0	68.9	108.6	117.0	121.0	105.5	129.0
60	101.6	169.6	57.7	57.5	70.0	74.5	116.4	125.0	129.7	113.5	138.0
64	109.3	182.6	62.5	62.7	76.0	80.3	124.6	134.0	139.0	121.8	147.0
68	117.3	195.0	67.6	68.3	82.0	86.3	133.2	144.0	148.6	130.6	157.0
72	125.7	208.1	72.9	74.2	89.0	92.8	142.2	153.0	158.7	139.7	168.0
76	134.5	221.7	78.4	80.3	95.0	99.4	151.5	164.0	169.3	149.3	179.0
80	143.6	235.3	84.2	86.8	102.0	106.4	161.2	174.0	180.3	159.4	190.0
84	153.2	250.8	90.2	93.6	109.0	113.7	171.4	185.0	191.9	169.8	202.0
88	163.2	266.3	96.5	100.7	117.0	121.2	181.9	197.0	203.9	180.8	214.0
92	173.7	282.5	103.1	108.2	125.0	129.1	192.9	209.9	216.6	192.2	227.0
96	184.6	299.3	110.0	116.1	133.0	137.4	204.3	222.0	229.8	204.1	240.0
100	195.9	317.2	117.2	124.3	142.0	146.0	216.2	235.0	243.5	216.6	254.0

104	207.7	335.2	124.7	132.9	*151.0*	*154.9*	228.5	*249.0*	257.9	229.5	*269.0*
108	220.0	354.3	132.4	142.8	*160.0*	*164.2*	241.3	*264.0*	272.9	243.0	*284.0*
112	232.8	374.2	140.5	151.3	*170.0*	*173.9*	254.6	*279.0*	288.6	257.0	*299.0*
116	246.1	394.8	148.9	161.1	*180.0*	*183.9*	268.4	*294.0*	304.9	271.6	*316.0*
120	259.9	417.7	157.7	171.3	*191.0*	*194.4*	282.7	*311.0*	321.9	286.8	*332.0*
124	274.3	438.7	166.7	182.0	*202.0*	*205.2*	297.6	*328.0*	339.7	302.6	*350.0*
128	289.1	462.0	176.2	193.1	*213.0*	*216.5*	312.9	*345.0*	358.2	319.0	*368.0*
132	304.6	486.2	185.9	204.7	*225.0*	*228.2*	328.9	*364.0*	377.6	336.0	*387.0*
136	320.6	511.4	196.1	216.8	*237.0*	*240.3*	345.4	*383.0*	397.7	353.6	*406.0*
140	337.3	539.0	206.6	229.4	*250.0*	*252.9*	362.6	*402.0*	418.7	372.0	*426.0*
144	354.5	564.8	217.5	242.4	*263.0*	*265.9*	380.4	*423.0*	440.6	390.9	*447.0*
148	372.3	593.8	228.8	256.0	*277.0*	*279.5*	398.9	*444.0*	462.0	410.6	*468.0*

Other values = Vapor psig (calculating superheat)
Values in bold italic = Liquid psig (calculating subcooling)

DESIGN PRESSURES FOR REFRIGERANT TESTING

Refrigerant	Name	Chemical Formula	Minimum Design Pressures (psig)		
			Low Side	High Side	
				Water or Evaporation Cooled	Air Cooled
R-11	Trichlorofluoromethane	CCl_3F	15	15	21
R-12	Dichlorodifluoromethane	CCl_2F_2	85	127	169
R-13	Chlorotrifluoromethane	$CClF_3$	521	547	547
R-13B1	Bromotrifluoromethane	$CBrF_3$	230	321	410
R-14	Tetrafluoromethane	CF_4	529	529	529
R-21	Dichlorofluoromethane	$CHCl_2F$	15	29	46
R-22	Chlorodifluoromethane	$CHClF_2$	144	211	278
R-30	Methylene chloride	CH_2Cl_2	15	15	15
R-40	Methyl chloride	CH_3Cl	72	112	151
R-113	Trichlorotrifluoroethane	CCl_2FCClF_2	15	15	15
R-114	Dichlorotetrafluoroethane	$CClF_2CClF_2$	18	35	53
R-115	Chloropentafluoroethane	$CClF_2CF_3$	123	181	238

4-24

DESIGN PRESSURES FOR REFRIGERANT TESTING (cont.)

Refrigerant	Name	Chemical Formula	Minimum Design Pressures (psig)		
			Low Side	High Side	
				Water or Evaporation Cooled	Air Cooled
R-170	Ethane	C_2H_6	618	695	695
R-290	Propane	C_3H_8	129	188	244
R-C318	Octafluorocyclobutane	C_4H_8	34	59	85
R-500	Dichlorodifluoromethane, 73.8% and ethylidene fluoride, 26.2%	CCl_2F_2/CH_3CHF_2	102	153	203
R-502	Chlorodifluoromethane, 48.8% and chloropentafluoroethane, 51.2%	$CHClF_2/CClF_2CF_2$	162	232	302
R-600	N-butane	C_4H_{10}	23	42	61
R-601	Isobutane	$CH(CH_3)_3$	39	63	88
R-611	Methyl formate	$HCOOCH_3$	15	15	15
R-717	Ammonia	NH_3	139	215	293
R-744	Carbon dioxide	CO_2	955	1058	1058
R-764	Sulfur dioxide	SO_2	45	78	115
R-1150	Ethylene	C_2H_4	732	732	732

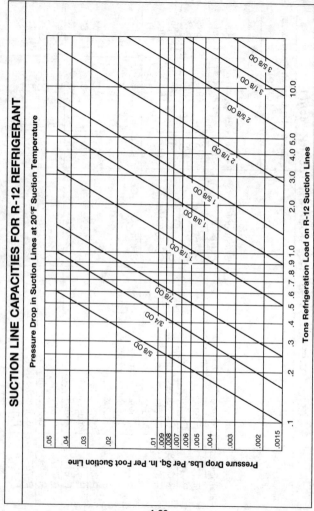

SUCTION LINE CAPACITIES FOR R-12 REFRIGERANT

Pressure Drop in Suction Lines at 20°F Suction Temperature

Pressure Drop Lbs. Per Sq. In. Per Foot Suction Line

Tons Refrigeration Load on R-12 Suction Lines

4-26

REFRIGERATION APPLICATIONS AND LOW-SIDE PRESSURE SETTINGS IN INCHES VACUUM

Application	R-12 (in. vacuum) Out	In	R-22 (in. vacuum) Out	In	R-502 (in. vacuum) Out	In
Freezer–open type	7	5	4	17	9	23
Freezer–closed type	1	8	11	22	17	29
Ice cube maker–flooded or dry type coil	4	17	16	37	22	47
Sweet water bath–soda fountain	21	29	43	56	52	67
Showcase–frost cycle	10	25	25	50	32	60
Showcase–defrost cycle	18	34	39	64	49	76
Beer, water, milk cooler	19	29	40	56	47	67
Walk-in cooler–defrost cycle	12	35	29	66	37	77
Ice-cream trucks hardening rooms	2	15	12	33	17	42
Vegetable display–defrost cycle	11	35	27	66	35	77
Eutectic brine tank ice cream truck	1	4	11	16	17	22
Reach-in cooler–defrost cycle	18	36	39	68	47	79
Beer coolers–blower dry type	15	34	33	64	42	76
Beer coolers–bare pipe dry type–frost cycle	12	27	29	53	37	64
Instantaneous beer coolers	12	29	29	56	37	67
Retail florist box–blower coil	26	42	51	77	61	88

COMPRESSOR REFRIGERANTS

Refrigerant	Type of Compressor	Application
R-11	Centrifugal	Large air conditioning systems ranging from 200 to 2,000 tons in capacity. Refrigerating systems for industrial process water and brine.
R-123	Centrifugal	Replacement for R-11.
R-12	Reciprocating, centrifugal, rotary	Large air conditioning and refrigeration systems. Small household refrigerators, water coolers, room and window air conditioners, automobile air conditioning.
R-134a	(Same as R-12)	Replacement for R-12.
R-401A	(Same as R-12)	Replacement for R-12.
R-22	Reciprocating, centrifugal, rotary	Residential and commercial air conditioning. Food-freezing plants, frozen-food storage, and display cases.
R-500	Reciprocating, centrifugal	Small home and commercial air conditioning equipment. Household refrigeration and commercial chillers.
R-502	Reciprocating	Frozen food and ice cream display cases, warehouses and food freezing plants. Medium-temperature display cases, truck refrigeration, and heat pumps.
R-407A	(Same as R-502)	Replacement for R-502.
R-407B	(Same as R-502)	Replacement for R-502.
R-507A	(Same as R-502)	Replacement for R-502.
R-503	Reciprocating	Low-temperature systems to $-130°F$ $(-90°C)$.
R-13	Reciprocating	Low-temperature systems to $-130°F$ $(-90°C)$ in cascade systems.
R-113	Centrifugal	Small to medium air conditioning systems. Industrial cooling, food freezing, and storage.

AVERAGE COMPRESSOR CAPACITIES (BTU/h)				
	Evaporating Temperatures (°F)			
	−30°		−15°	
	Condensing Temperatures (°F)			
HP	110°	120°	110°	120°
2	5,200	4,500	9,100	8,200
3	9,000	8,300	14,300	13,200
5	14,200	12,500	24,800	22,400
7½	25,000	20,000	31,000	28,000
10	31,000	26,000	43,600	44,800
15	42,600	37,500	74,400	67,200
20	56,000	44,700	82,000	71,000
25	70,000	56,000	96,000	85,000
30	80,000	67,000	116,500	102,500
40	94,000	75,000	155,000	135,000
50	122,000	100,000	188,500	159,500
60	168,000	134,000	240,000	220,000
70	196,000	156,000	272,000	239,000
75	210,000	167,000	291,000	256,000
80	224,000	178,000	310,000	273,000
90	252,000	201,000	349,000	307,000
100	280,000	223,000	388,000	341,000

AVERAGE COMPRESSOR CAPACITIES (BTU/h) *(cont.)*				
	Evaporating Temperatures (°F)			
	+20°		+40°	
	Condensing Temperatures (°F)			
HP	110°	120°	110°	120°
2	18,000	16,800	22,800	21,200
3	22,100	20,800	36,300	34,200
5	41,700	39,300	62,400	58,500
7½	53,000	48,000	87,000	81,700
10	81,000	75,000	120,000	112,000
15	111,000	102,000	171,600	160,000
20	154,000	142,000	235,000	218,000
25	188,000	174,000	283,000	263,000
30	225,000	210,000	349,000	324,000
40	325,000	306,000	439,000	406,000
50	375,000	350,000	585,000	550,000
60	450,000	420,000	710,000	670,000
70	571,000	534,000	800,000	742,000
75	582,000	542,000	855,000	795,000
80	622,000	578,000	900,000	842,000
90	750,000	700,000	1,027,000	955,000
100	777,000	723,000	1,170,000	1,100,000

SUCTION REFRIGERANT LINE SIZES PER COMPRESSOR CAPACITY					
	Length of Run (ft.)				
Compressor Capacity (Btu/h)	15	25	35	50	100
	Tube Dia.	Tube Dia.	Tube Dia.	Tube Dia.	Tube Dia.
18,500–20,000	5/8"	5/8"	5/8"	—	—
20,000–22,000	5/8"	5/8"	5/8"	—	—
22,000–24,000	5/8"	5/8"	5/8"	3/4"	3/4"
24,000–34,000	5/8"	5/8"	3/4"	3/4"	3/4"
38,000–40,000	3/4"	3/4"	3/4"	7/8"	7/8"
40,000–44,000	3/4"	7/8"	7/8"	7/8"	7/8"
44,000–51,000	7/8"	7/8"	7/8"	7/8"	7/8"
53,000–66,000	7/8"	7/8"	7/8"	1 1/8"	1 1/8"

Suction Line

Add 3 fluid ounces for each 10' of pipe over 35'.
Table presumes use of R-12.
Tube diameters are in inches.

LIQUID REFRIGERANT LINE SIZES PER COMPRESSOR CAPACITY

Compressor Capacity (BTU/h)	Length of Run (ft.)				
	15 Tube Dia.	25 Tube Dia.	35 Tube Dia.	50 Tube Dia.	100 Tube Dia.
18,500–20,000	5/16"	5/16"	5/16"	—	—
20,000–22,000	5/16"	5/16"	5/16"	—	—
22,000–24,000	5/16"	3/8"	3/8"	3/8"	3/8"
24,000–34,000	5/16"	3/8"	3/8"	3/8"	3/8"
38,000–40,000	5/16"	3/8"	3/8"	3/8"	3/8"
40,000–40,000	3/8"	3/8"	3/8"	3/8"	3/8"
44,000–51,000	3/8"	3/8"	3/8"	3/8"	3/8"
53,000–66,000	1/2"	1/2"	1/2"	1/2"	1/2"

Liquid Line

Add 3 fluid ounces for each 10' of pipe over 35'.
Table presumes use of R-12.
Tube diameters are in inches.

DISCHARGE REFRIGERANT LINE SIZES PER COMPRESSOR CAPACITY

Compressor Capacity (Btu/h)	Length of Run (ft.)				
	15 Tube Dia.	25 Tube Dia.	35 Tube Dia.	50 Tube Dia.	100 Tube Dia.
18,500–20,000	5/16"	3/8"	3/8"	—	—
20,000–22,000	3/8"	3/8"	3/8"	—	—
22,000–24,000	3/8"	3/8"	3/8"	1/2"	1/2"
24,000–34,000	3/8"	3/8"	1/2"	1/2"	1/2"
38,000–40,000	3/8"	1/2"	1/2"	1/2"	1/2"
40,000–44,000	3/8"	1/2"	1/2"	1/2"	1/2"
44,000–51,000	3/8"	1/2"	1/2"	1/2"	5/8"
53,000–66,000	1/2"	1/2"	5/8"	5/8"	3/4"

Discharge Line

Add 3 fluid ounces for each 10' of pipe over 35'.
Table presumes use of R-12.
Tube diameters are in inches.

4-33

RESISTANCE OF VALVES, ELBOWS, AND TEES OVER STRAIGHT LENGTHS OF PIPE

Type of Fitting	All Tube and Pipe Sizes with OD in Inches													
	1/4	3/8	1/2	5/8	3/4	7/8	1 1/8	1 3/8	1 5/8	2 1/8	2 5/8	3 1/8	3 5/8	4 1/8
	Amount of Feet to Be Added for Each Fitting													
Valve	1.5	1.5	2	2	2.5	3.0	4.0	5.0	6.0	7.5	9.0	11.0	13.0	15.0
Elbow (90°)	.75	.75	1	1	1.5	1.5	2	2.5	3	4.0	5.0	5.5	6.5	7.5
Tee	1.5	1.5	2	2	2.5	3.0	4.0	5.0	6.0	7.5	9.0	11.0	13.0	15.0

CAPILLARY TUBE LENGTH AND DIAMETER FOR REFRIGERANTS

R-12

Compressor Horsepower	Condenser Fan Type	Low Temperature		Medium Temperature		High Temperature	
		Length (in.)	ID (in.)	Length (in.)	ID (in.)	Length (in.)	ID (in.)
1/8	Fan	108	.028	84	.028	48	.028
1/8	Static	118	.028	92	.028	53	.028
1/6	Fan	120	.031	96	.031	72	.031
1/6	Static	132	.031	105	.031	79	.031
1/5	Fan	54	.031	36	.031	24	.031
1/5	Static	60	.031	39	.031	26	.031
1/4	Fan	43	.031	90	.040	60	.040
1/4	Static	47	.031	99	.040	66	.040
1/3	Fan	93	.040	72	.040	72	.040
1/2	Fan	96	.052	48	.052	90	.064
3/4	Fan	60	.052	92	.064	72	.064
1	Fan	132	.064	84	.064	54	.064
1 1/2	Fan	84	.064	60	.064	43	.064
2	Fan	55	.064	40	.064	26	.064

CAPILLARY TUBE LENGTH AND DIAMETER FOR REFRIGERANTS (cont.)

R-134A

Compressor Horsepower	Condenser Fan Type	Low Temperature		Medium Temperature		High Temperature	
		Length (in.)	ID (in.)	Length (in.)	ID (in.)	Length (in.)	ID (in.)
1/8	Fan	118	.028	96	.028	58	.028
1/8	Static	130	.028	106	.028	64	.028
1/6	Fan	132	.031	110	.031	86	.031
1/6	Static	144	.031	121	.031	95	.031
1/5	Fan	60	.031	41	.031	29	.031
1/5	Static	69	.031	45	.031	32	.031
1/4	Fan	47	.031	103	.040	72	.040
1/4	Static	52	.031	113	.040	79	.040
1/3	Fan	102	.040	83	.040	43	.040
1/2	Fan	105	.052	55	.052	108	.064
3/4	Fan	66	.052	106	.064	86	.064
1	Fan	144	.064	96	.064	65	.064
1½	Fan	92	.064	49	.064	52	.064
2	Fan	61	.064	46	.064	31	.064

R-22

HP							
⅕	Fan	65	.031	43	.031	29	.031
¼	Fan	52	.031	108	.040	72	.031
⅓	Fan	112	.040	86	.040	43	.040
½	Fan	115	.052	58	.052	108	.064
¾	Fan	72	.052	110	.064	86	.064
1	Fan	48	.052	101	.064	65	.064
1½	Fan	101	.064	72	.064	52	.064
2	Fan	66	.064	48	.064	31	.064

R-502

HP							
⅓	Fan	122	.040	95	.040	48	.040
½	Fan	127	.052	63	.052	119	.052
¾	Fan	79	.052	121	.064	94	.064
1	Fan	53	.064	119	.064	71	.064
1½	Fan	111	.064	79	.064	57	.064
2	Fan	73	.064	53	.064	34	.064

SATURATED VAPOR TEMPERATURE-PRESSURE CHART, FREON-12 AND FREON-22

Temp (°F)	Gauge Pressure (psi)		Temp (°F)	Gauge Pressure (psi)	
	Freon-12	Freon-22		Freon-12	Freon-22
−40	10.92*	0.61	10	14.65	32.93
−38	9.91*	1.42	12	15.86	34.88
−36	8.87*	2.27	14	17.10	36.89
−34	7.80*	3.15	16	18.38	38.96
−32	6.66*	4.07	18	19.70	41.09
−30	5.45*	5.02	20	21.05	43.28
−28	4.23*	6.01	22	22.45	45.53
−26	2.93*	7.03	24	23.88	47.85
−24	1.63*	8.09	26	25.37	50.24
−22	.24*	9.18	28	26.89	52.70
−20	.58	10.31	30	28.46	55.23
−18	1.31	11.48	32	30.07	57.83
−16	2.07	12.61	34	31.72	60.51
−14	2.85	13.94	36	33.43	63.27
−12	3.67	15.24	38	35.18	66.11
−10	4.50	16.59	40	36.98	69.02
−8	5.38	17.99	42	38.81	71.99
−6	6.28	19.44	44	40.70	75.04
−4	7.21	20.94	46	42.65	78.18
−2	8.17	22.49	48	44.65	81.40
0	9.17	24.09	50	46.69	84.70
2	10.19	25.73	52	48.79	88.10
4	11.26	27.44	54	50.93	91.5
6	12.35	29.21	56	53.14	95.1
8	13.48	31.04	58	55.40	98.8

*Inches vacuum.

SATURATED VAPOR TEMPERATURE-PRESSURE CHART, FREON-12 AND FREON-22 (cont.)

Temp (°F)	Gauge Pressure (psi)		Temp (°F)	Gauge Pressure (psi)	
	Freon-12	Freon-22		Freon-12	Freon-22
60	57.71	102.5	112	140.1	235.2
62	60.07	106.3	114	144.2	241.9
64	62.50	110.2	116	148.4	248.7
66	64.97	114.2	118	152.7	255.6
68	67.54	118.3	120	157.1	262.6
70	70.12	122.5	122	161.5	269.7
72	72.80	126.8	124	166.1	276.9
74	75.50	131.2	126	170.7	284.1
76	78.30	135.7	128	175.4	291.4
78	81.15	140.3	130	180.2	298.8
80	84.06	145.0	132	185.1	306.3
82	87.00	149.8	134	190.1	314.0
84	90.1	154.7	136	195.2	321.9
86	93.2	159.8	138	200.3	329.9
88	96.4	164.9	140	205.6	338.0
90	99.6	170.1	142	210.3	346.3
92	103.0	175.4	144	216.1	355.0
94	106.3	180.9	146	222.0	364.3
96	109.8	186.5	148	227.4	374.1
98	113.3	192.1	150	232.3	384.3
100	116.9	197.9	152	238.8	392.3
102	120.6	203.8	154	244.0	401.3
104	124.3	209.9	156	251.0	411.3
106	128.1	216.0	158	257.2	421.8
108	132.1	222.3	160	263.2	433.3
110	136.0	228.7			

CONDENSER CHART FREON-12

Typical head pressures and AIR OFF temperatures at different suction pressures and AIR ON temperatures.

Suction Pressure (psig)	Coil Temp. °f	AIR ON Condenser Temperatures (All Temps. IN°F)																	
		75°		80°		85°		90°		95°		100°		105°		110°		115°	
		Head Press. psig	Air Off Temp.	Head Press. psig	Air Off Temp.	Head Press. psig	Air Off Temp.	Head Press. psig	Air Off Temp.	Head Press. psig	Air Off Temp.	Head Press. psig	Air Off Temp.	Head Press. psig	Air Off Temp.	Head Press. psig	Air Off Temp.	Head Press. psig	Air Off Temp.
30	32	121	89	130	94	138	99	147	104	156	108	166	113	175	117	186	122	197	127
32	34	124	90	132	95	140	99	150	104	159	109	169	114	178	118	189	123	200	127
33	36	126	91	134	96	142	100	152	105	161	109	171	114	180	118	192	123	203	128
35	38	128	92	136	97	144	101	154	106	164	110	174	115	184	119	195	124	207	129
37	40	131	92	139	97	147	102	157	107	167	111	177	116	187	120	198	125	209	129
39	42	133	93	141	98	150	102	160	107	170	111	180	116	189	120	200	125	211	130
41	44	136	94	144	99	153	103	163	108	173	112	183	117	192	121	203	126	213	130
43	46	138	94	146	99	155	103	165	108	175	112	185	117	194	121	205	126	216	130
45	48	140	95	148	100	157	104	167	109	177	113	187	118	197	122	208	127	218	131
47	50	142	95	151	100	160	104	170	109	179	113	189	118	199	122	210	127	221	132
49	52	145	96	154	101	163	105	173	110	182	114	192	119	202	123	213	128	224	132
51	54	148	97	157	101	165	105	175	110	185	115	195	120	205	124	216	128	226	132
53	56	150	97	159	102	168	106	178	111	188	115	198	120	208	124	218	129	228	133

CONDENSER CHART FREON-22

Typical head pressures and AIR OFF temperatures at different suction pressures and AIR ON temperatures.

AIR ON Condenser Temperatures (All Temps. in °F)

Suction Pressure (psig)	Coil Temp. °f	75° Head Press. psig	75° Air Off Temp.	80° Head Press. psig	80° Air Off Temp.	85° Head Press. psig	85° Air Off Temp.	90° Head Press. psig	90° Air Off Temp.	95° Head Press. psig	95° Air Off Temp.	100° Head Press. psig	100° Air Off Temp.	105° Head Press. psig	105° Air Off Temp.	110° Head Press. psig	110° Air Off Temp.	115° Head Press. psig	115° Air Off Temp.
58	32	205	92	219	97	232	101	246	106	260	110	275	115	290	119	307	124	324	128
60	34	209	92	222	97	235	101	249	106	263	111	278	116	294	120	311	125	328	129
63	36	212	93	226	98	239	102	253	107	267	111	283	116	298	120	314	125	330	129
66	38	216	94	232	98	245	102	258	107	274	112	290	117	305	121	319	126	333	130
69	40	221	95	235	99	249	103	263	108	278	112	292	117	307	122	323	126	339	130
72	42	226	96	239	100	253	104	267	109	281	113	295	118	310	123	326	127	342	131
75	44	230	96	242	100	257	104	272	109	286	114	300	118	316	123	332	127	348	131
78	46	234	97	247	101	262	105	276	110	290	114	304	119	320	124	336	128	352	132
82	48	237	97	252	101	266	105	280	110	294	114	309	119	325	124	340	128	355	132
85	50	242	98	257	102	271	106	285	111	299	115	313	120	329	125	344	129	359	133
88	52	249	99	262	103	276	107	290	112	304	116	317	120	333	125	348	129	363	133
91	54	254	100	266	104	281	108	295	112	309	117	322	121	338	126	353	130	368	134
95	56	259	100	270	104	285	108	300	113	314	118	327	122	342	126	356	130	370	134

REFRIGERANT APPLICATIONS

Refrigerant	Boiling Point at Atmospheric Pressure, °F (°C)	Application
		High Temperature
R-113	118°F (48°C)	Low-capacity centrifugal packaged units for commercial and industrial air conditioning and chilling. Waste heat recovery in Organic Rankine Cycle engines.
R-11	75°F (24°C)	Centrifugal packaged units at higher system pressure and capacity than R-113. Secondary coolant in low-temperature systems and waste heat recovery.
R-123	82°F (28°C)	Replacement for R-11.
R-114	39°F (4°C)	High-capacity multistage centrifugal and rotary systems operating at intermediate pressure and displacement heat transfer in solar water heaters to reduce evaporator temperature in process chillers.
		Medium Temperature
R-12	−22°F (−30°C)	Formerly the most widely used for air conditioning and refrigeration.
R-134a	−16°F (−27°C)	Replacement for R-12 applications.

REFRIGERANT APPLICATIONS (cont.)

R-500	−28°F (−33°C)	Used in place of R-12 to increase capacity at the same compressor displacement.
R-22	−41°F (−41°C)	Air conditioners and heat pumps for residential and commercial applications and in refrigeration systems.
R-502	−50°F (−46°C)	Supermarket freezers and refrigerated cases. Operating at lower compressor discharge temperatures than R-22, it provides lower compression ratios and discharge temperatures and higher capacity.
R-507A	−52°F (−47°C)	Replacement to R-502.
R-13B1	−72°F (−58°C)	Medium- to low-temperature applications with one or two stages of compression.
R-116	−109°F (−78°C)	**Low Temperature** Specialty low-temperature applications.
R-13	−115°F (−82°C)	To produce evaporator temperatures as low as −100°F (−73°C) in the low-temperature stage of cascade refrigeration systems.
R-503	−128°F (−89°C)	To improve compressor capacity and low-temperature capability in the second stage of cascade systems that employ R-502, R-12, or R-22 in the first stage.
R-14	−198°F (−128°C)	Reciprocating compressors to produce evaporator temperatures down to −200°F (−129°C) in the third stage of triple cascade systems.

4-43

REFRIGERANT LUBRICANTS			
Refrigerant Type	**Appropriate Lubricant**		
R-11	—	AB	MO
R-12	POE	AB	MO
R-13	POE	AB	MO
R-22	POE	AB	MO
R-23	POE	—	—
R-123	—	AB	MO
R-124	POE	AB	MO
R-125	POE	—	—
R-134a	POE	—	—
R-176	POE	—	MO
R-401A	POE	AB	—
R-401B	POE	AB	—
R-401C	POE	AB	—
R-402A	POE	AB	MO
R-402B	POE	AB	MO
R-403B	POE	AB	MO
R-404A	POE	—	—
R-407A	POE	—	—
R-407B	POE	—	—
R-407C	POE	—	—
R-410A	POE	—	—
R-500	POE	AB	MO
R-502	POE	AB	MO
R-503	POE	AB	MO
R-507A	POE	—	—
R-717	—	—	MO
POE = Polyolester *AB = Alkylbenzene* *MO = Mineral oil*			

REFRIGERANT CYLINDERS AND ODP				
R Number	Chemical Class	ODP*	Cylinder Color	Similar To
R-11	CFC	1.0	Orange	—
R-12	CFC	1.0	White	—
R-22	HCFC	.05	Green	—
R-123	HCFC	.02	Light gray	R-11
R-134a	HFC	0.0	Sky blue	R-12
R-401a	HCFC	.05	Coral	R-12
R-402a	HCFC	.03	Sand	R-502
R-402b	HCFC	.02	Mustard	R-502
R-404a	HFC	0.0	Orange	R-502
R-407c	HFC	0.0	Med. brown	R-22
R-409a	HCFC	—	Tan	R-12
R-410a	HFC	0.0	Rose	—
R-416a	HCFC	—	—	R-12
R-500	CFC	.74	Yellow	—
R-502	HCFC	.28	Orchid	—
R-507	HFC	0.0	Teal	R-502
R-717	Ammonia	0.0	Silver	—
Any recovered refrigerant		—	Grey with yellow top	—
*ODP = Ozone depletion potential				

4-45

SPECIFIC HEAT OF VARIOUS LIQUIDS

Liquid Type	Temp. Range (°F)	Specific Heat
Alcohol, ethyl	32.0	0.548
Alcohol, methyl	5.0	0.59
Anilin	60.0	0.514
Benzol	105.0	0.423
$CaCl_2$ sp. gr. 1.14	5.0	0.764
$CaCl_2$ sp. gr. 1.20	−4.0	0.695
$CaCl_2$ sp. gr. 1.26	−4.0	0.651
Ethyl ether	32.0	0.529
Glycerine	59 to 120	0.576
NaCl plus 10 H_2O	64.0	0.791
NaCl plus 200 H_2O	64.0	0.978
Napthalene	185.0	0.396
Nitro benzole	84.0	0.362
Oils: Castor	—	0.471
Olive	44.0	0.471
Sesame	—	0.387
Turpentine	32.0	0.411
Petroleum	70 to 135	0.511
Sea water	64.0	0.938
(sp. gr. 1.0235)		
Toluol	150.0	0.490

SPECIFIC HEAT OF VARIOUS GASES

Gas Type	Temp. Range (°F)	Sp. Ht. at Constant Pressure	Sp. Ht. at Constant Volume
Acetone	79 to 230	0.3468	—
Air	−22 to 50	0.2377	0.168
Air	32 to 400	0.2375	—
Alcohol, C_2H_5OH	110 to 400	0.4534	0.399
Alcohol, CH_3OH	110 to 400	0.4580	—
Ammonia	73 to 212	0.5202	0.299
Benzene C_6H_6	94 to 235	0.2990	—
Carbon dioxide CO_2	−20 to 45	0.1843	0.171
Carbon monoxide CO	74 to 210	0.2425	0.176
Chlorine	61 to 700	0.1125	—
Chloroform $CHCl_3$	80 to 230	0.1441	—
Ether $C_4H_{10}O$	77 to 232	0.4280	—
Hydrochloric acid	50 to 212	0.1040	—
Hydrogen	54 to 300	3.4090	2.412
Methane CH_4	66 to 390	0.5929	—
Nitrogen	32 to 390	0.2438	—
Nitrous oxide	80 to 220	0.2126	—
Oxygen	50 to 400	0.2175	0.155
Sulfur dioxide	32	0.1544	—
Water vapor	32	0.4655	—
Water vapor	212	0.421	0.346

SPECIFIC HEAT OF VARIOUS SOLIDS

Material	Lb./ Cu. Ft.	Specific Heat	Material	Lb./ Cu. Ft.	Specific Heat
Asbestos	43.0	0.20	Cotton, loose	30.0	0.32
Ashes	43.0	0.20	Fats	58.0	0.46
Bakelite, laminated	86.0	0.35	Glass, common	164.0	0.199
Benzol	55.0	0.42	Glass, plate	161.0	0.161
Borax	—	0.24	Graphite	126.0	0.201
Bronze, phosphor	554.0	0.09	Gypsum, loose	70.0	0.26
Calcium carbonate	177.0	0.18	Ice −14°	57.5	0.53
Calcium sulfate	185.0	0.27	Litharge	—	0.21
Carborundum	195.0	0.16	Mica	—	0.10
Celluloid	94.0	0.37	Paper	58.0	0.324
Celluloid	90.0	0.36	Paraffin, 4° to 40°	—	0.377
Chalk	142.0	0.21	Paraffin, 32° to 68°	—	0.694
Coke	75.0	0.20	Plaster paris	103.0	1.14
Concrete, stone	147.0	0.19	Rubber	59.0	0.48
Concrete, cinder	105.0	0.18	Sugar	100.0	0.22
Cork	15.0	0.46	Sulfur	126.0	0.17
Corundum	247.0	0.20	Wood	44.0	0.373
Cotton, baled	93.0	0.32			

REFRIGERATION CONTROL

L1 ← 230 VAC → L2

Evaporator fan motor coil

Compressor motor coil

M

CM

High-pressure switch

HPC

LPC

Crankcase heater — CCH — Low-pressure switch

Control transformer

TS1 — LLS

Temperature switch — Liquid line solenoid

RECOMMENDED REFRIGERATION FIXTURE TEMPERATURE RANGES

Fixture (Cabinet)	Temp. (°F)	Temp. (°C)
Frozen food cabinet (closed)	−10 to −5	−23 to −21
Frozen food cabinet (open)	−7 to −2	−22 to −19
Dough retarding refrigerator	34 to 38	1 to 3
Retail market cooler	34 to 39	1 to 3
Restaurant storage cooling	35 to 39	2 to 3
Beverage precooler	35 to 40	2 to 4
Grocery refrigerator	35 to 40	2 to 4
Top display case (closed)	35 to 42	2 to 6
Dairy display case	36 to 39	2 to 3
Double display case	36 to 39	2 to 3
Delicatessen case	36 to 40	2 to 4
Restaurant service refrigerator	36 to 40	2 to 4
Back bar	37 to 40	3 to 4
Beverage cooler	37 to 40	3 to 4
Vegetable display Refrigerator (closed)	38 to 42	3 to 6
(open)	38 to 42	3 to 6
Florist storage case	38 to 45	3 to 7
Florist display refrigerator	40 to 50	4 to 10
Pastry display case	45 to 50	7 to 10
Candy case (storage)	58 to 65	15 to 18
Candy case (display)	60 to 65	16 to 18

TEMPERATURES FOR REFRIGERATION APPLICATIONS		
Application	**Temp. (°F)**	**Temp. (°C)**
Ice cream hardening	−25	−32
Ice cream storage	−20 to −10	−29 to −23
Freezer room	−15	−26
Locker room	−5 to 0	−21 to −18
Fresh meats	28 to 32	−2 to 0
Aging room	30 to 34	−1 to 1
Meats	30 to 34	−1 to 1
Poultry	30 to 34	−1 to 1
Curing room	32 to 36	0 to 2
Fur storage	33 to 37	0 to 3
Service	34 to 38	1 to 3
Chill room	35 to 39	2 to 3
Vegetables, fresh	36 to 42	2 to 6
Plants and flowers	38 to 50	3 to 10
Bananas	60 to 65	16 to 18

REFRIGERATION EQUIPMENT PRESSURE RANGES PER REFRIGERANT

Type of Case	Refrigerant 12		Refrigerant 22		Refrigerant 502		Approx. Temp. (°F)
	Suction Pressure (psig)	Head Pressure (psig)	Suction Pressure (psig)	Head Pressure (psig)	Suction Pressure (psig)	Head Pressure (psig)	
Walk-in meat cooler	12	150–175	–	–	–	–	34
Walk-in dairy cooler	12–14	150–175	–	–	–	–	36
Walk-in produce cooler	18–20	150–175	–	–	–	–	45
Walk-in freezer	0	130–150	5	225–250	15	280	0
Open type freezer	0	130–150	5	225–250	15	280	0
Ice cream case	5" Vacuum	120–150	0	200–225	5	260	–15
Open meat display case	6–9	140–160					32
Closed meat display case (forced air evaporator)	12	150–175					34
Closed meat case (gravity air evaporator)	6	130–150	–	–	–	–	34
Open produce display	16–18	150–175	–	–	–	–	45
Open dairy display	9	150–175	–	–	–	–	36
Beverage box	12	150–175	–	–	–	–	34
Reach-in cooler	12	150–175	–	–	–	–	34

TEMPERATURE–PRESSURE CONDITIONS IN THE REFRIGERATOR MECHANISM

Refrigerator Mechanism	R-12 Ambient Temp.	
	70°F	90°F
Refrigerator Evaporator EVAP Temperature at		
Start of cycle	15°F	15°F
Middle of cycle	5°F	5°F
End of cycle	0°F	0°F
Refrigerator Evaporator EVAP Pressure (psig) at		
Start of cycle	12	12
Middle of cycle	8	8
End of cycle	5	5
Refrigerator Condenser Temperature at		
Start of cycle	70°F	90°F
Middle of cycle	100°F	120°F
End of cycle	100°F	120°F
Refrigerator Condenser Pressure (psig) at		
Start of cycle	70	85
Middle of cycle	120	158
End of cycle	120	158

HEAD PRESSURES FOR REFRIGERANTS

Inlet Water Temperature (°F)

Refrigerant	50	55	60	65	70	75	80	85	90	95	100
					Head Pressure (psig)						
R-12	56	62	68	74	80	87	93	101	108	117	125
R-134a	55	61	68	76	84	92	101	110	120	131	142
R-22	95	104	113	123	133	144	155	158	180	194	208
R-717	98	108	119	130	140.5	152	164	177	191	205	220
R-500	71	78	86	94	103	112	121	131	142	153	165
R-502	116	126	137	148	160	173	186	200	214	230	246
R-404A	120	131	143	155	168	182	196	212	228	245	263

VENTILATION REQUIREMENTS FOR REFRIGERATION MACHINERY IN HOT AREAS

Compressor Rating		Total Air at 0.125 Static Pressure (CFM)	Free Area Intake Openings (sq. ft.)
HP	kW		
10	7.46	11,000	22
15	11.19	16,000	33
20	14.92	22,000	44
25	18.65	27,000	55
30	22.38	33,000	66
35	26.11	38,000	77
40	29.84	44,000	88
45	33.57	49,000	99
50	37.30	55,000	110

MOISTURE ABSORBANCY OF DESICCANTS

Desiccant (Drying Chemical)	Mesh (%)	Absorption from Liquid Percent of Weight of Desiccant (%)
Silica gel	8–20	16
Activated alumina	8–10	12
Synthetic Silicates	8–20	16

CAPACITIES OF AIR-COOLED CHILLERS

In Thousands of BTU per Hour

3-Ton Unit				4-Ton Unit				5-Ton Unit			
Air Temperature Entering Condenser (°F)											
90	95	100	105	90	95	100	105	90	95	100	105
36.76	36.28	35.10	33.64	49.01	48.37	47.37	44.85	61.26	60.46	58.50	56.07
36.72	36.18	34.92	32.65	48.96	48.24	47.13	43.53	61.20	60.30	58.20	54.42
36.65	36.04	34.50	31.32	48.86	48.04	46.57	41.76	61.08	60.06	57.50	52.20
36.54	35.82	33.72	29.60	48.72	47.76	45.51	39.45	60.90	59.70	56.20	49.32
36.36	35.53	32.30	27.43	48.48	47.37	43.60	36.57	60.60	59.22	53.84	45.72
36.00	35.03	30.60	23.40	48.00	46.70	41.30	31.20	60.00	58.38	51.00	39.00

COOLING CONTROL

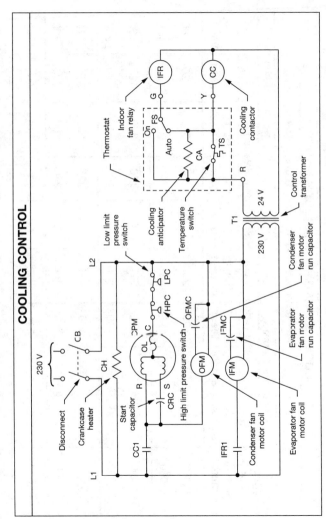

4-57

ROOM-COOLING REQUIREMENTS (BTU)

Approx. Btu Capacity Required	Type of Space Above Room Being Cooled								
	Occupied Room			Attic			Insulated Flat Roof		
	Area Being Cooled Has Exposed Walls Facing								
	North or East	South	West	North or East	South	West	North or East	South	West
6,000	400	200	100	200	100	64	250	120	80
7,000	490	250	125	235	130	97	295	155	105
8,000	580	300	150	270	160	130	340	190	130
10,000	750	440	390	340	270	200	470	340	240
12,000	920	580	470	410	320	225	550	375	275
13,000	1,000	660	550	450	350	250	600	400	300
16,000	1,290	970	790	570	480	390	750	650	540

ENERGY REQUIREMENT FORM FOR COOLING

Name: _____

Address: _____

Phone: _____

Space used for: _____

Interior Room Dimensions:

Length: _____   Width: _____   Height: _____

Windows:

No.: _____   Facing: _____   Size: _____ × _____

Window Loads

Sun exposed (interior shades)

West side:	_____ sq. ft. × 60 = _____ Btu/h _____ Watts
South side:	_____ sq. ft. × 40 = _____ Btu/h _____ Watts

Sun exposed (awnings)

West/south side:	_____ sq. ft. × 35 = _____ Btu/h _____ Watts
East, north, or shaded:	_____ sq. ft. × 15 = _____ Btu/h _____ Watts

Wall Load

South, west exposure:	_____ sq. ft. × 8 = _____ Btu/h _____ Watts
East, north exposure:	_____ sq. ft. × 5 = _____ Btu/h _____ Watts
Thin wall, all exposures:	_____ sq. ft. × 10 = _____ Btu/h _____ Watts
Interior walls:	_____ sq. ft. × 4 = _____ Btu/h _____ Watts
Interior glass partition:	_____ sq. ft. × 10 = _____ Btu/h _____ Watts

Floor Load

_____ sq. ft. × 3 = _____ Btu/h _____ Watts

Ceiling Load

Occupied above:	_____ sq. ft. × 3 = _____ Btu/h _____ Watts
Insulated roof:	_____ sq. ft. × 8 = _____ Btu/h _____ Watts
Uninsulated roof:	_____ sq. ft. × 20 = _____ Btu/h _____ Watts

Ventilation Load

_____ sq. ft. × 4 = _____ Btu/h _____ Watts

Occupancy Load

_____ sq. ft. × 400 = _____ Btu/h _____ Watts

Miscellaneous Loads

Electrical watts:	_____ × 3.4 = _____ Btu/h _____ Watts
Other:	= _____ Btu/h _____ Watts

	Total:	Btu/h	Watts

ETHYLENE GLYCOL PROPERTIES

% Glycol Solution	Temperature (°F)		Specific Heat	Specific Gravity	Equation Factor
	Freeze Point	Boiling Point			
0	+32	212	1.00	1.000	500
10	+26	214	0.97	1.012	491
20	+16	216	0.94	1.027	483
30	+4	220	0.89	1.040	463
40	−12	222	0.83	1.055	438
50	−34	225	0.78	1.067	416
60	−60	232	0.73	1.079	394
70	<−60	244	0.69	1.091	376
80	−49	258	0.64	1.101	352
90	−20	287	0.60	1.109	333
100	+10	287+	0.55	1.116	307

PROPYLENE GLYCOL PROPERTIES

% Glycol Solution	Temperature (°F)		Specific Heat	Specific Gravity	Equation Factor
	Freeze Point	Boiling Point			
0	+32	212	1.000	1.000	500
10	+26	212	0.980	1.008	494
20	+19	213	0.960	1.017	488
30	+8	216	0.935	1.026	480
40	−7	219	0.895	1.034	463
50	−28	222	0.850	1.041	442
60	<−60	225	0.805	1.046	421
70	<−60	230	0.750	1.048	393
80	<−60	230+	0.690	1.048	362
90	<−60	230+	0.645	1.045	337
100	<−60	230+	0.570	1.040	296

AIR CONDITIONING CONDENSATE PIPING

Parameter	AC Condensate Flow
Typical Range:	0.02–0.8 GPM/ton
Typical Average:	0.04 GPM/ton
Unitary packaged A/C equipment:	0.006 GPM/ton
Air handling units (100% outside air):	0.100 GPM/1,000 CFM
Air handling units (50% outdoor air):	0.065 GPM/1,000 CFM
Air handling units (25% outdoor air):	0.048 GPM/1,000 CFM
Air handling units (15% outdoor air):	0.041 GPM/1,000 CFM
Air handling units (0% outdoor air):	0.030 GPM/1,000 CFM

MINIMUM CONDENSATE PIPE SIZES

A/C (tons)	Minimum Drain size (in.)
0–20	1
21–40	1¼
41–60	1½
61–100	2
101–250	3
More than 250	4

ICE MACHINE SPECIFICATION RANGES

Ambient Temp. (°F)	Freeze Cycle			Harvest Cycle		
	Head Pressure (psig)	Suction Pressure (psig)	Cycle Time (min.)	Head Pressure (psig)	Suction Pressure (psig)	Cycle Time (min.)
AIR COOLED TYPE						
50	175–225	20–36	7–11	120–150	55–80	1.4–2.0
70	180–220	22–40	8–12	140–170	65–85	1.2–1.8
80	200–250	24–42	10–13	160–180	70–90	1.1–1.5
90	240–280	26–44	12–15	170–200	80–100	1.0–1.5
100	260–300	26–46	14–18	200–220	100–120	0.9–1.3
110	300–350	28–48	18–24	225–250	120–130	0.8–1.2
AIR COOLED (REMOTE) TYPE						
−20	170–190	24–38	7–10	150–170	70–90	1.0–1.6
50	170–190	24–38	8–11	150–170	70–90	1.0–1.6
70	170–190	24–38	8–12	150–170	70–90	1.0–1.5
80	180–220	24–38	8–13	160–180	80–100	1.0–1.5
90	200–240	24–40	9–14	170–190	80–100	0.9–1.4
100	230–280	26–42	10–15	190–220	90–110	0.8–1.3
110	260–320	26–42	12–17	220–250	110–130	0.8–1.3
120	280–340	28–44	14–19	240–270	120–140	0.8–1.3
WATER COOLED TYPE						
50	215–225	24–38	8–12	130–160	65–85	1.4–2.0
70	215–225	26–40	9–13	140–160	70–90	1.2–1.8
80	215–225	26–42	9–14	150–170	75–95	1.0–1.6
90	215–225	26–42	10–15	150–170	80–100	0.9–1.4
100	215–225	26–44	10–16	155–175	80–100	0.8–1.3
110	215–225	26–44	11–16	160–180	80–100	0.8–1.3

CHAPTER 5
Piping

In plumbing, pipe size is referred to as nominal pipe size (NPS) and is measured by the inside diameter (ID). In the heating, air conditioning, and refrigeration trades, pipe and tubing are referenced by their outside diameter (OD) measurement.

For example, a ¾" copper pipe in plumbing would be called a ⅞" copper pipe in the HVAC/R trade.

COMMON MEASUREMENT TERMS

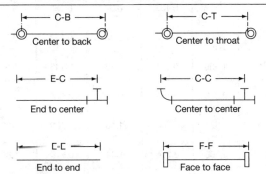

HEAVY FLUID SUPPORT MULTIPLIERS

For heavier fluids, multiply the support spacing on pages 5–4 and 5–13 by the following correction factors:

Specific Gravity of Fluid	1.00	1.10	1.20	1.40	1.60	2.00	2.50
Correction Factor	1.00	.98	.96	.93	.90	.85	.80

For insulated lines, reduce the spans by 70% of the given values.

PLASTIC PIPE DIMENSIONS AND WEIGHTS

SCHEDULE 40

Nominal Pipe Size (in.)	Outside Diameter (in.)	PVC		CPVC	
		Wall Thickness (in.)	Weight (lbs. per ft.)	Wall Thickness (in.)	Weight (lbs. per ft.)
¼	0.540	—	—	—	—
½	0.840	0.109	0.16	0.109	0.19
¾	1.050	0.113	0.22	0.113	0.25
1	1.315	0.133	0.32	0.133	0.38
1¼	1.660	0.140	0.43	0.140	0.51
1½	1.900	0.145	0.52	0.145	0.61
2	2.375	0.154	0.70	0.154	0.82
2½	2.875	0.203	1.10	0.203	1.29
3	3.500	0.216	1.44	0.216	1.69
4	4.500	0.237	2.05	0.237	2.33
6	6.625	0.280	3.61	0.280	4.10
8	8.625	0.322	5.45	—	—
10	10.75	0.365	7.91	—	—
12	12.75	0.406	10.35	—	—

PLASTIC PIPE DIMENSIONS AND WEIGHTS (cont.)

SCHEDULE 80

Nominal Pipe Size (in.)	Outside Diameter (in.)	PVC		CPVC	
		Wall Thickness (in.)	Weight (lbs. per ft.)	Wall Thickness (in.)	Weight (lbs. per ft.)
¼	0.540	0.119	0.10	0.119	0.12
½	0.840	0.147	0.21	0.147	0.24
¾	1.050	0.154	0.28	0.154	0.33
1	1.315	0.179	0.40	0.179	0.49
1¼	1.660	0.191	0.57	0.191	0.67
1½	1.900	0.200	0.69	0.200	0.81
2	2.375	0.218	0.95	0.218	1.09
2½	2.875	0.276	1.45	0.276	1.65
3	3.500	0.300	1.94	0.300	2.21
4	4.500	0.337	2.83	0.337	3.23
6	6.625	0.432	5.41	0.432	6.17
8	8.625	0.500	8.22	0.500	9.06
10	10.750	0.593	12.28	—	—
12	12.750	0.687	17.10	—	—

Nominal Pipe Size (in.)	Outside Diameter (in.)	PVDF		Polypropylene	
		Wall Thickness (in.)	Weight (lbs. per ft.)	Wall Thickness (in.)	Weight (lbs. per ft.)
½	0.840	0.147	0.24	0.147	0.14
¾	1.050	0.154	0.33	0.154	0.19
1	1.315	0.179	0.49	0.179	0.27
1¼	1.660	0.191	—	0.191	0.38
1½	1.900	0.200	0.81	0.200	0.45
2	2.375	0.218	1.13	0.218	0.62

PVC PIPE SUPPORT SPACING IN FEET

SCHEDULE 40

Nominal Pipe Size (in.)	Temperature (°F)				
	60	80	100	120	140
¼	3.75	3.50	3.00	2.50	2.00
½	4.25	4.00	3.50	3.00	2.50
¾	4.50	4.25	4.00	3.50	3.00
1	5.00	4.75	4.50	3.75	3.25
1¼	5.25	5.00	4.75	4.00	3.50
1½	5.50	5.25	5.00	4.25	3.75
2	6.00	5.50	5.00	4.50	4.00
2½	6.75	6.25	5.75	4.75	4.25
3	7.25	6.75	6.25	5.25	4.50
4	7.75	7.50	6.75	6.00	4.75
6	8.75	8.50	7.75	6.50	5.25
8	9.75	9.25	8.50	7.75	6.00
10	10.25	9.75	9.00	8.00	6.75
12	11.00	10.25	9.75	8.25	7.25

SCHEDULE 80

Nominal Pipe Size (in.)	Temperature (°F)				
	60	80	100	120	140
¼	4.25	4.00	3.75	3.00	2.75
½	4.50	4.25	4.00	3.75	3.00
¾	4.75	4.50	4.25	4.00	3.50
1	5.00	4.75	4.50	4.25	3.75
1¼	6.00	5.00	4.75	4.50	4.25
1½	6.50	5.50	5.25	5.00	4.75
2	6.75	5.75	5.50	5.25	5.00
2½	7.25	6.75	5.75	5.50	5.25
3	7.75	7.50	7.00	6.25	5.50
4	9.00	8.75	7.25	6.50	5.75
6	9.75	9.50	8.50	7.75	6.50
8	11.00	10.25	9.75	8.75	7.00
10	11.50	10.50	10.25	9.50	7.75
12	12.50	12.25	11.50	10.25	8.75

COPPER PIPE AND TUBING

Sweat fittings are measured by their inside diameter (ID) and compression fittings are measured by their outside diameter (OD). Always use a 50/50 solid core solder along with a high-quality flux when soldering sweat fittings. DO NOT use a rosin core type solder.

Type	Characteristics and Guidelines
K	Flexible copper tubing with a thicker wall than Type L and M. Required for all underground installations. Uses include plumbing, heating, steam, gas, and oil where thick-walled tubing is required. Can be used with sweat, flared, and compression fittings. Available in hard and soft types.
L	Standard copper tubing used for interior, aboveground applications including air conditioning, heating, steam, gas, and oil. Because of its flexibility, be very careful not to crimp the line when bending. Tools are available to make bending safer and easier. Sweat, compression, and flare fittings are available. DO NOT use compression fittings for gas lines. Available in hard and soft types.
M	Generally used with interior heating and pressure line applications. Wall thickness is less than Types K and L. Install with sweat fittings. Available in hard and soft types.

COPPER PIPE AND TUBING DIMENSIONS AND WEIGHTS

Nominal Pipe Size (in.)	Outside Diameter (in.), All Types	Wall Thickness (in.)			Inside Diameter (in.)			Pipe and Tube Weight (lbs. per ft.)		
		K	Type L	M	K	Type L	M	K	Type L	M
¼	.375	.035	.030	.025	.305	.315	.325	.145	.126	.106
⅜	.500	.049	.035	.025	.402	.430	.450	.269	.198	.145
½	.625	.049	.040	.028	.527	.545	.569	.344	.285	.204
⅝	.750	.049	.042	.030	.652	.666	.690	.418	.362	.263
¾	.875	.065	.045	.032	.745	.785	.811	.641	.455	.328
1	1.125	.065	.050	.035	.995	1.025	1.055	.839	.655	.465
1¼	1.375	.065	.055	.042	1.245	1.265	1.291	1.040	.884	.682
1½	1.625	.072	.060	.049	1.481	1.505	1.527	1.360	1.140	.940
2	2.125	.083	.070	.058	1.959	1.985	2.009	2.060	1.750	1.460
2½	2.625	.095	.080	.065	2.435	2.465	2.495	2.930	2.480	2.030
3	3.125	.109	.090	.072	2.907	2.945	2.981	4.000	3.330	2.680
3½	3.625	.120	.100	.083	3.385	3.425	3.459	5.120	4.290	3.580
4	4.125	.134	.110	.095	3.857	3.905	3.935	6.510	5.380	4.660
5	5.125	.160	.125	.109	4.805	4.875	4.907	9.670	7.610	6.660
6	6.125	.192	.140	.122	5.741	5.845	5.881	13.900	10.200	8.920
8	8.125	.271	.200	.170	7.583	7.725	7.785	25.900	19.300	16.500
10	10.125	.338	.250	.212	9.449	9.625	9.701	40.300	30.100	25.600
12	12.125	.405	.280	.254	11.315	11.565	11.617	57.800	40.400	36.700

COPPER PIPE AND TUBING LENGTHS IN FEET

Type	Drawn (hard)	Annealed (soft)
K	**Straight Lengths** Up to 8" diameter — 20' 10" diameter — 18' 12" diameter — 12'	**Straight Lengths** Up to 8" diameter — 20' 10" diameter — 18' 12" diameter — 12' **Coils** Up to 1" diameter — 60', 100' 1¼ and 1½" diameter — 60', 40' 2" diameter — 45'
L	**Straight Lengths** Up to 10" diameter — 20' 12" diameter — 18' N/A	**Straight Lengths** Up to 10" diameter — 20' 12" diameter — 18' **Coils** Up to 1" diameter — 60', 100' 1¼ and 1½" diameter — 60', 40' 2" diameter — 45'
M	**Straight Lengths** All diameters — 20'	N/A

ROLL GROOVE SPECIFICATIONS
FOR COPPER TUBING

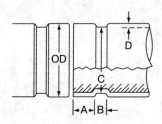

Nominal Copper Tube Size (in.)	OD Tube Distance Diameter (in.)	A Gasket Seat (in.)	B Groove Width (in.)	C Groove Diameter (in.)	D Groove Depth (in.)
2	2.125	.610	.300	2.029	.048
2½	2.625	.610	.300	2.525	.050
3	3.125	.610	.300	3.025	.050
4	4.125	.610	.300	4.019	.053
5	5.125	.610	.300	5.019	.053
6	6.125	.610	.300	5.999	.063

STANDARD CUT GROOVE SPECIFICATIONS FOR COPPER PIPE

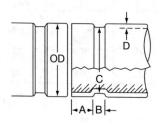

	OD	A	B	C	D
Nominal Pipe Size (in.)	Pipe Outside Diameter (in.)	Gasket Seal (in.)	Groove Width (in.)	Groove Diameter (in.)	Groove Depth (in.)
¾	1.050	.625	.313	.938	.056
1	1.315	.625	.313	1.189	.063
1¼	1.660	.625	.313	1.534	.063
1½	1.900	.625	.313	1.774	.063
2	2.375	.625	.313	2.249	.063
2½	2.875	.625	.313	2.719	.078
3	3.500	.625	.313	3.344	.078
4	4.500	.625	.375	4.344	.083
5	5.563	.625	.375	5.395	.084
6	6.625	.625	.375	6.455	.085
8	8.625	.750	.438	8.441	.092
10	10.750	.750	.500	10.562	.094
12	12.750	.750	.500	12.532	.109

STEEL PIPE DIMENSIONS AND WEIGHTS

SCHEDULE 40

Nominal Pipe Size (in.)	Outside Diameter (in.)	Wall Thickness (in.)	Inside Diameter (in.)	Pipe Weight (lbs. per ft.)
⅛	.405	.068	.269	.245
¼	.540	.088	.364	.425
⅜	.675	.091	.493	.568
½	.840	.109	.622	.851
¾	1.050	.113	.824	1.131
1	1.315	.133	1.049	1.679
1¼	1.660	.140	1.380	2.273
1½	1.900	.145	1.610	2.718
2	2.375	.154	2.067	3.653
2½	2.875	.203	2.469	5.793
3	3.500	.216	3.068	7.580
3½	4.000	.226	3.548	9.110
4	4.500	.237	4.026	10.790
5	5.563	.258	5.047	14.620
6	6.625	.280	6.065	18.970
8	8.625	.322	7.981	28.550
10	10.750	.365	10.020	40.480
12	12.750	.375	12.000	49.560

STEEL PIPE DIMENSIONS AND WEIGHTS *(cont.)*

SCHEDULE 80

Nominal Pipe Size (in.)	Outside Diameter (in.)	Wall Thickness (in.)	Inside Diameter (in.)	Pipe Weight (lbs. per ft.)
1/8	.405	.095	.215	.315
1/4	.540	.119	.302	.535
3/8	.675	.126	.423	.739
1/2	.840	.147	.546	1.088
3/4	1.050	.154	.742	1.474
1	1.315	.179	.957	2.172
1¼	1.660	.191	1.278	2.997
1½	1.900	.200	1.500	3.631
2	2.375	.218	1.939	5.022
2½	2.875	.276	2.323	7.661
3	3.500	.300	2.900	10.250
3½	4.000	.318	3.364	12.510
4	4.500	.337	3.826	14.980
5	5.563	.375	4.813	20.780
6	6.625	.432	5.761	28.570
8	8.625	.500	7.625	43.390
10	10.750	.500	9.750	54.740
12	12.750	.500	11.750	65.420

PIPE THREADING DIMENSIONS (NPT)

Nominal Pipe Size (in.)	Threads per Inch	Approximate Length of Thread (in.)	Approximate Number of Threads to Be Cut	Approximate Total Thread Makeup, Hand and Wrench (in.)
1/8	27	3/8	10	1/4
1/4	18	5/8	11	3/8
3/8	18	5/8	11	3/8
1/2	14	3/4	10	7/16
3/4	14	3/4	10	1/2
1	11 1/2	7/8	10	9/16
1 1/4	11 1/2	1	11	9/16
1 1/2	11 1/2	1	11	9/16
2	11 1/2	1	11	5/8
2 1/2	8	1 1/2	12	7/8
3	8	1 1/2	12	1
3 1/2	8	1 5/8	13	1 1/16
4	8	1 5/8	13	1 1/16
5	8	1 3/4	14	1 3/16
6	8	1 3/4	14	1 3/16
8	8	1 7/8	15	1 5/16
10	8	2	16	1 1/2
12	8	2 1/8	17	1 5/8

HORIZONTAL STEEL AND COPPER PIPE SUPPORT SPACING

Nominal Pipe Size (in.)	Maximum Hanger Spacing (ft.)				Maximum Rod Size (in.)
	Steel		Copper		
	Water Systems	Vapor Systems	Water Systems	Vapor Systems	
1/2	7	8	5	6	3/8
3/4	7	9	5	7	3/8
1	7	9	6	6	3/8
1 1/4	7	9	7	9	3/8
1 1/2	9	12	8	10	3/8
2	10	13	8	11	3/8
2 1/2	11	14	9	13	1/2
3	12	15	10	14	1/2
4	14	17	12	16	5/8
6	17	21	14	20	3/4
8	19	24	16	23	7/8
10	22	26	18	25	7/8
12	23	30	19	28	7/8

Nominal Pipe Size (in.)	Nominal Pipe Size (in.)												
	½	¾	1	1¼	1½	2	2½	3	4	6	8	10	12
	Minimun Centerline-to-Centerline Dimensions (in.)												
½	7.5	–	–	–	–	–	–	–	–	–	–	–	–
¾	8.0	8.0	–	–	–	–	–	–	–	–	–	–	–
1	8.0	8.5	8.5	–	–	–	–	–	–	–	–	–	–
1¼	8.5	8.5	8.5	9.0	–	–	–	–	–	–	–	–	–
1½	8.5	8.5	9.0	9.0	9.0	–	–	–	–	–	–	–	–
2	9.0	9.0	9.5	9.5	9.5	10.0	–	–	–	–	–	–	–
2½	10.0	10.0	10.5	10.5	10.5	11.0	12.0	–	–	–	–	–	–
3	10.0	10.5	10.5	11.0	11.0	11.5	12.5	12.5	–	–	–	–	–
4	11.5	11.5	12.0	12.0	12.0	12.5	13.5	14.0	15.0	–	–	–	–
5	12.0	12.0	12.5	12.5	12.5	13.0	14.0	14.5	15.5	–	–	–	–
6	12.5	12.5	13.0	13.0	13.0	13.5	14.5	14.5	16.0	17.0	–	–	–
8	13.5	14.0	14.0	14.5	14.5	15.0	16.0	16.0	17.5	18.5	19.5	–	–
10	15.0	15.0	15.5	15.5	15.5	16.0	17.0	17.5	18.5	19.5	21.0	22.0	–
12	16.5	16.5	17.0	17.5	17.0	17.0	18.5	19.0	20.0	21.0	22.5	23.5	25.0

Based on Schedule 40 and includes the outside dimensions for fittings, etc.

PIPE INSULATION THICKNESSES

Piping System Type	Nominal Pipe Sizes	Insulation Type and Thickness (in.)			
		A	B	C	D
Chilled Water— 40°F to 60°F	6" & smaller	1.0	1.5	2.0	1.5
	8" & larger	1.5	2.0	2.5	2.5
Chilled Water— 32°F to 40°F	1" & smaller	1.0	1.5	2.0	1.5
	1¼" to 6"	1.5	2.0	2.5	2.5
	8" & larger	2.0	2.5	3.5	3.0
Chilled Water— Below 32°F	2" & smaller	1.5	2.0	2.5	2.5
	2½" to 6"	2.0	2.5	3.5	3.0
	8" & larger	2.5	3.0	4.5	4.0
Condenser Water	All sizes	2.0	3.0	3.5	3.0
Heating Water—Low Temperature 100°F to 140°F	4" & smaller	1.0	1.5	2.0	1.5
	5" & larger	1.5	2.0	2.5	2.5
Heating Water—Low Temperature 141°F to 200°F	All sizes	1.5	2.0	2.5	2.5
Heating Water—Low Temperature 201°F to 250°F	2" & smaller	1.5	2.0	2.5	2.5
	2½" to 6"	2.0	2.5	3.5	3.0
	8" & larger	3.5	4.5	6.0	5.5
Heating Water— Medium Temperature 251°F to 350°F	1" & smaller	2.0	3.0	3.5	3.0
	1¼" to 4"	2.5	4.0	4.5	4.0
	5" & larger	3.5	5.0	6.0	5.5

PIPE INSULATION THICKNESSES *(cont.)*

Piping System Type	Nominal Pipe Sizes	Insulation Type and Thickness (in.)			
		A	B	C	D
Heating Water—High Temperature 351°F to 450°F	2" & smaller	2.5	3.5	4.5	4.0
	2½" to 4"	3.0	4.0	5.0	4.5
	5" & larger	3.5	4.5	6.0	5.5
Dual Temperature	All sizes	3.0	4.0	5.0	4.5
Heat Pump Loop	All sizes	2.5	3.5	4.5	4.0
Steam and Steam Condensate—Low Pressure	2" & smaller	1.5	2.0	2.5	2.5
	2½" to 6"	2.0	2.5	3.5	3.0
	8" & larger	3.5	4.5	6.0	5.5
Steam and Steam Condensate—Medium Pressure	1" & smaller	2.0	3.0	3.5	3.0
	2¼" to 4"	2.5	3.5	4.5	4.0
	5" & larger	3.5	4.5	6.0	5.5
Steam and Steam Condensate—High Pressure	2" & smaller	2.5	3.5	4.5	4.0
	2½" to 4"	3.0	4.5	5.0	4.5
	5" & larger	3.5	5.0	6.0	5.5
Refrigerant Suction and Liquid Lines	1" & smaller	1.0	1.5	2.0	1.5
	1¼" to 6"	1.5	2.0	2.5	2.5
	8" & larger	2.0	2.5	3.5	3.0
Refrigerant Hot Gas	All sizes	0.75	1.0	1.5	1.0
Air Conditioning Condensate	All sizes	0.5	0.5	1.0	0.75

LINEAR EXPANSION OF PIPING

Temp. (°F)	Inches of Expansion per 100 Ft. of Pipe			
	Copper	Steel	Cast Iron	Wrought Iron
−30	—	—	—	—
−20	.105	.072	.062	.073
−10	.211	.145	.124	.147
0	.316	.215	.186	.221
10	.428	.291	.251	.298
20	.541	.367	.317	.376
30	.654	.442	.383	.454
40	.767	.517	.449	.533
50	.880	.592	.515	.612
60	.993	.667	.581	.691
70	1.107	.742	.647	.770
80	1.221	.817	.713	.849
90	1.335	.892	.779	.928
100	1.449	.968	.845	1.007
110	1.565	1.048	.915	1.090
120	1.681	1.128	.985	1.174
130	1.797	1.208	1.056	1.258
140	1.913	1.287	1.127	1.342
150	2.029	1.366	1.198	1.426
160	2.145	1.445	1.269	1.510
170	2.261	1.524	1.340	1.594
180	2.377	1.603	1.411	1.678
190	2.494	1.682	1.482	1.762
200	2.611	1.761	1.553	1.846
210	2.727	1.843	1.626	1.931
220	2.843	1.925	1.699	2.016
230	2.959	2.008	1.773	2.101

LINEAR EXPANSION OF PIPING (cont.)

Temp. (°F)	Inches of Expansion per 100 Ft. of Pipe			
	Copper	Steel	Cast Iron	Wrought Iron
240	3.075	2.091	1.847	2.186
250	3.191	2.174	1.921	2.271
260	3.308	2.257	1.995	2.356
270	3.425	2.340	2.069	2.441
280	3.542	2.423	2.143	2.526
290	3.659	2.506	2.217	2.612
300	3.776	2.589	2.291	2.698
310	3.896	2.674	2.368	2.787
320	4.016	2.759	2.445	2.876
330	4.136	2.844	2.522	2.965
340	4.256	2.929	2.599	3.054
350	4.376	3.015	2.676	3.143
360	4.497	3.101	2.754	3.232
370	4.618	3.187	2.832	3.321
380	4.739	3.273	2.910	3.410
390	4.860	3.359	2.988	3.499
400	4.981	3.445	3.066	3.589

CHAPTER 6
Service

SERVICE CHECKLIST—BOILERS

System	Actions	Service Time Period
Steam heating high pressure more than 15 psi	**Pump and system:** Check feedwater and condensate pumps for proper operation and leaky packing. Examine traps, check valves, makeup float valves, expansion & condensate tank.	Daily
All steam and water discharges must be piped to a safe place.	**Low-water fuel cutoff:** Drain float chamber while boiler is running. This should interrupt the circuit and stop the burner.	Daily
	Burner operation: f the burner starts with a puff or operates roughly, repair immediately.	Daily
	Safety/relief valve: Pull try-lever to full open position with pressure on the boiler. Release try-lever to allow the valve to snap closed.	Monthly
	Water column or gauge glass: Open the drain valve quickly to void a small quantity of water. Water level should return quickly when the drain valve is closed.	Daily

SERVICE CHECKLIST—BOILERS (cont.)

System	Actions	Service Time Period
Steam heating low pressure 15 psi and less	**Pump and system:** Check feedwater and condensate pumps for proper operation and leaky packing. Examine traps, check valves, makeup float valves, expansion and condensate tank.	Daily
All steam and water discharges must be piped to a safe place.	**Low-water fuel cutoff:** Drain float chamber while boiler is running. This should interrupt the circuit and stop the burner.	Weekly
	Burner operation: If the burner starts with a puff or operates roughly, repair immediately.	Daily
	Safety/relief valve: Pull try-lever to full open position with pressure on the boiler. Release try-lever to allow the valve to snap closed.	Monthly
	Water column or gauge glass: Open the drain valve quickly to void a small quantity of water. Water level should return quickly when the drain valve is closed.	Weekly

SERVICE CHECKLIST – BOILERS *(cont.)*

System	Actions	Service Time Period
Hot water heating 160°F & 250 psi or less	**Pump and system:** Check feedwater pump for proper operation and leaky packing. Examine check valves, makeup float valves, and expansion tank.	Daily
All water discharge must be piped to a safe place.	**Low-water fuel cutoff** (if applicable): Drain float chamber while boiler is running. This should interrupt the circuit and stop the burner.	Monthly
	Burner operation: If the burner starts with a puff or operates roughly, repair immediately.	Daily
	Safety/relief valve: Pull try-lever to full open position with pressure on the boiler. Release try-lever to allow the valve to snap closed.	Monthly

SERVICE CHECKLIST—A/C AND REFRIGERATION

System	Actions	Service Time Period
Hermetic *Always deenergize electrical equipment before testing, cleaning, or performing maintenance.*	**Motors:** Take insulation resistance readings of motor windings. If less than one megohm, check for cause. Hermetic motor readings less than 30 megohms may indicate water in the system.	Annually
	Motor controls: Inspect starter contacts for deterioration, pitting, corrosion, etc.; check terminal connections for tightness; examine overload protection for adequate size and defects; determine that timing devices have correct operating sequence; check mechanical linkage.	Annually
	Fans: Check for broken, cracked, bent, or loose blades or hubs; check shaft and bearings; check belt tension and condition.	Annually
	Filters: Clean air filters serving the evaporator and the air-cooled condenser. If a water-cooled condenser is used, the water side must be kept clean.	Weekly

6-4

SERVICE CHECKLIST – A/C AND REFRIGERATION (cont.)

System	Actions	Service Time Period
Nonhermetic *Always deenergize electrical equipment before testing, cleaning, or performing maintenance.*	**Motors:** Take insulation resistance readings of motor windings. If less than one megohm, check for cause. Check air ventilation openings on open-type motors for obstruction. Check bearings on open-type motors for adequate and proper lubrication.	Annually
	Motor controls: Inspect starter contacts for deterioration, pitting, corrosion, etc.; check terminal connections for tightness; examine overload protection for adequate size and defects; determine that timing devices have correct operating sequence; check mechanical linkage for binding and looseness.	Annually
	Fans: Check for broken, cracked, bent, or loose blades or hubs; check shaft and bearings; check belt tension and condition.	Annually
	Filters: Clean air filters serving the evaporator and the air-cooled condenser. If a water-cooled condenser is used, the water side must be kept clean.	Weekly

6-5

SERVICE CHECKLIST — A/C AND REFRIGERATION (cont.)

System	Actions	Service Time Period
Nonhermetic (cont.) *Always deenergize electrical equipment before testing, cleaning, or performing maintenance.*	**Moisture indicator:** Determine any change in the indicator chemical color or the presence of gas bubbles in the liquid refrigerant.	**Weekly**
	Oil sight glass: Verify that sufficient oil is in the compressor crankcase. Oil leakage should not be tolerated. Any change in normal oil level should be investigated.	**Weekly**
	Temperature: In-operation temperature levels for the compressor suction and discharge should be established and recorded. Any unusual change in these temperatures should be investigated.	**Weekly**
	Pressure: Operating pressure levels should be established and recorded. Any unusual change in these pressures should be investigated.	**Weekly**

6-6

SERVICE CHECKLIST — ELECTRICAL/ELECTRONIC

System	Actions	Service Time Period
Electrical distribution system *Always deenergize electrical equipment before testing, cleaning, or performing maintenance.*	**Cool:** All vent and air circulation openings must be clear and operational. Verify that circuits are properly loaded and balanced.	**Monthly**
	Clean: Dust and/or dirt accumulations should be removed from the equipment and surroundings. Equipment should be thoroughly cleaned inside and outside. Space in switchrooms and switchgear enclosures should not be used for storing tools, supplies, or other material.	**Quarterly**
	Dry: Precautions should be taken to prevent steam, chemicals, moisture, or condensation from entering electrical enclosures.	**Monthly**
	Tight: Clean and tighten all loose parts and replace. Verify that moving parts do not bind and are free to operate.	**Annually**

6-7

SERVICE CHECKLIST—ELECTRICAL/ELECTRONIC (cont.)

System	Actions	Service Time Period
Electronic & computer equipment *Always deenergize electrical equipment before testing, cleaning, or performing maintenance.*	**Protection:** Computer electronic components should be protected by a surge suppression device. Verify that vital programs and records are stored remotely and/or in an approved safe located in a low-hazard area and protected by smoke detection and automatic sprinklers.	Monthly
	Heat: All vent and air circulation openings must be free from obstruction. Filters should be kept clean and sound, and the fans operable.	Monthly
	Smoke: Verify that detectors are installed and maintained. Verify that the actuation of any detector results in the sounding of alarms and the shutdown of air conditioning equipment.	Quarterly
	Dry: Precaution should be taken to prevent steam, chemicals, moisture, or condensation from entering equipment.	Monthly

SERVICE CHECKLIST—MECHANICAL/HOT WATER HEATERS

System	Actions	Service Time Period
Fans, blowers, & air induction louvers	**Fans & blowers:** Check for broken, cracked, bent, or loose blades or hubs; check shaft and bearings; check belt tension and condition.	Annually
	Vents & louvers: Inspect for damage and operation. Parts should be free of obstructions or blockage that would prevent proper intake for combustion air. Vents must not be blocked open.	Monthly
Deepwell pumps	**Motors:** Take insulation resistance readings of motor windings. If less than one megohm, check for cause. Check air ventilation opening on open-type motors for obstruction. Check bearings on open-type motors for adequate and proper lubrication.	Annually
	Motor controls: Inspect starter contacts for deterioration, pitting, corrosion, etc.; check terminal connections for tightness; examine overload protection for adequate size and defects; check mechanical linkage for binding and looseness.	Annually
Hot water heaters	**Relief valve:** Pull try-lever to full open position with pressure on the equipment. Release try-lever to allow the valve to snap closed. All discharges must be piped to a safe place.	Monthly

6-9

CENTRAL HEATING SERVICE CHECKLIST

1. Check air filter.
   If the filter is dirty, be sure to clean or replace.

2. Inspect fan and blower motor.
   Oil both if not self-lubricated.

3. Inspect belt.
   If the belt is frayed or cracked, replace it.

4. Clean pilot burner and light.
   It should have a clear blue flame.

5. Check safety pilot.
   Be sure it is operating properly.

6. Check the operation of the solenoid valve.
   Be sure it will open and close properly.

7. Check the differential.
   Make changes on the heat anticipator if needed.

8. Check the operation of the fan control.

9. Check and be sure the limit switch will operate.

10. Observe the heater through several cycles.

OIL BURNER STARTUP PROCEDURES

1. Check oil level in storage tank.

2. Open valve in oil supply line to burner.

3. Bleed air from fuel pump. One-pipe systems must be bled. Two-pipe systems will usually bleed themselves.

4. Turn on the electricity.

5. Place a can under the bleed port to catch purged oil. Loosen the bleed plug and start the burner. Allow burner to run until a solid stream of oil is purged from the port. Turn off burner and screw in bleed port.

6. Set thermostat to call for heat.

7. If burner does not start, reset primary safety control and reset oil burner motor overload.

8. After the oil burner is started, make necessary adjustments.

9. Check operation of the draft control.

10. Check operation of the primary safety control.

11. Clean or replace the air filters.

12. Adjust or replace the fan belt.

13. Lubricate all bearings.

14. Check the fan control settings for accuracy.

15. Check the limit control setting for accuracy.

GAS HEATING STARTUP PROCEDURES

1. Close off all gas valves.

2. Wait approximately 5 min. for any unburned gas to escape from combustion area.

3. Open the gas cock in the gas line.

4. Turn off–on–pilot knob to pilot position.

5. Strike match and hold flame by the pilot burner.

6. Depress pilot knob.

7. Hold pilot knob in for approximately 1 min. after lit.

8. Release pilot knob.

9. Turn thermostat down below room temperature or turn off electrical power to furnace.

10. Turn gas valve knob to on position.

11. Turn on electrical power or raise temperature setting of thermostat above room temperature.

12. Check flame on main burners. Adjust primary air shutter if there is any yellow in the flame. If burner cannot be properly adjusted, close off the gas valves, remove and clean the burners. Reinstall burners, complete the above steps, and adjust the primary air.

GAS HEATING STARTUP PROCEDURES *(cont.)*

13. Check for carbon monoxide in the flue gases inside the draft diverter. Make any adjustments indicated.

14. Check the fan on and off temperature, and the limit off temperature with a thermometer inserted into the circulating air stream as close as possible to the sensing element.

15. Check the heat anticipator.

16. Turn off the electrical power to furnace after fan has stopped running.

17. Clean or replace the air filter.

18. Check condition of all bearings. Replace or repair as required.

19. Lubricate all bearings requiring service.

20. Check belt condition and tension. Replace or adjust as required.

21. Check calibration of thermostat. Calibrate as required.

22. Clean and vacuum furnace and area around furnace.

23. Replace all covers.

HEAT PUMP STARTUP PROCEDURES

1. Verify that electricity has been on to the outdoor unit for 24 hr.

2. Check the outdoor coil and clean if necessary.

3. Check all bearings; make necessary repairs.

4. Lubricate all bearings requiring service.

5. Check condition and tension of all belts. Replace or adjust as required.

6. Check condition of compressor contactor contacts. Replace if necessary.

7. Check tightness of all electrical connections. Tighten loose connections.

8. Check and repair burnt or frayed wiring.

9. Set thermostat for cooling, and lower temperature setting below room temperature.

10. Check temperature rise on outdoor coil.

11. Check suction and discharge pressures.

12. Check refrigerant charge in system. If short, repair leak and add refrigerant.

13. Check the amperage all motors.

14. Check temperature drop on indoor coil. If more than 24°F (13.33°C), check for dirty indoor coil. If less than 18°F (10°C), check for inefficient compressor or refrigerant shortage on cooling. On heating, the temperature rise should be approximately 30°F (16.67°C).

15. Check thermostat calibration, Calibrate if necessary.

16. Clean or replace air filter.

17. Clean and vacuum indoor unit and area around unit.

18. Replace covers.

ELECTRIC HEATING STARTUP PROCEDURES

1. Turn off all electric power to unit.

2. Visually check for any burnt or bad wiring and replace as required.

3. Tighten all electrical connections.

4. Turn on electric power.

5. Set thermostat well above room temperature to ensure that all stages are demanding.

6. Check voltage drop on all elements after all relays are closed.

7. Check amperage to all elements. Replace any bad elements.

8. Check setting of heat anticipator. Adjust as required.

9. Set thermostat temperature selector below room temperature.

10. After fan has stopped, turn off all electricity.

11. Check the condition of all bearings.

12. Lubricate all bearings requiring service.

13. Check condition and tension of belt. Replace or adjust as required.

14. Replace or clean air filter.

15. Clean and vacuum furnace and area around furnace.

16. Replace all covers.

17. Turn on electricity.

REFRIGERATION STARTUP PROCEDURES

1. Check condenser; clean if necessary.

2. Check all bearings; make necessary repairs.

3. Lubricate bearings.

4. Check condition and tension of all belts. Replace or adjust as required.

5. Check condition of compressor contactor contacts. Replace if necessary.

6. Check tightness of all electrical connections.

7. Check and repair burnt or frayed wiring.

8. Start unit and check suction and discharge pressures.

9. Check compressor oil level.

10. Check refrigerant charge. If short, repair leak and add refrigerant.

11. Check amperage at all motors.

12. Check temperature drop on evaporator.

13. Clean and vacuum indoor unit and area around unit.

14. Replace covers.

15. Allow unit to operate for 24 hr. and check the fixture temperature to see if it corresponds to the thermostat setting.

AIR CONDITIONING STARTUP CHECKLIST

Compressors

☐ Energize the crankcase heaters 8 hr. before start-up and before taking insulation resistance readings of hermetic motor windings. Crankcase heaters should be left energized for the rest of the season to prevent refrigerant migration to the crankcase.

☐ Test the lubricating oil for color and acidity; check crankcase oil level.

Motors

☐ Check the air passages of open motors for cleanliness and obstructions.

☐ Check the condition and lubricate bearings.

☐ Take insulation resistance readings. If less than one megohm resistance, check for cause. Repair or replace.

Motor Controls

☐ Inspect starter contacts for deterioration from short cycling, arcing, or corrosion.

☐ Check terminal connections for tightness.

☐ Examine the overload protection for defects and for proper size.

☐ Check mechanical linkages for binding and excessive looseness.

☐ Check timing devices for correct operating sequence.

Operating and Safety Controls

☐ Verify controls are properly calibrated and in working order, including thermostatic controls, oil pressure safety switches, and flow switches.

AIR CONDITIONING STARTUP CHECKLIST *(cont.)*

Refrigerant Circuits

☐ Be sure the circuit is equipped with a moisture indicator. If moisture is indicated, install new liquid line filter/drier cores. Determine and correct the source of the moisture.

☐ Check the expansion valve for proper operation and superheat settings over full range of operation.

Condensers and Evaporators

☐ Verify proper cleaning of heat transfer surfaces has been completed.

☐ Cooling towers: Check baffles for tightness and soundness. Clean the baffles, sump, and the spray nozzles. Check the makeup water valve for proper operation.

Pumps

☐ Check the bearings, packings, shaft couplings, and seals. Lubricate bearings.

Fans

☐ Check for broken, cracked, bent, or loose blades. Check hubs, fan shaft, and bearings.

☐ Check belt condition and tension.

☐ Replace air filters.

Piping

☐ Check all piping supports for signs of distress.

☐ Check for external damage and excessive vibration.

RECIPROCATING A/C AND REFRIGERATION LOG

Man.	No.	Size
Loc.	Tech.	Date

Preseason Maintenance and Startup Log

Motors—Open/Hermetic

☐ Take insulation resistance readings of motor windings. If the readings indicate less than one megohm resistance, check for the cause. Hermetic Motors Readings less than 30 megohms may indicate moisture in the system or refrigerant in the motor/compressor.

☐ Check air ventilation openings on open-type motors for obstruction.

☐ Check bearings on open-type motors for adequate and proper lubrication.

Fans

☐ Check for broken, cracked, bent, or loose blades and hubs.

☐ Check fan shaft and bearings.

☐ Check belt tension and condition.

Refrigerant Circuit

☐ Be sure that liquid line is equipped with a moisture indicator.

☐ If moisture is indicated, dehydrate the system. Determine and correct cause.

Motor Controls

☐ Inspect starter contacts for deterioration—pitting corrosion.

☐ Check terminal connections for tightness.

☐ Examine overload protection for adequate size and defects.

☐ Determine that timing devices have correct operating sequence.

☐ Check mechanical linkage.

RECIPROCATING A/C AND REFRIGERATION LOG (cont.)

Man.	No.	Size
Loc.	Tech.	Date

Preseason Maintenance and Startup Log

Operating and Safety Controls

☐ Test thermostatic controls.

☐ Test oil pressure differential switches and high pressure cut out.

☐ Examine flow switch by removing and checking for corrosion and proper linkage operation.

☐ Determine that all controls are properly calibrated and in good working condition.

☐ Check thermostatic expansion valve for proper superheat.

Pumps

☐ Check the condition of bearings, packing, shaft coupling, and seals.

Condenser
Air/Smell and Tube/Evaporative

☐ Clean heat transfer surfaces.

☐ Cooling towers: Baffles should be tight, sound, and clean. The sump, spray nozzles, and overflow drain should be clean. The makeup water valve should be checked for proper operation.

Compressors

☐ Crankcase heaters energized 24 hr. before startup.

☐ Crankcase oil at normal level in sight glass.

☐ Lubricating oil tested for acidity and color.

☐ Examine valves for signs of wear, cracking, and fatigue.

RECIPROCATING A/C AND REFRIGERATION LOG *(cont.)*

Man.	No.		Size						
Loc.	Tech.		Date						

Log This Information on a Weekly Basis

Item	Operating Conditions	Operating Standards	Cleaning Dates				
			1	2	3	4	5
Moisture indicator Sight glass	This device should be observed to determine any change in the indicator chemical color or the presence of gas bubbles in the liquid refrigerant.	Color					
Oil sight glass	Observe this glass to establish that sufficient oil is in the compressor crankcase. Oil leakage should not be tolerated.	Level					
Temperature	In operation temperature levels for the compressor suction and discharge should be established and recorded.	Discharge temp. / Suction temp.					
Pressure	The operating pressure levels should be established and recorded.	Discharge pressure / Suction pressure					
Filters	Regular cleaning of the air filters serving the evaporator and the air-cooled condenser is critical. If a water-cooled condenser is used, the water side must be kept clean.	Evaporator filter / Condenser filter					

6-21

CENTRIFUGAL A/C AND REFRIGERATION LOG

Man.		No.				Size		
Loc.		Tech.				Date		
		Compressor						
			Bearing		Oil			
Date or Time	Outside Temp.	Position Capac. Indicator	Temp.	Temp.	Level	Temp. Reser voir	Temp. Leaving Cooler	Press.

Compressor:
After 40,000 operating hr. or 5 years, a compressor should be disassembled. Impeller(s) should be examined for rubbing, grooves, and cracks and cleaned, tested, and balanced. Guide vanes, linkage, and bushings examined for lost motion, wear, and sticking stems and tested. The main shaft, pinion, and gears should be tested.

CENTRIFUGAL A/C AND REFRIGERATION LOG (*cont.*)

Man.				No.					Size	
Loc.				Tech.					Date	

| Date or Time | Outside Temp. | Compressor | | | | Chiller | | | | |
| | | Motor | | Gear oil | | Refrigerant | | | Water Temp | |
		Amps	Volts	Press.	Temp.	Level	Press.	Temp.	In	Out

Tubes:
1. The water side surfaces should be cleaned annually.
2. Eddy current analysis should be performed on condenser tubes within 3 to 5 years of service, and on evaporator tubes within 5 to 7 years of service.

CENTRIFUGAL A/C AND REFRIGERATION LOG (*cont.*)

Man.						No.			Size	
Loc.						Tech.			Date	

Date or Time	Outside Temp.	Condenser				Purge		Refrig. Tubes	
		Refrigerant		Water Temp		Freq. of Oper.	Amt. Water Removed or Rem.	Tubes cleaned	Eddy current tested
		Press	Temp.	In	Out				

Oil/Refrigerant:
When oil or refrigerant is added record date and amount. Also Indicate leak tests, repairs, and adjustments. Oil samples should be tested.

Purge Unit:
Should be overhauled at least once a year.

CENTRIFUGAL A/C AND REFRIGERATION LOG (*cont.*)

Man.			No.				Size		
Loc.			Tech.				Date		

Date or Time	Outside Temp.	Oil Press. Safety Contl.		Low Refrig. Press. Temp. Safety Contl.		Low Chilled Water Temp. Control		Flow and Pressure Differential Contl.	
		Setting	Cond.	Setting	Cond.	Setting	Cond.	Chilled Water Cond.	Condenser Water Cond.

Controls:
All safety and operating controls should be tested annually, calibrated to design conditions, and their set points recorded. Defective safety devices and controls should be replaced.

6-25

ABSORPTION A/C AND REFRIGERATION LOG

Man.		No.		Size	
Loc.		Tech.		Date	

Date or Time	Outside and AMB. Temp.	Absolute Pressure	Water Temperature			
			Chilled		Absorber	
			In	Out	In	Out

Tubes:
1. The water side surfaces should be cleaned annually.
2. Eddy current analysis should be performed on condenser tubes within 3 to 5 years of service and on evaporator tubes within 5 to 7 years of service.

ABSORPTION A/C AND REFRIGERATION LOG (*cont.*)

Man.				No.			Size	
Loc.				Tech.			Date	

Date or Time	Outside and AMB. Temp.	Water Temp. Condenser		Generator Water Temp. or Steam Pressure	Capacity Control Valve Position	Solution Level in Absorber
		In	Out			

Lithium Bromide Solution:
An analysis should be made twice a year by a testing laboratory.

ABSORPTION A/C AND REFRIGERATION LOG (*cont.*)

Man.				No.			Size	
Loc.				Tech.			Date	

Date or Time	Outside and AMB. Temp.	Purge Unit Operation		Solution Analysis		Chemical Added Internally		
		Time	Ok?	Date	Ok?	Date	Type	Amount

Canned Motor Pump Unit:
Overhaul units after 3 years of normal service. Examine, clean or replace seals, bearings, cooling, and seal water passages. Other parts such as pump impeller, motor stator, and rotor should be examined for defects and proper repairs or replacement made.

ABSORPTION A/C AND REFRIGERATION LOG (*cont.*)

Man.								Size	
Loc.			Tech.					Date	

Date or Time	Outside and AMB. Temp.	Refrig. Tubes		Low Refrig. Temp.		Low Chilled Water Temp.		Flow Switches	
		Cleaned	Eddy Current Tested	Setting	Condition	Setting	Condition	Chiller Water Condition	Condenser Water Condition

Controls:
All safety and operating controls should be tested annually, calibrated to design conditions, and their set points recorded. Defective safety devices and controls should be replaced.

CHILLER LOG

Man.		No.		Size	
Loc.		Tech.		Date	

	Evaporator					
Time of Reading	Water Temp In	Water Temp Out	Chill Water GPM	EVAP Press.	Refrig. Temp	Refrig. Level
12:00 am						
2:00 am						
4:00 am						
6:00 am						
8:00 am						
10:00 am						
12:00 pm						
2:00 pm						
4:00 pm						
6:00 pm						
8:00 pm						
10:00 pm						

6-30

CHILLER LOG (cont.)

Man.			No.		Size	
Loc.			Tech.		Date	

	Condenser				
Time of Reading	Water Temp In	Water Temp Out	Condenser GPM	Condenser Pressure	Refrigerant Temp
12:00 am					
2:00 am					
4:00 am					
6:00 am					
8:00 am					
10:00 am					
12:00 pm					
2:00 pm					
4:00 pm					
6:00 pm					
8:00 pm					
10:00 pm					

CHILLER LOG (*cont.*)						
Man.		**No.**			**Size**	
Loc.		**Tech.**			**Date**	
	Compressor					
Time of Reading	Motor AMP Average	Volts	Oil Temp.	Oil Press.	Oil Level	Bearing Temp.
12:00 am						
2:00 am						
4:00 am						
6:00 am						
8:00 am						
10:00 am						
12:00 pm						
2:00 pm						
4:00 pm						
6:00 pm						
8:00 pm						
10:00 pm						

CHILLER LOG (cont.)

Man.			No.		Size	
Loc.			Tech.		Date	

Time of Day	Miscellaneous					
	Start up Time	Shutdown Time	Ambient Temp.	Relative Humidity	Tower Water Cond./TDS	Makeup Water Cond./TDS
12:00 am						
2:00 am						
4:00 am						
6:00 am						
8:00 am						
10:00 am						
12:00 pm						
2:00 pm						
4:00 pm						
6:00 pm						
8:00 pm						
10:00 pm						

HEATING BOILER LOG

Man.		No.		Year
Loc.		Tech.		Date

	Safety/Relief Valve Tested				
Log Month	**Week (✔)**				
	1	**2**	**3**	**4**	**5**
January					
February					
March					
April					
May					
June					
July					
August					
September					
October					
November					
December					

Test Instructions

Safety/Relief Valve: Pull try-lever to full open position with pressure on the boiler. Release try-lever; allow the valve to snap closed. All discharges must be piped to a safe place.

HEATING BOILER LOG (*cont.*)

Man.		No.		Year	
Loc.		Tech.		Date	

Log Month	Water Column or Gauge Glass Drained				
	Week (✔)				
	1	2	3	4	5
January					
February					
March					
April					
May					
June					
July					
August					
September					
October					
November					
December					

Test Instructions

Water Column or Gauge Glass (steam systems only): Open the drain valve quickly to void a small quantity of water. Water level should return quickly when the drain valve is closed. All discharges must be piped to a safe place.

HEATING BOILER LOG (cont.)

Man.		No.		Year
Loc.		Tech.		Date

Log Month	Low Water Fuel Cutoff Tested				
	Week (✔)				
	1	2	3	4	5
January					
February					
March					
April					
May					
June					
July					
August					
September					
October					
November					
December					

Test Instructions

Low Water Fuel Cutoff: Drain float chamber while the boiler is running. This should interrupt the circuit and stop the burner. All discharges must be piped to a safe place.

Servicing: The low water cutoff should be dismantled for a complete overhaul by competent service personnel at least annually. The internal and external mechanism, including linkage contacts, mercury bulbs, floats, and wiring should be checked. *Record the service dates.*

Service Dates					

HEATING BOILER LOG (*cont.*)

Man.		No.		Year
Loc.		Tech.		Date

Log Month	Circulating/Return Pump and System Checked				
	Week (✔)				
	1	2	3	4	5
January					
February					
March					
April					
May					
June					
July					
August					
September					
October					
November					
December					

Test Instructions

Pump and System: Check pump for proper operation and leaky packing. Examine traps, check valves, makeup float valves, expansion or condensate tank, and other parts of the system.

HEATING BOILER LOG (*cont.*)

Man.		No.		Year
Loc.		Tech.		Date

Log Month	Burner Operation Checked				
	Week (✔)				
	1	2	3	4	5
January					
February					
March					
April					
May					
June					
July					
August					
September					
October					
November					
December					

Test Instructions

Burner Operation: If the burner starts with a puff or operates roughly, repair immediately.

Servicing

Stoker, Oil, or Gas Burner and Controls: Stoker, oil, or gas burner and all operating and protective controls should be thoroughly checked at least once every three months by a competent service organization. *Record the service dates*

Service Dates					

AVERAGE LIFE EXPECTANCY OF HEATING AND COOLING EQUIPMENT IN YEARS

Residential Equipment

Central air conditioning	13 years
Central system heat pump	12 years
Window air conditioning	12 years
Chest freezer	17 years
Portable electric heater	10 years
Domestic refrigerator	15 years
Residential water heater	11 years

Commercial Equipment

Packaged A/C with gas heat	15 years
Packaged heat pump	12 years
Reach-in refrigerator	16 years
Reach-in freezer	17 years
Ice machine	11 years
Walk-in cooler	20 years
Walk-in freezer	20 years
Commercial water heater	12 years
Supermarket refrigerated display cases	18 years

SYSTEM CHECKLIST

Loc._____ Address:_____

Tech._____ Date:_____ Other:_____

MECHANICAL ROOMS

- Clean and dry? _____ Trash or chemicals? _____
- Items in need of attention _____

_____

MAJOR MECHANICAL EQUIPMENT

- Preventive maintenance plan in use? _____

Control System

- Type _____
- System operation _____
- Date of last calibration _____

Boilers

- Rated Btu input _____Condition _____
- Combustion air: is there at least 1 sq. in. free area per 2,000 Btu input? _____
- Fuel or combustion odors _____

Cooling Tower

- Clean? _____
- Leaks or overflow? _____
- Slime or algae growth? _____
- Eliminator performance _____
- Biocide treatment working? (list type) _____
- Spill containment plan implemented? _____
- Dirt separator working? _____

SYSTEM CHECKLIST (cont.)

Loc._____ Address:_____
Tech._____ Date:_____ Other:_____

Chillers
- Refrigerant leaks?_____

- Evidence of condensation problems? _____

- Waste oil and refrigerant properly stored and disposed of?

_____

Air Handling Unit
- Unit identification _____

- Area served _____

Outdoor Air Intake, Mixing Plenum, and Damper
- Outdoor air intake location _____

- Nearby contaminant sources? (describe) _____

_____

- Bird screen in place and unobstructed? _____

- Design total cfm _____ Outdoor air (O.A.) ctm _____

  Date last tested and balanced _____

- Minimum % O.A. (damper setting)_____

  Minimum cfm O.A. $\frac{\text{(total cfm} \times \text{minimum \% O.A.)}}{100}$ = _____

- Current O.A. damper setting (date, time, and HVAC operating mode) _____

- Damper control sequence (describe) _____

_____

- Condition of dampers and controls (date) _____

SYSTEM CHECKLIST (*cont.*)

Loc._____ Address:_____
Tech._____ Date:_____ Other:_____

Fans
- Control sequence _____

  _____

- Condition (note date) _____

- Indicated temperatures   Supply air _____ Mixed air _____
                                Return air _____ Outdoor air ____

- Actual temperatures   Supply air _____ Mixed air _____
                             Return air _____ Outdoor air ____

Coils
- Heating fluid discharge temerature _____ ΔT _____

  Cooling fluid discharge temperature _____ ΔT _____

- Controls (describe) _____

  _____

- Condition (note date) _____

Humidifler
- Type _____

- If biocide is used, note type _____

- Condition (no overflow, drains trapped, all nozzles working?)

  _____

- No slime, visible growth, or mineral deposits? _____

SYSTEM CHECKLIST (cont.)

Loc._____ Address:_____
Tech._____ Date:_____ Other:_____

DISTRIBUTION SYSTEM

Zone/ Room	System Type	Supply Air	
		Ducted/ Unducted	CFM

Return Air		Power Exhaust		
Ducted/ Unducted	CFM	CFM	Control	Serves (e.g., toilet)

Condition of Distribution System and Terminal Equipment

- Adequate access for maintenance? _____
- Ducts and coils clean and obstructed? _____
- Air paths unobstructed? Supply _____ Return _____
  Transfer _____ Exhaust _____ Makeup _____
- Note locations of blocked air paths, diffusers, or grilles _____
  _____
- Unintentional openings into plenums? _____
- Controls operating properly? _____
- Air volume correct? _____
- Drain pans clean? _____ Growth or odors? _____

SYSTEM CHECKLIST (*cont.*)

Loc._____ Address:_____
Tech._____ Date:_____ Other:_____

Filters

Location	Type/Rating	Size

Date Last Changed	Condition

OCCUPIED SPACE

Thermostat Types _____

Zone/Room	Thermostat Location	Controlled Device

Setpoints		Measured Temperature	Day/Time
Summer	Winter		

SYSTEM CHECKLIST (*cont.*)

Loc._____ Address:_____
Tech._____ Date:_____ Other:_____

Humidistats/Dehumidistats Type _____

Zone/ Room	Humidistat/ Dehumidistat Location	Controlled Device

Setpoints (%RH)	Measured Temperature	Day/ Time

- Potential problems (location) _____
  _____

- Thermal comfort or air circulation (drafts, obstructed airflow, stagnant air, etc.)
  _____
  _____

- Malfunctioning equipment _____

- Major sources of odors or contaminants
  _____
  _____

HOW TO MAINTAIN AND SERVICE
AIR CONDITIONING SYSTEMS

- Is the filter clean?

- Does the filter fit the filter rack well and is the rack well sealed?

- Are the supply and return registers and grilles open and unobstructed?

- Is the condenser coil clean and unobstructed?

- Is the condensing unit location acceptable?

- Is the condensing unit level?

- Is the evaporator coil clean?

- Is the evaporator condensate drain and drain line clean and unobstructed?

- Check for air bypassing the evaporator coil.

- Is the blower wheel clean?

- Is the blower motor clean?

- Inspect blower motor mounting and vibration isolators.

- Oil the blower motor if necessary. Do not overoil.

- Check blower belt and adjust or replace as necessary.

- Is the thermostat properly located, level, and clean?

- Are all power and low-voltage wiring connections clean and tight?

- Inspect start and run capacitors for signs of overheating.

HOW TO MAINTAIN AND SERVICE
AIR CONDITIONING SYSTEMS (*cont.*)

- Is the condensing unit contactor (contacts) in good condition?

- Is the system properly charged and operating with acceptable superheat and subcooling?

- Is the temperature differential of the evaporator airflow between 16 and 20°F?

- Does the condenser air temperature rise exceed 15°?

- Is the compressor crankcase heater tightly fitted to the compressor and working?

- Is the system operating with acceptable voltage plus or minus 10% of the rated voltage for the system?

- If the condensing unit uses a fused disconnect, are the fuses the correct size (amperage)?

- Inspect ductwork for leaking, disconnected joints, and damage.

- Check for vibrating or rubbing copper tube and piping.

- Repair or replace suction line insulation as necessary.

- Add wire ties to loose or hanging control wiring.

- Operate system and verify the proper operation of controls in all modes of operation.

A/C SERVICING—PRESSURES

Approximate Operating Pressures for Comfort Air Conditioning Systems at Various Outside Air Temperatures for R-22 Refrigerant

Outside Air Temp. (°F)	Low Side Pressure Range (psig)	High Side Pressure Range (psig)
70	60–65	170–200
80	60–70	180–225
90	65–75	200–240
100	70–78	240–300

- Lower than normal low side pressures can be caused by lack of sufficient airflow, dirty air filter, dirty indoor coil, lack of refrigerant charge, or refrigerant leak.

- Higher than normal low side pressures can be caused by a refrigerant overcharge, lack of sufficient outdoor airflow, dirty outdoor air coil, air in the refrigerant system, abnormally high indoor air temperature, or bad compressor valves.

- Lower than normal high side pressure can be caused by refrigerant undercharge, refrigerant leak, or bad compressor valves.

- Higher than normal high side pressure can be caused by a refrigerant overcharge, lack of sufficient airflow over the outdoor coil, dirty outdoor coil, or air in the refrigerant system.

REFRIGERANT AND OIL RETROFIT

Retrofit guidelines vary depending upon the type of refrigerant, type and design of equipment, and manufacturer. For specific retrofit instructions for a particular piece of equipment, contact the manufacturer.

1. Determine what replacement refrigerant is the best replacement for the system type.

2. Determine which refrigerant oil is most compatible with the retrofit refrigerant.

3. Recover the system refrigerant.

4. Remove the oil from the compressor.

5. Drain oil from all low spots in the system. This includes oil in the bottom of accumulators or oil separators, if present. Low spots in suction lines can be drained by drilling a small hole in the low spot. The hole can be brazed over later.

6. Charge the compressor crankcase with the new oil.

7. Replace the filter drier.

8. Evacuate the system to 500 microns or use the triple evacuation method.

9. Charge the system with the new refrigerant and operate the system.

10. Make any system adjustments as necessary. (Thermostatic expansion valve, etc.)

11. Operate the system and record the operating temperatures and pressures. Leave a copy of the results with the equipment. Label the system with the new refrigerant and oil charge now in the system and include the date of the retrofit.

REFRIGERANT RECOVERY CYLINDER PRESSURE SERVICE RATING REQUIREMENTS

- R-410a recovery cylinders must be service rated for at least 400 psig.
- DOT 4BA 400 & DOT 4BW 400 are acceptable cylinders.
- Dedicated cylinders and equipment for 410a must be clearly marked "For 410a Service Only."
- Never allow a cylinder of 410a to get warmer than 125°F.
- Never fill a recovery cylinder over 80% of capacity.

ALTERNATIVE REFRIGERANTS AND RECOMMENDED LUBRICANTS

R-11 Alternative	Lubricant
R-123	MO
R-12 Alternatives	**Lubricant**
R-134a	Ester
R-401a	Ester or AB
R-401b	Ester of AB
R-406a	MO, AB or Ester oils
R-502 Alternatives	**Lubricant**
R-408a	MO, AB or Ester oils
R-404a	Ester
R-507	Ester
R-402a	MO or AB
R-402b	MO or AB
R-403a	MO or AB
R-403b	MO or AB
R-407a	POE-Ester
R-22 Alternatives	**Lubricant**
R-407c	Ester
R-410a*	Ester

*R-410a is not a direct replacement or drop-in refrigerant. It must only be used in systems originally designed for 410a.

PROCEDURES FOR REFRIGERANT RECLAIMING

Procedure for Recovering Refrigerant into Cylinders <u>Only</u>

1. Visually inspect the cylinder to be filled. Strictly follow all DOT requirements for inspection.
2. Place the cylinder on a scale. Note empty weight of cylinder to determine maximum gross weight.
3. Connect transfer hoses to the cylinder. Make certain they are leak free. If at all possible, change hoses when recovering different types of refrigerants to avoid contamination by mixing refrigerants.
4. Open the cylinder outlets, begin the transfer, following manufacturer's instructions for the recovery unit.
5. **Do not leave the cylinder unattended.** Watch the scale closely. **Do not overfill.** Do not exceed the gross weight limit. Do not fill more than 80% by volume. (It is illegal to transport an overfilled cylinder.)
6. When the scale reaches the gross weight limit—stop the transfer process. Tightly close all valves and other outlets.
7. Disconnect the transfer hose. **Avoid contact with liquid refrigerant/oil mixtures.** Immediately replace all caps and other cylinder closures.
8. Weigh the cylinder. Record the weight on **all** appropriate forms and on the cylinder **hang tag.**
9. Completely fill out the cylinder **hang tag** attached to the cylinder. **Be sure the hang tag indicates the correct refrigerant in the cylinder.** *It is illegal to transport a cylinder without correctly identifying the contents,* including an empty cylinder.

REPLACING A BURNED COMPRESSOR

1. Recover the burned refrigerant into a suitable recovery container dedicated for the purpose and clearly marked as such.

2. Recover the refrigerant through one or more filter driers in series with the recovery machine.

3. If the bad compressor is brazed in the lines do not unbraze or unsweat the lines. Cut the lines with a tubing cutter and remove the compressor.

4. Inspect the suction line to determine if carbon is deposited inside the line. (The inside is normally clean copper.) If carbon exists, consider replacing the suction line.

5. Drill holes in any low spots in the system to drain out any residual oil. The oil may be highly acidic.

6. Purge the system with nitrogen.

7. Replace the compressor. Perform all brazing with a dry nitrogen bleed.

REPLACING A BURNED COMPRESSOR (*cont.*)

8. Replace the liquid line filter drier with one larger than the existing drier.

9. Add an acid core suction line filter drier.

10. Leak test the system with 20% R-22 backed up with 150 psig of nitrogen.

11. Leave the leak test charge in the system for at least 30 min., then bleed the charge.

12. Evacuate the system to 1500 microns.

13. Break the vacuum with dry nitrogen and pressurize to 100 to 150 psig. Allow this charge to remain in the system for at least 30 min. before bleeding the charge.

14. Evacuate the system to approximately 700 microns and repeat step 13.

15. Evacuate the system to 400 microns and hold it at 400 microns or less for at least 1 hr.

FOLLOW UP AFTER BURNOUT REPLACEMENT

1. Allow the new compressor to operate for 8 to 12 hr.

2. Take an acid test and record the results before going to the next step.

3. Recover the refrigerant and filter it through a set of new filter driers.

4. Remove and replace the compressor oil with new fresh oil.

5. Remove and replace both the liquid line and suction line filter driers.

6. Perform a vapor acid test on the recovered refrigerant. If the refrigerant shows acid do not use it. If clear of acid it may be recharged into the system.

7. Allow the system to operate for 24 to 48 hr. and repeat step 6.

8. Take an acid test before recovering the refrigerant and keep track of the results.

9. Allow the system to operate for 7 to 10 days, then take another acid test.

10. If the oil or refrigerant shows acid, perform the same oil change and filter replacement again.

11. When the system comes up perfectly clean, replace the filter driers one last time.

EVACUATION AND DEHYDRATION

Evacuation and Dehydration Notes

1. Do not use a converted refrigerant compressor as a vacuum pump; instead, use a two-stage deep vacuum pump made for HVAC & refrigeration evacuation.

2. Use special vacuum pump oil in vacuum pumps.

3. Change vacuum pump oil after every use on wet systems or every 10 hr. of use.

4. An evacuation is complete when a vacuum of 500 microns has been reached.

5. Once a 500-micron vacuum has been achieved, valve off the pump while leaving the vacuum indicator gauge in the system. If the vacuum pressure rises but levels off between 1,000 and 2,000 microns, the system is leak-tight but is still too wet. Continue the evacuation.

6. An evacuation that reaches 500 microns but does not rise above 1,000 microns indicates a tight and dry system.

EVACUATION PRESSURES AND TEMPERATURES TO VAPORIZE WATER OUT OF A SYSTEM		
Microns	**in. Hg**	**Boiling Temp. (°F)**
762,000	0 (Atmospheric Press)	212
500,000	10.24	192
200,000	22.05	152
100,000	25.98	125
50,000	27.95	101
30,000	28.74	84
20,000	29.13	72
15,000	29.33	63
10,000	29.53	52
6,000	29.69	39
4,000	29.76	29
2,000	29.84	15
1,000	29.88	1
500	29.90	−12
300	29.91	−21
200	29.91	−28
150	29.92	−33
100	29.92	−40
50	29.92	−50
0	29.921	—

ESTIMATED EVACUATION TIME TO REACH A 500-MICRON VACUUM

5 cu. ft. internal volume; air at approximately 40% relative humidity.

Pump Size	Line Diameter (6' long)	Time to Reach 500 Microns
1 CFM	¼" ID	80 min.
2 CFM	¼" ID	60 min.
3 CFM	¼" ID	56 min.
4 CFM	¼" ID	49 min.
5 CFM	¼" ID	45 min.
6 CFM	¼" ID	44 min.
1 CFM	⅜" ID	50 min.
2 CFM	⅜" ID	30 min.
3 CFM	⅜" ID	23 min.
4 CFM	⅜" ID	20 min.
5 CFM	⅜" ID	16 min.
6 CFM	⅜" ID	15 min.

REQUIRED EVACUATION VACUUM LEVELS

| Appliance Type | Inches Hg. Vacuum using Eqt. Manufactured | |
	Before Nov. 15, 1993	*After* Nov. 15, 1993
R-22 appliance on component containing less than 200 lb. of refrigerant	0	0
R-22 appliance on component containing 200 lb. or more of refrigerant	4	10
Other high-pressure appliance on component containing less than 200 lb. of refrigerant (R-12, -500, -502, -114)	4	10
Other high-pressure appliance containing 200 lb. or more of refrigerant (R-12, -500, -502, -114)	4	15
Very high-pressure appliance (R-13, -503)	0	0
Low-pressure appliance (R-11, R-123)	25	29 (25 mm Hg. Absolute)

NUMBER OF FLUSHES AND TYPICAL MINERAL OIL RESIDUAL LEVELS

Flush Number	Starting Levels				
	50%	40%	30%	20%	10%
1st	25%	16%	9%	4%	1%
2nd	12.5%	6.4%	2.7%	.8%	.1%
3rd	6.25%	2.56%	.8%	.16%	.01%
4th	3.1%	1.0%	.25%	.03%	0%

ADDING OIL TO A CLOSED SYSTEM

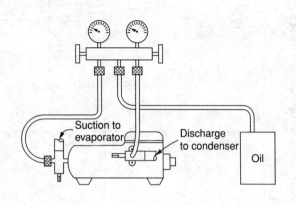

Suction to evaporator

Discharge to condenser

Oil

ADDING OIL TO A SYSTEM WITH AN OIL PUMP

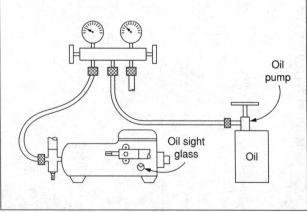

Oil pump

Oil sight glass

Oil

LIQUID LEVEL INDICATOR PARTS

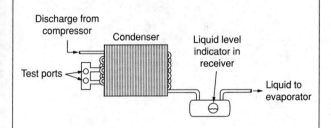

Discharge from compressor

Condenser

Test ports

Liquid level indicator in receiver

Liquid to evaporator

CHECKING REFRIGERANT CHARGE USING A SIGHT GLASS

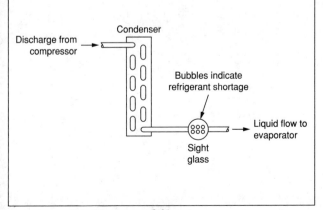

Condenser

Discharge from compressor

Bubbles indicate refrigerant shortage

Liquid flow to evaporator

Sight glass

CONNECTIONS—LOADING A CHARGING CYLINDER

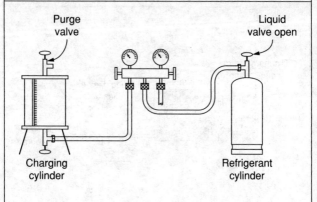

Purge valve

Liquid valve open

Charging cylinder

Refrigerant cylinder

CONNECTIONS—EVACUATING A CHARGING CYLINDER

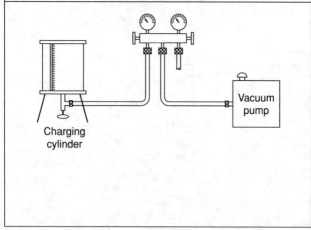

Charging cylinder

Vacuum pump

CONNECTIONS FOR LIQUID CHARGING A SYSTEM

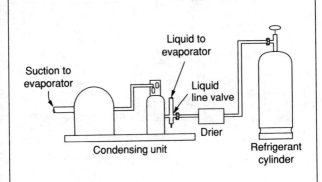

Liquid to evaporator

Liquid line valve

Suction to evaporator

Drier

Condensing unit

Refrigerant cylinder

LIQUID DRIER IN CHARGING LINE

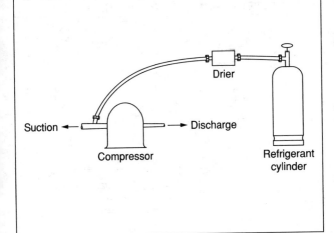

Drier

Suction ←

→ Discharge

Compressor

Refrigerant cylinder

CONNECTIONS FOR REMOVING REFRIGERANT WITH A RECOVERY UNIT

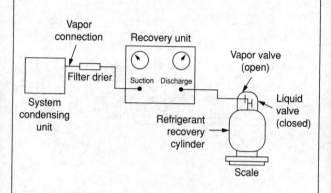

CLOSED-CIRCUIT THERMOCOUPLE TEST

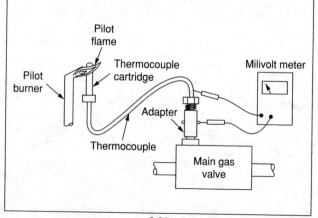

SUCTION LINE CLEANUP FILTER DRIER

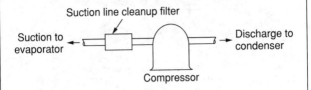

Suction line cleanup filter

Suction to evaporator ←

Discharge to condenser →

Compressor

LIQUID LINE CLEANUP DRIER

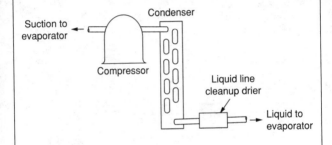

Condenser

Suction to evaporator ←

Compressor

Liquid line cleanup drier

Liquid to evaporator →

VAPOR RECOVERY CONNECTIONS

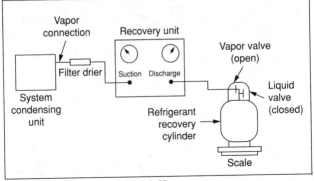

Vapor connection

Recovery unit

Vapor valve (open)

Liquid valve (closed)

Filter drier

Suction   Discharge

System condensing unit

Refrigerant recovery cylinder

Scale

6-65

RECOVERY/RECYCLE UNIT SCHEMATIC USING SINGLE PASS FILTER DRIER

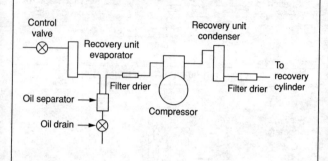

LIQUID RECOVERY CONNECTIONS

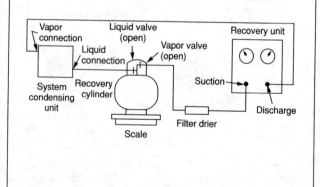

RECOVERY/RECYCLE UNIT SCHEMATIC USING AN OIL SEPARATOR

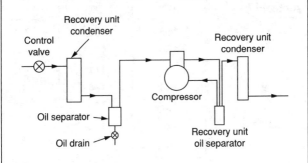

RECOVERY/RECYCLE UNIT SCHEMATIC USING MULTIPLE PASS FILTER DRIERS

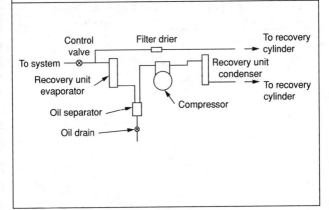

RECOMMENDED COIL PRESSURES

Indoor Coil Pressure (psig)

Outdoor Ambient (°F)	Indoor Temp (°F)		
	60	70	80
37	168	190	211
32	158	180	200
27	150	171	191
22	143	164	184
17	137	159	178
12	133	154	174
7	129	151	170
2	126	147	167
−3	123	145	164
−8	123	143	162

Outdoor Coil Pressure

Outdoor Ambient (°F)	Outdoor Coil Pressures (psig)
62	57
57	51
52	46
47	41
42	38
37	34
32	30
30	25
25	20

APPROXIMATE SETTINGS, LOW PRESSURE CONTROL (R-12)

Type of Equipment	Cut-in Setting (psig)	Cut-out Setting (psig)
Walk-in meat cooler	38	12
Walk-in dairy cooler	38	14
Walk-in produce cooler	38 to 40	16
Walk-in florist cooler	40 to 42	18
Open meat display case	26	9
Closed meat display case (gravity air evaporator)	24	4
Closed meat display case (forced air evaporator)	32	12
Open dairy display case	30	9
Open produce display case	38	16
Medium temperature reach-in	38	12
Beverage box (soft drinks, beer, etc.)	38	12
Air conditioner (central) (safety cut-out only)	120	54

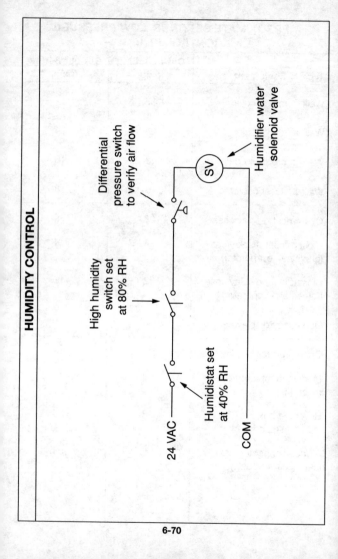

HUMIDITY CONTROL

Differential pressure switch to verify air flow

Humidifier water solenoid valve

SV

High humidity switch set at 80% RH

Humidistat set at 40% RH

24 VAC

COM

6-70

CHAPTER 7
Troubleshooting

OIL HEATING SYSTEMS		
Problem	**Analysis**	**Correction**
Burner will not start	Thermostat off or set too low	Turn thermostat on; set to higher temperature
	Burner motor overload tripped	Push motor overload reset button
	Primary control off on safety	Reset safety switch lever
	Dirty thermostat contacts	Clean thermostat contacts
	Bad thermostat	Replace thermostat
	Blown fuse or tripped breaker	Replace fuse or reset breaker
	Disconnect switch open	Close switch
	Shorted flame detector circuit	Replace flame detector
	Shorted flame detector leads	Separate and insulate leads
	Flame detector exposed to direct light	Protect detector from light
	Faculty friction clutch	Replace element or control
	Hot contacts stuck	Replace element or control
	Dirty cold contacts	Clean contacts
	Flame detector bimetal carboned	Clean bimetal
	Loose connection or broken wire on flame detector	Repair connection or replace wire
	Low line voltage or power failure	Notify power company
	Limit control open	Set limit control to 200°F then jumper control terminals; if burner starts, replace the control

OIL HEATING SYSTEMS *(cont.)*

Problem	Analysis	Correction
Burner will not start *(cont.)*	Open electric circuit to limit control	Repair or replace wiring
	Defective internal primary control circuit	Replace control
	Dirty burner relay contacts in primary control	Clean contacts
	Defective burner motor	Replace burner motor
	Binding burner blower wheel	Turn off power and rotate blower by hand
	Seized fuel pump	Turn power off and rotate blower by hand; replace fuel pump
Burner starts and fires but short cycles (does not heat enough)	Thermostat in warm draft	Relocate thermostat
	Heat anticipator set wrong	Correct anticipator setting
	Vibration at thermostat	Correct vibration or relocate thermostat
	Limit control set too low	Reset limit to 200°F
	Dirty air filter	Clean or replace filter
	Low or fluctuating voltage	Notify power company; check circuit
	Loose wiring connection	Repair connection
	Furnace blower running too slow	Speed up blower to obtain an 85°F to 95°F temperature
	Defective blower motor bearings	Replace motor
	Defective blower bearings	Replace bearings
	Dirty furnace blower wheel	Clean blower wheel
	Wrong blower motor rotation	Change rotation or replace motor
	Return air restricted	Clear restriction
Burner starts and fires, then locks out on safety	Too little primary air; long dirty flame	Increase combustion air
	Too much primary air; short lean flame	Reduce combustion air
	Unbalanced flame	Replace nozzle

OIL HEATING SYSTEMS (cont.)

Problem	Analysis	Correction
Burner starts and fires, then locks out on safety (cont.)	Too little or restricted draft	Correct draft or remove restriction
	Excessive draft	Adjust barometric damper
	Dirty flame detector bimetal element	Clean element
	Faulty flame detector friction clutch	Replace flame detector control
	Welded or shorted cold contacts in flame detector	Replace flame detector control
	Air leaking into flue pipe around flame detector mount	Seal air leaks
	Dirty flame detector cad cell face	Clean cad cell face
	Loose or defective flame detector cad cell wires	Repair or replace cad cell holder and wires
	Faulty flame detector cad cell; resistance exceeds 1,500 ohms	Replace cad cell
	Defective primary control circuit	Replace primary control
Burner starts, but no flame is established	Oil tank empty	Contact oil distributor
	Oil tank shut-off valve closed	Open valve
	Water in oil tank	Remove water
	Air leak in oil supply line	Repair leak
	Oil filter plugged	Install new filter
	Oil pump strainer plugged	Clean strainer
	Restricted oil line	Repair or replace line
	Excessive combustion air	Adjust air supply damper
	Excessive vent draft	Adjust barometric damper to .030" to .035" water column
	Off-center spray from nozzle	Replace nozzle
	Nozzle strainer plugged	Replace nozzle
	Nozzle orifice plugged	Replace nozzle
	Faulty oil pump	Replace pump
	Low fuel pressure	Adjust pressure to desired pressure

OIL HEATING SYSTEMS *(cont.)*

Problem	Analysis	Correction
Burner starts, but no flame is established *(cont.)*	Faulty pump coupling	Replace coupling
	Faulty transformer	Replace transformer
	No or weak ignition spark	Properly ground transformer case
	Dirty or shorted ignition electrodes	Clean electrodes
	Improper position or gap of ignition electrodes	Correctly position and reset electrode gap
	Cracked or burned lead insulation	Replace electrode leads
	Loose or disconnected electrode leads	Repair or replace leads
	Defective electrode lead insulators	Replace electrodes
	Oil pump or blower overloading motor	Remove overload condition
	Faulty oil pump motor	Replace motor
	Low voltage	Notify power company
Burner starts and fires but loses flame and locks out on safety	Dirty face and cad cell	Clean cad cell face
	Faulty cad cell; resistance exceeds 1500 ohms	Replace cad cell
	Loose or defective cad cell wires	Repair or replace wires
	Stack control bimetal dirty	Clean bimetal element
	Faulty friction clutch in stack control	Replace stack control
	Air leaking into vent pipe around stack control mount	Seal air leaks
	Defective stack control cold contacts	Replace stack control
	Too much combustion air	Adjust combustion air damper
	Too little combustion air	Adjust combustion air damper
	Unbalanced flame	Replace nozzle
	Excessive vent draft	Adjust barometric damper
	Too little vent draft	Adjust barometric damper
	Vent restricted	Clear restriction
	Oil pump loses prime	Prime pump at bleed port

OIL HEATING SYSTEMS (cont.)

Problem	Analysis	Correction
Burner starts and fires but loses flame and locks out on safety *(cont.)*	Air leak in oil supply line	Repair leaks
	Partially plugged nozzle	Replace nozzle
	Partially plugged nozzle strainer	Replace nozzle
	Water in oil storage tank	Remove water from storage tank
	Oil too heavy	Change to Number 1 oil
	Plugged fuel pump strainer	Clean strainer or replace pump
	Restricted oil line	Clear restriction
Too much heat; burner runs continuously	Defective thermostat	Repair or replace thermostat
	Shorted thermostat wires	Repair or replace wires
	Thermostat in cold location	Relocate thermostat
	Thermostat not level	Level thermostat
	Defective primary control	Replace control
Too little heat; burner runs continuously	Too much combustion air	Reduce combustion air
	Air leaking into heat exchanger	Repair leaks
	Fxcessive vent draft	Adjust barometric damper
	Wrong burner head adjustment	Correct burner head setting
	Plugged heat exchanger	Clean heat exchanger and adjust burner
	Too little combustion air	Increase combustion air
	Insufficient vent draft	Adjust barometric damper
	Insufficient indoor air	Speed blower to obtain 85°F to 95°F temperature rise
	Dirty indoor blower	Clean blower
	Dirty furnace filter	Clean or replace filter
	Partially plugged nozzle	Replace nozzle
	Nozzle too small	Replace with larger nozzle
	Low oil pressure	Increase to proper pressure

ELECTRIC HEATING SYSTEMS

Problem	Analysis	Correction
Unit will not run	Blown fuse	Replace fuse and correct cause
	Burned transformer	Replace transformer and correct cause
	Thermostat not calling for heat	Set thermostat
	Defective thermostat	Replace thermostat
	Defective heating relay	Replace relay
Fan will not run	Burned fan motor	Repair or replace fan motor
	Broken fan belt	Replace fan belt
	Burned contacts in fan relay	Replace fan relay
	Defective fan control	Replace fan control
	Defective wiring or connections	Repair wiring or connections
Fan motor hums but will not start	Defective fan motor bearings	Replace bearings or fan motor
	Defective fan motor starting switch	Repair starting switch or replace motor
	Defective starting capacitor	Replace capacitor
	Burned start windings in motor	Repair or replace motor
	Defective blower bearings	Replace bearings
	Loose wiring connections in motor starting circuit	Repair wiring
Fan motor cycles	Defective motor bearings	Replace bearings or motor
	Defective blower bearings	Replace blower bearings
	Defective run capacitor	Replace capacitor
	Defective fan control	Replace fan control
	Defective fan relay	Replace relay
	Defective motor windings	Repair or replace motor
Fan blows cold air	Defective heat sequencing relays	Replace relays
	Burned heat elements	Replace elements
	Loose wiring connections	Repair wiring
	Defective thermostat	Replace
	Fan set to on position	Set to auto position
	Defective fan control	Replace

ELECTRIC HEATING SYSTEMS *(cont.)*

Problem	Analysis	Correction
Not enough heat	Dirty air filters	Clean or replace filters
	Unit too small	Install more elements
	Too little air flow through furnace	Increase air flow; remove restrictions
	Thermostat heat anticipator not properly set	Reset anticipator
	Defective fan motor	Repair or replace fan motor
	Air conditioning evaporator dirty	Clean evaporator
	Thermostat not properly located	Relocate thermostat
	Thermostat set too low	Set thermostat
	Thermostat out of calibration	Calibrate thermostat
	Air ducts not insulated	Insulate ducts
	Burned elements	Replace
	Defective heat sequencing relays	Replace
	Defective thermostat	Replace
	Defective element limits	Replace limits
	Outdoor thermostat set too low	Reset thermostat
Too much heat	Unit too large	Reduce Btu input
	Thermostat heat anticipator not properly set	Set heat anticipator
	Thermostat not properly located	Relocate thermostat
	Thermostat set too high	Set thermostat
	Thermostat out of calibration	Calibrate thermostat
High humidity in building	Cooking	Vent cookstove
	Bathing	Vent bathroom
	Humidity	Increase temperature rise through furnace

ELECTRIC HEATING SYSTEMS *(cont.)*

Problem	Analysis	Correction
Blown element limits	Shorted heating element	Replace element; correct cause
	Dirty filters	Clean or replace
	Dirty blower	Clean blower
	Broken or slipping fan belt	Adjust or replace belt
	Defective blower motor	Replace motor
	Not enough air through furnace	Remove restriction
	Loose electrical connections	Repair connections
High operating costs	Unit too small	Increase number of elements
	Dirty air filters	Clean or replace filters
	Dirty air conditioning evaporator	Clean evaporator
	Air ducts not insulated	Insulate ducts
	Thermostat in wrong location	Relocate thermostat
	Dirty blower	Clean blower
	Defective thermostat	Replace thermostat
	Fan belt slipping	Replace or adjust fan belt
	Low or high voltage	Notify power company
	Thermostat setting too high	Lower setting

GAS HEATING SYSTEMS

Problem	Analysis	Correction
Unit will not run	Blown fuse	Replace fuse and correct cause
	Burned transformer	Replace transformer and correct cause
	Thermostat not calling for heat	Set thermostat
	Defective wiring or connections	Repair wiring or connections
Fan will not run	Burned fan motor	Repair or replace motor
	Broken fan belt	Replace fan belt

GAS HEATING SYSTEMS (cont.)

Problem	Analysis	Correction
Fan will not run *(cont.)*	Burned contacts in fan relay	Replace fan relay
	Defective fan control	Replace fan control
	Defective wiring or connections	Repair wiring or connections
Fan motor hums but will not start	Defective bearings in fan motor	Replace bearings or fan motor
	Defective starting switch in fan motor	Repair starting switch or replace motor
	Defective starting capacitor	Replace capacitor
	Burned start winding in motor	Repair or replace motor
	Defective blower bearings	Replace bearings
	Loose wiring connections in motor starting circuit	Repair wiring
Fan motor cycles	Defective motor bearings	Replace bearings or fan motor
	Defective blower bearings	Replace blower bearings
	Defective run capacitor	Replace run capacitor
	Defective fan control	Replace fan control
	Return air too cool	Allow air to warm
	Fan control differential too close	Adjust fan control
	Fan control off setting too high	Adjust fan control
	Too much air flow through furnace	Reduce air flow
	Defective motor windings	Repair or replace motor
	Fan control on setting too low	Adjust control
Pilot not burning properly or is out	Faulty thermocouple	Replace
	Dirty or corroded thermocouple connection	Clean connection
	Gas supply turned off	Restore gas supply
	Pilot burner orifice dirty	Clean orifice
	Thermocouple not installed in flame properly	Properly install thermocouple
	Drafts affecting pilot flame	Shield pilot from drafts
	Defective pilot safety device	Replace pilot safety device

GAS HEATING SYSTEMS (cont.)

Problem	Analysis	Correction
Fan cycles while main burner stays on	Wrong size orifices in burners	Replace orifices with proper size
	Low manifold gas pressure	Increase gas pressure
	Too much air flowing through furnace	Reduce air flow
	Too cold return air	Allow air to warm
Main burner cycles while blower stays on	Dirty air filters	Clean or replace filters
	Wrong size orifices in burners	Replace orifices with proper size
	High manifold gas pressure	Reduce gas pressure
	Faulty limit control	Replace
	Too little air flow through furnace	Increase air flow; clear restrictions
Not enough heat	Dirty air filters	Clean or replace
	Wrong size orifices in burners	Replace orifices with proper size
	Low manifold gas pressure	Increase gas pressure
	Too little air flow through furnace	Increase air flow; clear restrictions
	Thermostat heat anticipator not properly set	Set heat anticipator
	Defective fan motor	Replace or repair
	Unit too small	Replace unit with larger size
	Air conditioning evaporator dirty	Clean evaporator
	Thermostat not properly located	Move thermostat
	Thermostat out of calibration	Calibrate thermostat
Too much heat	Unit too large	Reduce Btu input; replace unit
	Thermostat heat anticipator not properly set	Set anticipator
	Thermostat not properly located	Move thermostat
	Thermostat out of calibration	Calibrate thermostat
Pilot burning; main gas valve will not operate	Blown fuse	Replace fuse and check for cause
	Defective gas valve	Replace gas valve

GAS HEATING SYSTEMS (cont.)

Problem	Analysis	Correction
Pilot burning; main gas valve will not operate (cont.)	Burned transformer	Replace transformer and check for cause
	Burned thermostat heat anticipator	Replace thermostat
	Bad thermostat	Replace thermostat
	Bad electrical connections	Repair connections
	Broken thermostat wire	Repair broken wire
Delayed ignition of main burner	Poor flame travel to the burner	Correct flame travel
	Poor flame distribution over the burner	Correct flame distribution
	Low manifold gas pressure	Adjust gas pressure
	Defective step-opening regulator	Adjust or replace regulator
Roll-out on main burner ignition	Restricted heat exchanger	Clear restrictions
	Quick opening main gas valve	Install surge arrestor
Flame flashback	Low manifold gas pressure	Adjust manifold gas pressure
	Extremely small main burner flame	Adjust primary air
	Distorted burner or carry-over wing slots	Repair burner or carry-over wing slots
	Defective main burner orifice	Replace orifices
	Orifice misaligned	Replace orifices
	Erratic gas valve operation	Replace valve
	Dirty burner	Clean burners
	Improper gas–air mixture	Be sure that proper gas is being used
	Unstable gas supply pressure	Install a two-stage pressure regulator
Resonance (loud, rumbling noise)	Excess primary air to main burner	Adjust primary air
	Defective orifice spud	Replace orifice spud
	Dirty orifice spuds	Clean orifices
Yellow flame	Too little primary air	Adjust primary air
	Dirty orifice spud	Clean spuds
	Orifice spuds misaligned	Align orifice spuds

GAS HEATING SYSTEMS *(cont.)*

Problem	Analysis	Correction
Yellow flame *(cont.)*	Restricted heat exchanger	Clean heat exchanger
	Poor vent operation	Correct venting
Floating main burner flame	Air blowing into heat exchanger	Check for defective heat exchanger
	Restricted heat exchanger	Clean heat exchanger
	Negative pressure in furnace room	Increase air supply to room
Main burner flame too large	Orifices too large	Replace orifices
	Excessive manifold gas pressure	Adjust pressure regulator
	Defective gas pressure regulator	Replace regulator
	Wrong type of gas being used	Install changeover kit
Main burner flame too small	Dirty orifice spuds	Clean orifice spuds
	Low manifold gas pressure	Adjust pressure regulator
	Orifices too small	Replace orifices
	Wrong type of gas being used	Install changeover kit
Odor in building	Vent not operating properly	Correct venting problem
	Poor ventilation	Check flame conditions and correct
High operating costs	Unit too small	Install proper size unit
	Dirty air filters	Clean or replace filters
	Dirty air conditioning evaporator	Clean evaporator
	Air ducts not insulated	Insulate ducts
	Thermostat in wrong location	Relocate thermostat
	Dirty blower	Clean blower

OIL-FIRED FURNACES

Problem	Analysis	Correction
Change in size of fire	Cold oil	Adjust pressure
	Low pressure	Adjust at pump
	Dirty nozzle	Clean or replace
	Plugged strainer	Clean

OIL-FIRED FURNACES *(cont.)*

Problem	Analysis	Correction
Burner motor does not start	Contact dirty or open on primary relay	Clean or replace relay
	Fuse burned out	Replace
	Primary relay off on safety	Push reset button
	Defective thermostat	Replace
	Motor stuck or burned out or overload protector out	Replace if burned out
	Limit control open	Check setting and correct
	Relay transformer burned out	Replace relay
Oil spray but no ignition	Improper spacing	Reset
	Faulty relay	Replace
	Dirty electrodes	Clean or replace
	Loose connection	Tighten
	Cracked porcelain	Replace
	Dead transformer	Replace
Noisy operation	Air in oil line	Bleed oil line; look for leaks
	Pump noise	Continued running sometimes works in gears. If not, replace.
	Blower noisy	Oil bearing. Tighten shaft collars; adjust belt tension; align and tighten pulleys; position rubber isolators.
	Hum vibration	Isolate pipes from structural members
	Loose coupling	Tighten setscrews
	Combustion noise	Adjust noise
	Bad coupling alignment	Loosen fuel unit or motor
	Furnace too small	Check heat loss to be sure furnace properly sized
	Burner noisy	Check mounting and position; adjust air
Burner will not run continuously	Poor flame due to too much air; too little oil	Check nozzle, air adjustment, oil pressure, and size of nozzle
	Control wired wrong	Check and rewire
	Lockout timing too short	Replace primary control
	Water or air in oil	Look for leak in supply

OIL-FIRED FURNACES (cont.)

Problem	Analysis	Correction
No oil flow	Frozen pump shaft	Replace
	Oil level below intake line in the supply tank	Fill tank
	Clogged nozzle	Clean or replace
	Slipping or broken coupling	Tighten or replace coupling
	Clogged strainer	Remove and clean strainer
	Air leak in intake line	Tighten fittings and plugs; check valves
	Restricted intake line (high vacuum reading)	Replace kinked tubing; check valves, filters
	Two-pipe system air bound	Check bypass plug
	Single-pipe system air bound	Loosen gauge port and drain oil until foam is gone
Pulsation	Dirty or improperly set electrodes	Clean and reset; wire primary control for continuous ignition
	Air adjustment	Readjust air
	Too much oil impingement	Check nozzle and pump pressure; check nozzle size and angle and position of drawer assembly
	Pressure over fire	Correct draft to 0.02" W.C. negative
Short cycling of fan	Input too low	Check burner input
	Temperature rise too low due to excessive speed of blower	Slow blower down and check ventilation
	Fan control setting	Set lower turn on (115°F)
Short cycling on limit control	Temperature rise too high due to blower running too slow	Increase blower speed
	Fan control setting too high	Reset lower (115°F)
	Input too high	Check burner input
	Control out of position	Place cad cell in proper position
	Limit setting low	Reset to maximum
	Temperature rise too high due to restricted returns or outlets	Open dampers or add additional outlets or returns

OIL-FIRED FURNACES *(cont.)*

Problem	Analysis	Correction
High fuel consumption	Flue loss too great	Measure CO_2 and flue-gas temperature; if loss is more than 25%, reset air, check input, and speed up blower. Check static pressure in return and outlet plenum and correct to recommended values.
	Input too high	Check burner input
Not heating	Insufficient air circulating	Speed up blower. Check size and location of ducts and outlets. Set fan control and blower for continuous air circulation.
	Low input	Check nozzles and input

GAS-FIRED FURNACES

Problem	Analysis	Correction
Noisy flame	Burr in orifice	Remove burr or replace orifice
	Too much primary air	Adjust air shutters
	Noisy pilot	Reduce pilot gas
Flame too large	Defective regulator	Replace
	Pressure regulator set too high	Reset using manometer
	Burner orifice too large	Replace with correct size
Floating flame	Insufficient primary air	Increase primary air supply
	Blocked venting	Clean
Yellow-tip flame	Clogged burner ports	Clean ports
	Too little primary air	Adjust air shutters
	Misaligned orifices	Realign
	Clogged draft hood	Clean
Delayed ignition	Improper pilot location	Reposition pilot
	Burner ports clogged	Clean ports
	Low pressure	Adjust pressure regulator
	Pilot flame too small	Check orifice; clean; increase pilot gas

GAS-FIRED FURNACES (cont.)

Problem	Analysis	Correction
Burner will not turn off	Defective or sticking automatic valve	Clean or replace
	Limit switch maladjusted	Replace
	Poor thermostat location	Relocate
	Short circuit	Check operation at valve; check for short and correct
Rapid fan cycling	Blower speed too high	Readjust to lower speed
	Fan switch differential too low	Readjust or replace
Blower will not stop	Shorts	Check wiring and correct
	Manual fan on	Switch to automatic
	Fan switch defective	Replace
Noisy blower and motor	Belt tension improper	Readjust (usually allow 1" slack)
	Bearings dry	Lubricate
	Belt rubbing	Reposition
	Fan blades loose	Replace or tighten
	Defective belt	Replace
	Pulleys out of alignment	Realign
Rapid burner cycling	Excessive anticipation	Adjust thermostat anticipatory for longer cycles
	Poor thermostat location	Relocate
	Clogged filters	Clean or replace
	Limit setting too low	Readjust or replace limit
Blower will not run	Loose wiring	Check and tighten
	Power not on	Check power switch; check fuses and replace if necessary
	Defective motor overload	Replace motor protector or motor
	Fan control adjustment	Readjust or replace
Not enough heat	Lamp or some other heat source too close to thermostat	Move heat source away from thermostat
	Fan speed too low	Check motor and fan belt and tighten if too loose
	Thermostat set too low	Raise setting
	Thermostat out of calibration	Recalibrate or replace

GAS-FIRED FURNACES *(cont.)*

Problem	Analysis	Correction
Not enough heat *(cont.)*	Thermostat improperly located	Relocate thermostat
	Dirty air filter	Clean or replace
	Limit set too low	Reset or replace
Burner will not turn on	Defective wiring	Check connections; tighten and repair shorts
	Pilot flame too large or too small	Readjust
	Too much draft	Shield pilot
	Defective automatic valve	Replace
	Dirt in pilot orifice	Clean
	Defective thermostat	Check for switch closure and repair or replace
	Improper thermocouple	Properly position thermopile
	Defective thermocouple	Replace
Too much heat	Short in wiring	Locate and correct
	Bypass open	Close bypass
	Thermostat set too high	Lower setting
	Valve sticks open	Replace valve
	Thermostat out of calibration	Recalibrate or replace
	Thermostat in draft or on cold wall	Relocate thermostat to sense average temperature

COAL-FIRED FURNACES

Problem	Analysis	Correction
Inadequate fire	Dirty flue	Clean flue
	Grate clogged with slate and clinkers	Dump grate and rebuild fire; dislodge by poking gently with poker
	Dirty furnace	Clean furnace
	Insufficient draft	Clean ash pit and remove obstruction to primary air supply; adjust damper control
	Poor-quality fuel	Replace with better-quality fuel

COAL-FIRED FURNACES (cont.)

Problem	Analysis	Correction
Noisy blower and motor	Belt tension improper	Readjust to allow 1" slack
	Defective belt	Replace
	Bearings dry	Lubricate
	Belt rubbing	Reposition
	Fan blades loose	Replace or tighten
	Pulleys out of alignment	Realign
Blower won't run	Fan control adjustment too high	Readjustment or replace
	Power not on	Check power switch; check fuses and replace if necessary
	Defective motor overload, protector, or motor	Replace motor
	Loose wiring	Check and tighten
Stoker operates continuously	Fire out	Rebuild fire
	Controls out of adjustment	Readjust controls or call local sales representative for service
	Dirty furnace	Clean furnace
	Dirty fire	Clean fire
Coal stoker stops	Dirty fire	Clean fire
	Power off	Check main power switch, fuses, and correct
	Obstruction in feed screw	Remove obstruction
Blower won't stop	Fan switch defective	Replace
	Manual fan on	Switch to automatic
	Short in wiring	Check wiring and correct
Stoker motor fails to start	Overload	Push reset button on transmission; push reset button on stoker
	Limit control contacts open	Let furnace cool off
	No electrical power	Check main power switch, fuses, and correct
	Blown fuses	Replace fuses
Not enough heat	Lamp or other heat source too near to the thermostat	Remove heat source
	Thermostat set too low	Raise setting
	Thermostat out of calibration	Recalibrate or replace

COAL-FIRED FURNACES *(cont.)*

Problem	Analysis	Correction
Not enough heat *(cont.)*	Limit set too low	Reset or replace
	Dirty air filter	Clean or replace
	Thermostat improperly located	Relocate thermostat
	Fan speed too slow	Check motor and fan belt, and tighten if too loose
Excessive coal in firebox	Stoker windbox full of siftings	Clean out windbox
	Stoker feeding too much coal	Reduce coal feed rate
	Accumulation of clinkers in fire	Clean fire
	Insufficient air	Open manual draft
Excessive fuel use	Improper draft	Adjust dampers for correct rate of combustion
	Dirty flue	Clean flue
	Dirty furnace	Clean furnace
Rapid fan cycling	Blower speed too high	Readjust to lower speed
	Fan switch differential too low	Readjust or replace

OIL BURNERS

Problem	Analysis	Correction
No oil flow at nozzle	Oil level below intake line in supply tank	Fill tank with oil
	Clogged strainer or filter	Remove and clean strainer Replace filter element
	Clogged nozzle	Replace nozzle
	Air leak in intake line	Tighten all fittings in intake line. Tighten unused intake port plug. Check filter cover and gasket.
	Restricted intake line (high vacuum reading)	Replace any kinked tubing and check any valves in intake line.
	A two-pipe system that becomes airbound	Check for and insert bypass plug. Make sure return line is below oil level in tank.

OIL BURNERS (cont.)

Problem	Analysis	Correction
No oil flow at nozzle *(cont.)*	A single-pipe system that becomes airbound	Loosen gauge port plug or easy flow valve and bleed oil for 15 sec. after foam is gone in bleed hose. Check intake line fittings for tightness. Check all pump plugs for tightness.
	Slipping or broken coupling	Tighten or replace coupling
	Rotation of motor and fuel unit is not the same as indicated by arrow on pad at top of unit	Install fuel unit with correct rotation
	Frozen pump shaft	Return unit to approved service station or factory for repair. Check for water and dirt in tank.
Oil leak	Loosen plugs or fittings	Dope with good-quality thread sealer. Retighten.
	Seal leaking	Replace fuel unit
	Leak at pressure adjust screw or nozzle plug	Washer may be damaged. Replace the washer or O-ring.
	Blown seal (single-pipe system)	Check to see if bypass plug has been left in unit. Replace fuel unit.
	Blown seal (two-pipe system)	Check for kinked tubing or other obstructions in return line. Replace fuel unit.
	Cover	Tighten cover screws or replace damaged gasket
Noisy operation	Air in inlet line	Check all connections. Use only good flare fittings.
	Bad coupling alignment	Loosen fuel-unit mounting screws slightly and shift fuel unit in different positions until noise is eliminated. Retighten mounting screws.
	Tank hum on two-pipe system and inside tank	Install return-line hum eliminator in return line

OIL BURNERS (cont.)

Problem	Analysis	Correction
Improper nozzle cutoff	To determine the cause of improper cutoff, insert a pressure gauge in the nozzle port of the fuel unit. After a minute of operation, shut burner down. If the pressure drops from normal operating pressure and stabilizes, the fuel unit is operating properly and air is the cause of improper cutoff. If, however, the pressure drops to 0 psi, fuel unit should be replaced.	
	Filter leaks	Check face of cover and gasket for damage
	Air leak in intake line	Tighten intake fittings. Tighten unused intake port and return plug.
	Strainer cover loose	Tighten 4 screws on cover
	Air pocket between cutoff valve and nozzle	Run burner stopping and starting unit until smoke and afterfire disappear
	Partially clogged nozzle strainer	Clean strainer or change nozzle
	Leak at nozzle adapter	Change nozzle and adapter
Pulsating pressure	Partially clogged strainer or filter	Remove and clean strainer. Replace filter element.
	Air leak in intake line	Tighten all fittings
	Air leaking around cover	Be sure strainer cover screws are tightened securely. Check for damaged cover gasket
Low oil pressure	Defective gauge	Check gauge against master gauge or other gauge
	Nozzle capacity is greater than fuel unit capacity	Replace fuel unit with unit of correct capacity

OIL BURNER PRIMARY CONTROL

Problem	Analysis	Correction
Repeated safety shutdown	Slow combustion thermostat response	Move combustion thermostat to better location. Adjust for more efficient burner flame. Clean surface of cad cell.
	Low line voltage	Check wiring and rewire if necessary. Contact local power company.
	Short cycling of burner	Clean filters. Reset or replace differential or auxiliary controls. Repair or replace faulty auxiliary control. Set thermostat heat anticipation at higher amp valve. Clean holding circuit contacts.
	High resistance in combustion thermostat circuit	Replace combustion thermostat
	High resistance in thermostat or operating control circuit	Check circuit and correct cause
	Short circuit in combustion thermostat cable	Repair cable or replace combustion thermostat
Relay will not pull in	No power. Open power circuit	Repair, replace, or reset fuses, line switch, limit control, auxiliary controls
	Combustion thermostat open	Repair or replace combustion thermostat
	Open thermostat circuit	With power to relay, momentarily short thermostat terminals on relay. If burner starts, check wiring.
	Ignition timer contacts open	Clean magnet
	Open circuit in relay coil	Replace relay

GAS BURNERS

Problem	Cause
Pilot does not light	Air in gas line
	Blocked pilot orifice
	High or low gas pressure
	Flame runner improperly located
Pilot goes out frequently during standby, or safety switch needs frequent resetting	Restriction in pilot gas line
	Low gas pressure
	Blocked pilot orifice
	Poor draft condition
	Loose thermocouple connection on 100% shutoff
	Defective thermocouple or pilot safety switch
	Draft tube set into or flush with inner wall of combustion chamber
Pilot goes out when motor starts	Restriction in pilot gas line
	Excessive pressure drop when main gas valve opens
	High or low gas pressure
Long, yellow flame	Air shutter not open enough
	Air openings or blower wheel clogged
	Too much input
Motor does not run	Burned-out fuse or current off
	Motor burned out
	Thermostat or limit defective or improperly set
	Relay or transformer defective
	Improper wiring
	Tight motor bearings from lack of oil
Motor running but no flame	Pilot out
	Thermocouple not generating sufficient voltage
	Pilot safety switch needs to be reset
	Very low or no gas pressure
	Motor running too slow
Short, noisy burner flame	Pressure regulator set too low
	Vent in regulator plugged
	Air shutter open too wide
	Too much pressure drop in gas line
	Defective regulator
Main gas valve does not close when blower stops	Defective valve
	Obstruction on valve seat
Regulator vent leaking gas	Hole in diaphragm

HEAT PUMPS—COOLING CYCLE

Problem	Analysis	Correction
No cooling, but compressor runs continuously	Defective compressor valves	Replace valves and valve plate or compressor
	Low refrigerant	Repair leak and recharge
	Defective reversing valve	Replace
	Air or noncondensables	Remove noncondensables
	Wrong superheat setting on indoor expansion valve	Adjust setting
	Loose thermal bulb on indoor expansion valve	Tighten thermal bulb
	Dirty indoor coil	Clean coil
	Dirty indoor filters	Clean or replace filters
	Indoor blower belt slipping	Replace or adjust belt
	Restriction in refrigerant system	Locate and remove restriction
Too much cooling; compressor runs continuously	Faulty wiring	Repair wiring
	Faulty thermostat	Replace thermostat
	Wrong thermostat location	Relocate thermostat
Liquid refrigerant flooding compressor (TXV system)	Wrong superheat setting on indoor expansion valve	Adjust superheat
	Loose thermal bulb on indoor expansion valve	Tighten thermal bulb
	Faulty indoor expansion valve	Replace expansion valve
	Defective indoor check valve	Replace check valve
	Refrigerant overcharge	Remove overcharge
Liquid refrigerant flooding compressor (capillary tube system)	Refrigerant overcharge	Remove overcharge
	Dirty indoor filter	Clean or replace filter
	Dirty indoor coil	Clean coil
	Indoor blower belt slipping	Replace or adjust belt
	Indoor check valve defective	Replace check valve

HEAT PUMPS—HEATING CYCLE

Problem	Analysis	Correction
No heating, but compressor runs continuously	Low refrigerant	Repair leak and recharge
	Compressor valves defective	Replace valves and valve plate or compressor
	Leaking reversing valve	Replace reversing valve
	Defective defrost control, time clock, or relay	Replace defrost control, time clock, or relay
Too much heat; compressor runs continuously	Faulty controls	Repair wiring
	Faulty thermostat	Replace thermostat
	Wrong thermostat location	Relocate thermostat
Compressor cycles on low pressure control at end of defrost cycle	Defective reversing valve	Replace
	Defective power element on indoor expansion valve	Replace power element
	Shortage of refrigerant	Repair leak; recharge
Unit runs in cooling cycle but pumps down in cooling cycle	Faulty outdoor expansion valve	Clean or replace expansion valve
	Defective power element on outdoor expansion valve	Replace power element
	Defective reversing valve	Replace
	Dirty outdoor coil	Clean coil
	Belt slipping on outdoor blower	Replace or adjust belt
	Defective indoor check valve	Replace check valve
	Restriction in refrigerant circuit	Locate and remove restriction
Defrost cycle will not terminate	Low refrigerant	Repair leak; recharge
	Defrost control out of adjustment	Adjust control
	Defective defrost control, time clock, or relay	Replace defrost control, time clock, or relay
	Defective reversing valve	Replace
	Defective compressor valves	Replace valves and valve plate or compressor
	Faulty electrical	Repair wiring
Defrost cycle initiates without ice on coil	Low refrigerant	Repair leak and recharge
	Defrost control out of adjustment	Adjust control

HEAT PUMPS—HEATING CYCLE (cont.)

Problem	Analysis	Correction
Defrost cycle initiates without ice on coil (cont.)	Defective defrost control, time clock, or relay	Replace defrost control, time clock, or relay
	Defrost control sensing element not making proper contact	Improve contact
	Outdoor coil dirty	Clean coil
	Outdoor fan belt slipping	Replace belt or adjust
Reversing valve will not shift	Defective reversing valve	Replace reversing valve
	Defective compressor valves	Replace valves and valve plate or compressor
	Faulty fan relay on either indoor or outdoor section	Replace
	Burned transformer	Replace
Indoor blower off with auxiliary heat on	Defective indoor fan relay	Replace
	Defective indoor fan motor	Repair or replace motor
	Faulty wiring or loose terminals	Repair wiring or terminals
	Faulty thermostat	Replace
Outdoor blower runs during defrost cycle	Faulty outdoor fan relay	Replace
Compressor short cycles on defrost control	Low refrigerant	Repair leak and recharge
	Defrost control out of adjustment	Adjust defrost control
	Defective defrost control, time clock, or relay	Replace defrost control, time clock, or relay
	Defective power element on outdoor expansion valve	Replace power element
	Fan belt slipping on outdoor blower	Replace or adjust belt
Excessive ice buildup on indoor coil	Defective defrost relay	Replace defrost relay
	Defective compressor valves	Replace valves and valve plate or compressor
	Low refrigerant	Repair leak and recharge
	Defrost control out of adjustment	Adjust defrost control

HEAT PUMPS—HEATING CYCLE (cont.)

Problem	Analysis	Correction
Excessive ice buildup on indoor coil *(cont.)*	Defrost control sensing element not making proper contact	Improve contact
	Defective defrost control, time clock, or relay	Replace control, time clock, or relay
	Defective reversing valve	Replace reversing valve
	Wrong superheat setting on outdoor expansion valve	Adjust superheat
	Defective power element on outdoor expansion valve	Replace power element
	Plugged outdoor expansion valve	Clean or replace expansion valve
Ice buildup on lower section of outdoor coil	Defective defrost relay	Replace defrost relay
	Defective compressor valves	Replace valves and valve plate or compressor
	Low refrigerant	Repair leak and recharge
	Defrost control out of adjustment	Adjust defrost control
	Defrost sensing element not making proper contact	Improve contact
	Defective reversing valve	Replace reversing valve
	Wrong superheat setting on outdoor expansion valve	Adjust superheat
Liquid refrigerant flooding compressor on heating cycle	Wrong superheat setting on outdoor expansion valve	Adjust superheat
	Outdoor expansion valve thermal bulb not making proper contact	Improve contact
	Outdoor expansion valve stuck open	Clean or replace expansion valve
	Leaking outdoor check valve	Replace check valve
	Refrigerant overcharge	Remove overcharge
	Defective outdoor check valve	Replace check valve
Excessive operating costs	Low refrigerant	Repair leak and recharge

HEAT PUMPS—HEATING CYCLE (cont.)

Problem	Analysis	Correction
Excessive operating costs (cont.)	Defective reversing valve	Replace reversing valve
	Defrost control out of adjustment	Adjust control
	Refrigerant overcharge	Remove overcharge
	Dirty indoor or outdoor coil	Clean coil
	Blower belt slipping on indoor or outdoor blower	Replace or adjust belt
	Dirty indoor air filters	Clean or replace filters
	Wrong thermostat location	Relocate thermostat
	Ducts not insulated	Insulate ducts
	Wrong size unit	Replace with proper size
	Outdoor thermostat not controlling auxiliary heat	Adjust, relocate, or provide shield

HEAT PUMPS—COOLING OR HEATING CYCLE

Problem	Analysis	Correction
Compressor hums but will not start	Fuse	Replace fuse and correct cause
	Faulty wiring	Repair
	Loose electrical terminals	Tighten connections
	Compressor overloaded	Locate and remove overload
	Faulty starting capacitor	Replace capacitor
	Faulty starting relay	Replace relay
	Burned compressor motor	Replace compressor
	Defective compressor bearings	Replace bearings or compressor
	Stuck compressor	Replace compressor
Compressor cycling on overload	Low voltage	Determine cause; repair
	Loose electrical terminals	Repair terminals
	Single-phasing of 3-phase power	Replace fuse or repair wiring; notify power company
	Defective contactor contacts	Replace contacts or contactor
	Defective compressor overload	Replace overload

HEAT PUMPS—COOLING OR HEATING CYCLE (cont.)

Problem	Analysis	Correction
Compressor cycling on overload (cont.)	Compressor overloaded	Locate and remove overload
	Defective start capacitor	Replace capacitor
	Defective run capacitor	Replace capacitor
	Defective starting relay	Replace starting relay
	Refrigerant overcharge	Remove overcharge
	Defective compressor bearings	Replace bearings or compressor
	High head pressure	Remove noncondensables from system
	Defective reversing valve	Replace reversing valve
Compressor off on high pressure control	Refrigerant overcharge	Remove overcharge
	Control out of adjustment	Adjust control
	Defective indoor fan motor	Repair or replace
	Defective outdoor fan motor	Repair or replace
	Defective fan relay on either indoor or outdoor section	Repair or replace
	Too long defrost cycle	Replace time clock, defrost relay, or termination thermostat
	Defective reversing valve	Replace reversing valve
	Blower belt slipping on indoor or outdoor coil	Adjust or replace belt
	Indoor or outdoor coil dirty	Clean proper coil
	Dirty indoor air filters	Replace or clean filters
	Air bypassing indoor or outdoor coil	Prevent air bypassing
	Air volume too low over indoor or outdoor coil	Increase indoor ductwork or remove restriction from coils
	Auxilliary heat strips ahead of indoor coil	Locate heat strips downstream of indoor coil
Compressor cycles on low pressure control	Low refrigerant	Repair leak and recharge
	Low suction pressure	Increase load
	Defective expansion valve	Repair or replace expansion valve
	Dirty indoor or outdoor coil	Clean coil

HEAT PUMPS—COOLING OR HEATING CYCLE (cont.)

Problem	Analysis	Correction
Compressor cycles on low pressure control *(cont.)*	Slipping blower belt	Replace or adjust blower belt
	Dirty air indoor filter	Clean or replace filter
	Ductwork restriction	Increase ductwork
	Liquid drier or suction strainer restricted	Replace drier or strainer
	Defrost thermostat element loose or making poor contact	Tighten or increase contact
	Air temperature too low for evaporation	Relocate unit or provide adequate air temperature
	Defrost cycle too long	Replace time clock, defrost relay, or termination thermostat
	Defective evaporator fan motor	Repair or replace fan motor or relay
Outdoor fan runs, but compressor will not	Faulty wiring, loose connections	Repair wiring or connections
	Defective starting capacitor	Replace
	Defective starting relay	Replace
	Defective run capacitor	Replace
	Shorted or grounded compressor motor	Replace
	Stuck compressor	Replace
	Compressor overloaded	Determine and remove overload
	Defective contactor contacts	Replace contactor or contacts
	Single-phasing of 3-phase power	Locate problem and repair or contact power company
	Low voltage	Locate and correct cause
Outdoor fan motor will not start	Faulty electrical wiring or loose connections	Repair wiring or connections
	Defective outdoor fan motor	Repair or replace motor
	Defective outdoor fan relay	Replace fan relay
	Defective defrost control, timer, or relay	Replace control, timer, or relay
Outdoor section does not run	Blown fuse	Replace fuse; correct fault

HEAT PUMPS—COOLING OR HEATING CYCLE (cont.)

Problem	Analysis	Correction
Outdoor section does not run (cont.)	Faulty electrical wiring or loose terminals	Repair wiring or terminals
	Compressor overloaded	Determine overload and correct
	Defective transformer	Replace
	Burned contactor coil	Replace
	Compressor overload open	Determine cause and correct
	High pressure control open	Determine cause and correct
	Low pressure control open	Determine cause and correct
	Thermostat off	Turn thermostat on and set
Indoor blower will not run	Blown fuse	Replace fuse and correct cause
	Faulty electrical wiring or loose connections	Repair wiring or connections
	Burned transformer	Replace
	Indoor fan relay defective	Replace
	Faulty indoor fan motor	Repair or replace motor
	Faulty thermostat	Replace
Indoor coil iced over	Dirty filters	Clean or replace filters
	Dirty coil	Clean coil
	Blower fan belt slipping	Replace or adjust belt
	Outdoor check valve sticking closed	Replace check valve
	Defective indoor expansion valve	Clean or replace expansion valve
	Low indoor air temperature	Increase temperature
	Low refrigerant	Repair leak and recharge
Noisy compressor	Low oil level in compressor	Determine reason for loss of oil and correct. Replace oil.
	Defective suction or discharge valves	Replace valves and plate or compressor
	Loose hold-down bolts	Tighten
	Broken internal springs	Replace compressor
	Inoperative check valves	Repair or replace check valve
	Loose thermal bulb on indoor expansion valve	Tighten thermal bulb

HEAT PUMPS—COOLING OR HEATING CYCLE (cont.)

Problem	Analysis	Correction
Noisy compressor *(cont.)*	Improper superheat setting on indoor expansion valve	Adjust superheat
	Stuck open indoor expansion valve	Clean or replace valve
Compressor loses oil	Low refrigerant	Repair leak and recharge
	Low suction pressure	Increase load on evaporator
	Restriction in refrigerant circuit	Remove restriction
	Indoor expansion valve stuck open	Clean or replace expansion valve
Unit operates normally in one cycle, but high suction pressure on other cycle	Leaking check valve	Replace
	Loose thermal bulb on outdoor or indoor expansion valve	Tighten thermal bulb
	Leaking reversing valve	Replace
	Expansion valve stuck open on indoor or outdoor	Repair or replace, expansion valve
Unit pumps down in cool or defrost cycle but operates normally in heat cycle	Defective reversing valve	Replace
	Defective power element on indoor expansion valve	Replace power element
	Restriction in refrigerant circuit	Locate and remove restriction
	Clogged indoor expansion valve	Clean or replace expansion valve
	Check valve in outdoor section sticking closed	Replace
Head pressure high	Overcharge of refrigerant	Remove overcharge
	Air or noncondensables in system	Remove noncondensables
	High air temperature supplied to condenser	Reduce air temperature
	Dirty indoor or outdoor coil	Clean coil
	Dirty indoor air filters	Clean or replace filters
	Indoor or outdoor blower belt slipping	Replace or adjust blower belt
	Air bypassing indoor or outdoor coil	Prevent air bypassing

HEAT PUMPS—COOLING OR
HEATING CYCLE (cont.)

Problem	Analysis	Correction
Suction pressure high	Defective compressor suction valves	Replace valves and valve plate or compressor
	Excessive load on cooling	Determine cause; correct
	Leaking reversing valve	Replace reversing valve
	Leaking check valves	Replace check valve
	Indoor or outdoor expansion valve stuck open	Clean or replace expansion valve
	Loose thermal bulb on indoor or outdoor expansion valve	Tighten bulb
Suction pressure low	Low refrigerant	Repair leak; recharge
	Blower belt slipping on indoor or outdoor blower	Replace belt or adjust
	Dirty indoor air filters	Clean or replace
	Defective check valves	Replace check valves
	Restriction in refrigerant circuit	Locate and remove restriction
	Ductwork small or restricted	Repair or replace ductwork
	Defective expansion valve power element on indoor or outdoor coil	Replace power element
	Clogged indoor or outdoor expansion valve	Clean or replace valve
	Wrong superheat setting on indoor or outdoor expansion valve	Adjust superheat setting
	Dirty indoor or outdoor coil	Clean coil
	Bad contactor contacts	Replace contactor or contacts
	Low refrigerant charge	Repair leak and recharge

SELF-CONTAINED AIR CONDITIONERS

Problem	Analysis	Correction
Unit fails to start	Starting switch off	Place starting switch in start position
	Power supply off	Check voltage at connection terminals
	Reset button out	Push reset button
	Loose connection in wiring	Check external and internal wiring connections
	Valves closed	See that all valves are opened
Motor hums but fails to start	Motor single-phasing on 3-phase circuits	Test for blown fuse and overload tripout
	Belts too tight	See that the motor is full floating on trunnion base. See that the belts are in the pulley groove and not binding.
	No oil in bearings. Bearings tight from lack of lubrication	Use proper oil for motor
Noisy compressor	Too much vibration in unit	Check for point of vibration in setup
	Bearing knock	Liquid in crankcase
	Slugging oil	Low suction pressure
	Oil level low in crankcase	Pump down system and add oil if too low
Unit fails to cool	Thermostat set wrong	Check thermostat setting
	Coil frosted	Dirty filters; restricted airflow through unit may be caused by some obstruction at air grille. Fan not operating. Attempting to operate unit at too low a coil temperature.
	Fan not running	Check fan-motor electrical circuit. Also determine if fan blade and motor shaft revolve freely.
Unit runs continuously but no cooling	Shortage of refrigerant	Check liquid-refrigerant level. Test for leaks, repair, and add refrigerant to proper level.
Unit cycles too often	Thermostat differential too close	Check differential setting of thermostat and adjust setting

SELF-CONTAINED AIR CONDITIONERS *(cont.)*

Problem	Analysis	Correction
Thermostat jumpered (system works)	Thermostat contacts dirty	Clean and replace
	Thermostat set point too high	Reset or replace
	Thermostat damaged	Replace thermostat
	Break in thermostat circuit	Locate and correct
Unit vibrates	Belts jerking	Motor not floating freely
	Not setting level	Level all sides
	Shipping bolts not removed	Remove all shipping bolts and steel bandings
	Unit suspension springs not balanced	Adjust unit suspension until unit ceases vibrating
Condensate leaks	Drain lines not properly installed	Drainpipe sizes, proper fall in drain line, traps, and possible obstruction from foreign matter should be checked. Organic slime formation in pan and drain lines and sometimes present on evaporator fins. This formation is largely biological and usually complex in nature. Different localities produce different types. It is largely a local problem to combat and should be observed as such. Periodic cleaning will tend to reduce the trouble but will not eliminate it totally. Filtering air thoroughly will also help, but at times some capacity must be sacrificed when doing this.
Room temperature overshoots thermostat setting (too cold)	Thermostat not mounted level (mercury-switch types)	Remount thermostat in level position
	Thermostat not properly calibrated	Recalibrate or replace
	Thermostat exposed to heat source	Move thermostat to better location
	Thermostat set point too low	Reset
	System sized improperly	Determine correct sizing and make system adjustments

SELF-CONTAINED AIR CONDITIONERS (cont.)

Problem	Analysis	Correction
Room thermostat does not reach setting (too warm)	System sized improperly	Determine correct sizing and make system adjustments
	Thermostat not mounted level (mercury-switch types)	Remount thermostat in level position
	Thermostat not properly calibrated	Recalibrate or replace
	Thermostat subject to draft	Wiring hole may not be plugged. Move thermostat to better position.
	Thermostat set point too high	Reset
	Thermostat damaged	Replace thermostat
System cycles too often	Thermostat differential too small	Reset or replace thermostat
	Thermostat exposed to heat source	Relocate thermostat
	Thermostat subject to vibrations	Remount thermostat in location free from vibrations
	Thermostat heating element improperly set	Reset or replace thermostat
	Thermostat exposed to cold draft	Remount in better location
System does not cycle often enough (burner operates too long)	System sized improperly	Determine correct sizing and make system adjustments
	Thermostat not exposed to circulating air	Remount in better location
	Contacts dirty	Clean or replace
	Thermostat differential too great	Reset or replace thermostat
	Thermostat heating element improperly set	Reset or replace thermostat
	Thermostat set too high	Reset
Room temperature swings excessively	Thermostat not exposed to circulating air	Remount in better position
	System sized improperly	Determine correct sizing and make system adjustments
	Thermostat exposed to heat source	Remount in better position

SELF-CONTAINED AIR CONDITIONERS *(cont.)*

Problem	Analysis	Correction
Burner fails to stop	Thermostat in cold location	Relocate in better location
	Thermostat set too high	Reset
	Thermostat out of adjustment	Recalibrate or replace thermostat
	Defective thermostat	Replace thermostat
	Thermostat contacts stuck	Correct

AIR CONDITIONING SYSTEMS

Problem	Analysis	Correction
Unit will not run	Blown line fuse	Replace; check for cause
	Thermostat not demanding	Turn on thermostat and set temperature
	Blown transformer fuse	Replace; check for cause
	Burned out transformer	Replace transformer
	Faulty wiring, loose connections	Repair wiring or connections
Outdoor unit will not run	Blown fuse to outdoor unit	Replace fuse and check for cause
	Thermostat set too high	Reset thermostat
	Burned out contactor coil	Replace
	Burned contactor contacts	Replace
	Compressor overload open	Determine overload and correct
	Faulty wiring or connections	Repair wiring or connections
Compressor will not start	Defective or dirty contactor contacts on controls	Replace or clean contacts
	Bad starting capacitor	Replace
	Bad starting relay	Replace
	Bad run capacitor	Replace
	Burned out compressor motor	Repair motor or replace compressor
	Switch open	Close switch
	Solenoid valve closed	Examine holding coil; if burned out or defective, replace

AIR CONDITIONING SYSTEMS *(cont.)*

Problem	Analysis	Correction
Compressor will not start *(cont.)*	Compressor not unloading for start	Check oil level and oil pressure. Check capacity control mechanism. Check unloader pistons.
	Control circuit not functioning properly	Locate open control and determine cause
	Compressor stuck	Determine location and cause. Repair or replace compressor.
	Overload tripped or fuse blown	Reset overload, replace fuses, and examine for cause of condition and correct
	No charge of gas in system operated on low pressure control	With no gas in system, there is insufficient pressure to throw in low-pressure control. Recharge system; correct leak.
	Out on safety controls, high-pressure cutout, oil pressure safety switch	Discharge service valve may be partially closed. Check oil level. Reset control button. Check for faulty wiring.
Outdoor fan motor will not start	Faulty wiring or loose connections	Repair wiring or connections
	Burned out fan motor	Replace
	Bad fan motor bearings	Replace bearings or motor
Compressor hums but will not run	Bad starting capacitor	Replace
	Bad starting relay	Replace
	Burned out compressor motor	Repair or replace
	Stuck compressor	Replace compressor
	Defective contactor contacts	Replace
	Three-phase compressor single-phasing	Replace fuse or reset circuit breaker
	Low voltage	Check circuit; call utility
Compressor starts, but motor will not get off starting winding	Running capacitor shorted	Check by disconnecting running capacitor
	Low line voltage	Bring up voltage
	Tight compressor	Check oil level. Check binding.
	Starting and running windings shorted	Check resistance. Replace starter if defective.

AIR CONDITIONING SYSTEMS (cont.)

Problem	Analysis	Correction
Compressor starts, but motor will not get off starting winding (cont.)	High discharge pressure	Check discharge shutoff valve. Check pressure.
	Defective relay	Check operation manually. Replace relay if defective.
	Starting capacitor weak	Check capacitance. Replace if low.
	Improperly wired	Check wiring against diagram
Compressor cycling on overload	Bad starting capacitor	Replace
	Bad starting relay	Replace
	Bad running capacitor	Replace
	Weak overload	Replace
	Bad contactor contacts	Replace
	Low voltage	Check circuit; call utility
	Burned compressor motor	Repair or replace compressor
	Refrigerant overcharge	Remove overcharge
	Low refrigerant	Repair leak recharge
	High suction pressure	Reduce load or repair compressor
	Air or noncondensables in system	Remove air or noncondensables
Compressor off on high-pressure control	Refrigerant overcharge	Remove overcharge
	Dirty outdoor coil	Clean coil
	Slipping outdoor fan belt	Replace or adjust fan belt
	Air or noncondensables in system	Remove air or noncondensables
Compressor cycling or off on low-pressure control	Shortage of refrigerant	Repair leak; recharge
	Dirty or defective expansion valve	Clean or replace expansion valve
	Defective expansion valve power element	Replace power element
	Dirty indoor filters	Clean or replace filters
	Dirty indoor coil	Clean coil
	Indoor blower belt slipping	Replace or adjust fan belt
	Restriction in refrigerant system	Locate and remove restrictions

7-39

AIR CONDITIONING SYSTEMS *(cont.)*

Problem	Analysis	Correction
Compressor short-cycles	Coils on evaporator clogged with frost or dirt	Clean or defrost coils
	Low head pressure	Check and adjust water-regulating valve. Check refrigerating charge.
	Control set at too close a differential	Check control settings for proper application, clean, and adjust or replace
	Fans not running on evaporator	Check all electrical connections, fuses, thermal overloads, and thrown switches
	Discharge valve leaks slightly	Replace valve plate
	Thermal bulb on expansion valve has lost charge	Detach thermal bulb from suction line and hold in the palm of one hand, with the other hand gripping the suction line. If flooding through is observed, bulb has lost charge. If no flooding through is noticed, replace expansion valve.
	Leaky liquid line solenoid valve	Repair or replace
Noisy compressor	Defective compressor valves	Replace valves and valve plate
	Wrong expansion valve superheat setting	Adjust superheat setting
	Expansion valve stuck	Repair or replace expansion valve
	Poor contact of expansion valve thermal bulb	Improve contact
	Overcharge of refrigerant (cap tube system)	Remove overcharge
	Slugging due to flooding back of refrigerant	Check thermal-bulb location and fastening. Reset expansion valve. Trap suction line so refrigerant will not flood back on the off cycle.

AIR CONDITIONING SYSTEMS *(cont.)*

Problem	Analysis	Correction
Noisy compressor *(cont.)*	Worn parts such as pistons, piston pins, or connecting rods	Determine location of cause. Repair or replace compressor.
	Restriction in line	Be sure discharge valve is wide open, or install muffler
	Too light a foundation, or loose foundation bolts	Check size of foundation. Check shim. Tighten all bolts.
	Too much oil in circulation, causing hydraulic knocks	Check oil level. Remove excess, if any. Also check for oil at refrigerant test cock.
Insufficient oil pressure	Low oil level	Check level and add oil if low
	Broken or loose oil line	Check and correct
	Strainer pickup tube loose, split, or has defective flare	Check and correct defect if necessary
	Defective oil pump	Replace pump
	Plug out of crankshaft	Inspect. Install plug if missing.
	Excessive oil foaming	Adjust expansion valve. Check crankcase heater setting or use of pumpdown cycle.
Compressor loses oil	Shortage of refrigerant	Repair leak; recharge system with refrigerant and oil
	Expansion valves stuck open	Repair or replace expansion valve
	Restriction in refrigeration system	Locate restriction and remove
	Refrigerant piping improperly sized	Resize piping
Oil leaves compressor	High crankcase pressure due to worn rings	Replace rings and liners if worn
	Short-cycling	Check control setting for proper job conditions. Repair or replace faulty control.
	Wet suction causing oil foaming	Adjust expansion valves and check thermal bulbs for proper mounting. Check refrigerant piping.

AIR CONDITIONING SYSTEMS *(cont.)*

Problem	Analysis	Correction
Oil does not return to crankcase	Separator valve not functioning properly	Check oil-separator operation
	Insufficient charge of refrigerant	Test for leaks and add refrigerant
	Refrigerant piping laid out incorrectly	Check and correct if necessary
	High crankcase pressure	Check piston ring blowby. Check oil-separator operation.
No cooling, but compressor runs continuously	Shortage of refrigerant	Repair leak and recharge
	Defective compressor valves	Replace valves and valve plate or compressor
	High suction pressure (air or noncondensables in system)	Remove air or noncondensables
	Wrong expansion valve superheat setting	Adjust superheat setting
	Dirty or defective expansion valve	Repair or replace
	Indoor coil dirty	Clean coil
	Indoor air filter dirty	Clean or replace filter
	Indoor blower belt slipping	Replace or adjust fan belt
	Restriction in refrigerant circuit	Locate and remove
	Dirty outdoor coil	Clean coil
Fixture temperature too high	Compressor inefficient	Check valves and pistons
	Cooling coils too small	Add surface or replace
	Expansion valve too small	Raise suction pressure with larger valve
	Refrigerant shortage	Repair leak and recharge
	Iced or dirty coil	Defrost or clean
	Unit too small	Add unit or replace
	Expansion valve set too high	Lower setting
	Restricted or small gas lines	Clear restriction or increase line size
	Expansion valve or strainer plugged	Clean or replace
	Control set too high	Reset control

AIR CONDITIONING SYSTEMS *(cont.)*

Problem	Analysis	Correction
Too much cooling; compressor runs continuously	Thermostat setting low	Reset thermostat
	Thermostat in wrong location	Relocate thermostat
	Faulty wiring	Repair wiring
	Control contacts frozen	Clean points or replace control
	Service load too great	Keep doors closed
	Unit too small	Add unit or replace
	Dirty condenser	Clean condenser
	Defective insulation	Correct or replace
	Shortage of gas	Repair leak and recharge
	Location too warm	Change to cooler location
	Air in system	Purge
	Iced or plugged coil	Defrost or clean
	Plugged expansion valve or retainer	Clean or replace
	Compressor inefficient	Check valves and pistons
Running cycle too long	Unit too small for application	Capacity must by increased by increasing speed (if belt drive), or replace compressor
	Leaky valve in compressor	Repair or replace
	Control contacts stuck	Check, clean, repair, or replace
	Insufficient refrigerant charge	Test for leaks and add refrigerant
	Excessive head pressure due to overcharge, air in condenser, or dirty condenser	Remove excess refrigerant, purge system, clean condenser
Liquid refrigerant flooding compressor—cap tube system	Refrigerant overcharge	Remove overcharge
	Indoor coil dirty	Clean coil
	Indoor fan belt slipping	Replace or adjust belt
	Indoor air filters dirty	Clean or replace filters
	Indoor fan not running	Repair on replace
Liquid refrigerant flooding compressor—expansion valve system	Expansion valve superheat setting wrong	Adjust superheat setting
	Expansion valve stuck open	Repair or replace expansion valve

AIR CONDITIONING SYSTEMS (cont.)

Problem	Analysis	Correction
Liquid refrigerant flooding compressor— expansion valve system (cont.)	Expansion valve thermal bulb loose	Improve contact
	Refrigerant overcharge	Remove overcharge
	Too cold indoor air temperature	Raise thermostat setting
High head pressure	Refrigerant overcharge	Remove overcharge
	High ambient temperature	Provide cooler air to condenser
	Air or noncondensables in system	Purge air or noncondensables
	Excessive loading	Reduce load
	Dirty outdoor coil and/or condenser	Clean coil
	Outdoor fan motor not running	Repair or replace
	Outdoor fan belt slipping	Replace or adjust fan belt
	Discharge service valve not fully open	Open full
	Water flow restricted or too warm on watercooled unit	Check and adjust water-regulating valve
	Restriction in discharge line	Determine cause and correct
	Fouled water-cooled condenser	Clean tubes
	Inadequate air circulation in air-cooled or evaporative condenser	Remove obstruction. Clean condenser.
Low head pressure	Leaky discharge valve or discharge to suction relief valve leaking	Repair or replace valves and valve plate or compressor
	Liquid refrigerant flooding back from evaporator	Check operation of expansion valve. Also check fastenings and insulation of thermal bulb.
	Low air temperature over outdoor coil	Provide warmer air
	Low refrigerant charge	Check for leak. Repair and add refrigerant.
	Too much water flowing through condenser	Adjust water-regulating valve

AIR CONDITIONING SYSTEMS *(cont.)*

Problem	Analysis	Correction
High suction pressure	Defective compressor valves	Replace valves and valve plate
	Leaking suction valve	Repair or replace valve
	Refrigerant overcharge	Remove overcharge
	High head pressure	See "High head pressure"
	Return air temperature high	Provide cooler air
	Excessive load on evaporator	Check air infiltration and insulation
	Expansion valve stuck open or leaks	Clean or replace valve
Low suction pressure	Low refrigerant	Repair leak and recharge
	Low return air temperature	Reset thermostat higher
	Wrong expansion valve superheat setting	Readjust superheat setting
	Dirty or defective expansion valve	Clean or replace expansion valve
	Defective expansion valve power element	Replace power element
	Indoor blower belt slipping	Replace or adjust belt
	Indoor blower not running	Repair or replace
	Restriction in refrigerant circuit	Locate restriction and remove
	Indoor air filters dirty	Clean or replace filter
	Indoor coil dirty	Clean coil
	Restricted cap tube	Replace cap tube
	Evaporator coil frosted	Defrost coil
	Dirty suction screens or suction-stop valve partially closed	Clean screens. Check suction valve and open wide.
	Light load on evaporator	Check capacity against load. Reduce capacity if necessary.
	Restriction in liquid line (clogged strainer or dehydrator, etc.) causing liquid to flash and restrict flow	Clean strainer. Correct restriction.

AIR CONDITIONING SYSTEMS (cont.)

Problem	Analysis	Correction
Low suction pressure *(cont.)*	Excessive pressure drop through evaporator, with clogged external equalizer	Be sure external equalizer blows clear and not plugged
	Too much oil circulating in system, reducing heat transfer in evaporator	Check for excess oil in system. Remove excess oil.
	Low head pressure	Check and adjust water-regulating valve. Check refrigerating charge.
Indoor blower not running	Blown fuse	Replace fuse; correct cause
	Indoor fan relay defective	Replace
	Burned indoor motor	Replace
	Broken belt	Replace
	Faulty wiring or loose connections	Repair wiring or connections
Indoor coil icing	Shortage of refrigerant	Repair leak and recharge
	Low return air temperature	Raise thermostat setting
	Indoor blower not running	Repair or replace
	Indoor blower belt slipping	Replace or adjust belt
	Restriction in refrigerant system	Locate and remove restriction
	Indoor air filter dirty	Clean or replace filters
	Dirty indoor coil	Clean coil
	Dirty or defective expansion valve	Clean or replace valve
High operating costs	Defective compressor valves	Replace valves and valve plate or compressor
	Shortage of refrigerant	Repair leak and recharge
	Refrigerant overcharge	Remove overcharge
	Dirty outdoor coil	Clean coil
	Dirty indoor coil	Clean coil
	Dirty indoor air filters	Clean or replace
	Thermostat in wrong location	Relocate thermostat
	Air ducts not insulated	Insulate ducts
	Unit too small	Install proper size unit

AIR CONDITIONING SYSTEMS (cont.)

Problem	Analysis	Correction
High operating costs (cont.)	Indoor or outdoor fan belt slipping	Replace or adjust belt
	Outdoor thermostat set too high	Adjust thermostat

COMPRESSORS

Problem	Analysis	Correction
Compressor will not start—no hum	Line disconnect switch open	Close start or disconnect switch
	Fuse removed or blown	Replace fuse
	Overload protector tripped	Repair or replace
	Control stuck in open position	Repair or replace control
	Control off due to cold location	Relocate control
	Wiring improper or loose	Check wiring against diagram; repair
Compressor will not start, trips overload	Improperly wired	Check wiring against diagram; repair
	Low voltage to unit	Check circuit; call utility
	Starting capacitor defective	Replace
	Relay failing to close	Repair or replace
	Compressor motor has a winding open or shorted	Replace compressor
	Internal mechanical trouble in compressor	Replace compressor
	Liquid refrigerant in compressor	Add crankcase heater and/or accumulator
Compressor starts but does not switch off of start winding	Improperly wired	Check wiring against diagram
	Low voltage to unit	Check circuit; call utility
	Relay failing to open	Repair or replace
	Run capacitor defective	Determine reason and replace
	Excessively high discharge pressure	Check discharge shut-off valve, possible overcharge, or insufficient cooling on condenser

COMPRESSORS (cont.)

Problem	Analysis	Correction
Compressor starts but does not switch off of start winding—*cont.*	Compressor motor has a winding open or shorted	Replace compressor
	Internal mechanical trouble in compressor	Replace compressor
Compressor starts and runs but short cycles on overload protector	Excess current passing through overload protector	Check circuit; repair
	Low voltage to unit (or unbalanced if 3-phase)	Check circuit; repair
	Overload protector defective	Check; replace
	Run capacitor defective	Replace
	Excessive discharge pressure	Check ventilation, restrictions in system
	Suction pressure too high	Check for design error. Use stronger unit
	Compressor too hot; return gas hot	Check refrigerant charge (fix leak); add if necessary
	Compressor motor has a winding shorted	Replace compressor
Unit runs OK but short cycles on	Overload protector.	Check size; replace
	Thermostat	Widen differential
	High-pressure cut-out due to insufficient air or water supply, overcharge, air in system	Check air or water supply to condenser; correct if necessary. Reduce refrigerant charge. Purge.
	Low-pressure cut-out due to liquid line solenoid leaking, compressor valve leak, undercharge, or restriction in expansion device	Replace. Replace. Fix leak; add refrigerant. Replace device.
Unit operates long or continuously	Low refrigerant	Fix leak; add charge
	Control contacts stuck or frozen closed	Clean contacts or replace
	Space has excessive load or poor insulation	Determine fault and correct
	System inadequate to handle load	Replace with larger system
	Evaporator coil iced	Defrost

COMPRESSORS *(cont.)*

Problem	Analysis	Correction
Unit operates long or continuously *(con't)*	Restriction in refrigeration system	Determine location and remove
	Dirty condenser	Clean condenser
	Filter dirty	Clean or replace
Start capacitor open, shorted, or blown	Relay contacts not operating properly	Clean contacts or replace relay if necessary
	Prolonged operation on start cycle due to low voltage to unit, improper relay, or starting load too high	Determine reason and correct. Replace. Correct by using pump down arrangement if necessary.
	Excessive short cycling	Determine reason for short cycling and correct
	Improper capacitor	Determine correct size and replace
Run capacitor open, shorted, or blown.	Improper capacitor	Determine correct size and replace
	Excessively high line voltage (110% of rated max.)	Determine reason and correct
Relay defective or burned out	Incorrect relay	Check and replace
	Incorrect mounting angle	Remount relay in correct position
	Line voltage too high or too low	Determine reason and correct
	Excessive short cycling	Determine reason and correct
	Relay being influenced by loose vibrating mounting	Remount rigidly
	Incorrect run capacitor	Replace with proper capacitor
Space temperature too high	Control setting too high	Reset control
	Expansion valve too small	Use larger valve
	Cooling coils too small	Add surface or replace
	Inadequate air circulation	Improve air movement
Suction line frosted or sweating	Expansion valve passing excess refrigerant or is oversized	Readjust valve or replace with smaller valve
	Expansion valve stuck open	Clean valve; replace if necessary

COMPRESSORS (cont.)

Problem	Analysis	Correction
Suction line frosted or sweating (cont.)	Evaporator fan not running	Determine reason and correct
	Overcharge of refrigerant	Correct charge
Liquid line frosted or sweating	Restriction in dehydrator or strainer	Replace part
	Liquid shut-off (king valve) partially closed	Open valve fully
Unit noisy	Loose parts or mountings	Find and tighten
	Tubing rattle	Reform to be free of contact
	Bent fan blade causing vibration	Replace blade
	Fan motor bearings worn	Replace motor
Compressor will not start	No power to motor	Check power at fuses; replace if necessary
		Check starter contacts, connections, overloads, and timer (if part winding start)
		Reset or repair as necessary
		Check power at motor terminals
		Repair wiring if damaged
	Control circuit is open	Check high pressure, oil failure, and low pressure switches. Repair.
		Check control circuit fuses; replace
		Check wiring for open circuit
Motor hums but does not start	Low voltage to motor	Check incoming power; call power company
	Motor shorted	Repair or replace
	Single phase failure in the 3-phase power supply	Check power wiring
	Compressor is seized due to damage or liquid	Remove belts or coupling. Manually turn crankshaft to check compressor.
	Compressor is not unloaded	Check unloader system

COMPRESSORS (cont.)

Problem	Analysis	Correction
Compressor starts, but motor cycles off on overloads	Compressor has liquid or oil in cylinders	Check compressor crankcase temperature
		Throttle suction stop valve on compressor to clear cylinders
	Suction pressure is too high	Unload compressor when starting. Use internal unloaders if present. Install external bypass unloader.
	Bearings are tight	Lubricate bearings
	Motor is running on single-phase power	Check power lines, fuses, starter, motor, to determine where open circuit has occurred
Compressor starts but short cycles automatically	Low refrigerant charge	Check and add if necessary
	Driers plugged or saturated with moisture	Replace cores
	Refrigerant feed control is defective	Repair or replace
	No load	To prevent short cycling, if objectionable, install pump-down circuit, anti-recycle timer or false load system
	Unit is too large for load	Reduce compressor speed. Install false load system
	Suction strainer blocked or restricted	Check and clean or replace as necessary
Motor is noisy or erratic	Motor bearing failure or winding failure	Check and repair as needed
	If electronic starter, check calibration on control elements	Adjust as necessary
Compressor runs continuously but does not keep up with the load	Load is too high	Speed up compressor or add compressor capacity
		Reduce load
	Refrigerant metering device is underfeeding, causing the compressor to run at too low a suction pressure	Check and repair liquid feed problems
		Check discharge pressure and increase if low

COMPRESSORS (cont.)

Problem	Analysis	Correction
Compressor runs continuously but does not keep up with the load (cont.)	Faulty control circuit, may be low pressure control or capacity controls	Check and repair
	Compressor may have broken valve plates	Check compressor for condition of parts. Check compressor discharge temperature.
	Thermostat control is defective and keeps unit running	Check temperatures of product or space and compare with thermostat control. Replace or readjust thermostat.
	Defrost system on evaporator not working properly	Check and repair as needed
	Suction bags in strainers are dirty and restrict gas flow	Clean or remove
	Hot gas bypass or false load valve struck	Check and repair or replace
Compressor loses excessive amount of oil	High suction superheat causes oil to vaporize	Insulate suction lines. Adjust expansion valves to proper superheat.
	Too low of an operating level in chiller will keep oil in vessel	Raise liquid level in flooded evaporator (R-12 systems only)
	Oil not returning from separator	Make sure all valves are open.
		Check float mechanism and clean orifice.
		Check and clean return line
	Oil separator is too small	Check selection
	Broken valves cause excessive heat in compressor and vaporization of oil	Repair compressor
	Slugging of compressor with liquid refrigerant that causes excessive foam in the crankcase	Dry up suction gas to compressor by repairing evaporator
		Refrigerant feed controls are overfeeding
		Check suction trap level controls
		Install a refrigerant liquid transfer system to return liquid to high side

COMPRESSORS *(cont.)*

Problem	Analysis	Correction
Noisy compressor operation	Loose flywheel or coupling	Tighten
	Coupling not properly aligned	Check; align if required
	Loose belts	Align and tighten per specs
		Check sheeve grooves
	Poor foundation or mounting	Tighten mounting bolts, grout base, or install heavier foundation
	Check compressor with stethoscope if noise is internal	Open, inspect, and repair as necessary
	Check for liquid or oil slugging	Eliminate liquid from suction mains
		Check crankcase oil level
Low evaporator capacity	Inadequate refrigerant feed to evaporators	Clean strainers and driers
		Check expansion valve superheat setting
		Check for excessive pressure drop due to change in elevation, too small of lines (suction and liquid lines). A heat exchanger may correct this.
		Check expansion valve size
	Expansion valve bulb in a trap	Change piping or bulb location to correct
	Oil in evaporator	Warm the evaporator drain oil and install an oil trap to collect oil
	Evaporator surface fouled	Clean
	Air or product velocity is too low	Increase to rated velocity
		Check defrost time
		Check method of defrost
	Brine flow through evaporator may be restricted	Check circulating pumps
		Check process piping for restriction
Discharge pressure too high	Air in condenser	Purge noncondensibles
	Condenser tubes fouled	Clean
	Water flow is inadequate	Check water supply and pump

COMPRESSORS (cont.)

Problem	Analysis	Correction
Discharge pressure too high (cont.)	Water flow is inadequate (cont.)	Check control valve
		Check water temperature
	Air flow is restricted	Check and clean coils, eliminators, and dampers
	Liquid refrigerant backed up in condenser	Find source of restriction and clear
		If system is overcharged, remove refrigerant as required
		Check to make sure equalizer (vent) line is properly installed and sized
	Spray nozzles on evaporator condensers plugged	Clean
Discharge pressure too low	Ambient air is too cold	Install a fan cycling control system
	Water quantity not being regulated properly through condenser	Install or repair water regulating valve
	Refrigerant level low	Check for liquid seal, add refrigerant if necessary
	Evaporator condenser fan and water switches are improperly set	Reset condenser controls
Suction pressure too low	Light load condition	Shut off some compressors
		Unload compressors
		Slow down RPM of compressor
		Check process flows
	Short of refrigerant	Add if necessary
	Evaporators not getting enough refrigerant	Discharge pressure too low. Increase to maintain adequate refrigerant flow.
		Check liquid feed lines for adequate refrigerant supply
		Check liquid line drier
	Refrigerant metering controls are too small	Check superheat or liquid level and correct as indicated

COMPRESSORS *(cont.)*

Problem	Analysis	Correction
Suction pressure too high	Low compressor capacity	Check compressors for possible internal damage
		Check system load
		Add more compressor capacity

REFRIGERATION SYSTEMS

Problem	Analysis	Correction
Compressor fails to start (no hum)	Power failure	Contact power company
	Disconnect switch open	Check circuits; close switch
	Fuse blown	Determine cause; replace fuse
	Burned-out compressor motor	Replace
	Motor starter	Repair or replace
	Control circuit open	Locate cause and repair
	Oil failure control	Reset and check control
	Overload protector tripped	Check overload
	Thermostat setting too high	Set to lower temperature
	Low-pressure control open	Reset and check pressures
	High-pressure control open	Reset and check pressures
	Loose wiring	Repair wiring
Compressor will not start (hums and trips overload protector)	Improperly wired	Rewire unit
	Low voltage to unit	Determine reason and correct
	Bad starting capacitor	Determine reason and replace
	Starting relay open	Determine reason and replace starting relay
	Burned-out compressor motor	Replace compressor motor
	Mechanical problems in compressor	Replace compressor

REFRIGERATION SYSTEMS (cont.)

Problem	Analysis	Correction
Compressor will not start (hums and trips overload protector) (cont.)	Liquid refrigerant in compressor crankcase	Install crankcase heater
	Bad run capacitor	Determine reason and replace capacitor
	Unequalized pressures on PSC motor	Allow pressures to equalize or install a hard-start kit
Compressor will run but remains on start winding	Improperly wired	Rewire unit
	Low voltage to unit	Determine reason and correct
	Starting relay does not open	Determine cause and replace starting relay
	Bad running capacitor	Determine reason and replace
	High discharge pressure	Open compressor discharge service valve; purge possible overcharge of refrigerant
	Open or shorted motor winding	Replace compressor
	Mechanical trouble in compressor	Replace compressor
	Defective overload protector	Replace overload protector
Compressor starts and runs but short cycles	Defective overload protector	Replace overload protector
	Low voltage to unit	Determine reason and correct
	Defective run capacitor	Determine reason and replace
	High discharge pressure	Open compressor discharge service valve. Purge possible overcharge of refrigerant.
	Suction pressure too low	Properly charge system with refrigerant. Increase load on evaporator.
	Suction pressure too high	Reduce air flow over evaporator. Purge overcharge of refrigerant. Replace compressor valves.

REFRIGERATION SYSTEMS (cont.)

Problem	Analysis	Correction
Compressor starts and runs but short cycles *(cont.)*	Compressor too hot	Properly charge system with refrigerant
	Shorted motor winding	Replace compressor
	Dirty or iced evaporator	Increase air flow over evaporator. Replace broken belt. Replace defective fan motor.
	Low-pressure control differential set too close	Readjust differential
	High-pressure control differential set too close	Readjust or replace control
	Condenser water regulating valve inoperative	Clean and repair or replace
	Condenser water temperature too high	Clean and repair water pump, piping spray nozzles, and coil
	Erratic thermostat	Relocate or replace thermostat
Unit operates excessively	Short of refrigerant	Repair leak and recharge unit
	Thermostat contacts stuck closed	Clean contacts or replace thermostat
	Excessive load	Check heat load; replace unit; replace insulation
	Evaporator coil iced	Defrost unit and check operation
	Restriction in refrigerant system	Locate and remove
	Dirty condenser	Clean condenser
	Restricted air over evaporator	Determine cause and correct
	Inefficient compressor	Check compressor valves and repair
Compressor loses oil	Traps in hot gas or suction lines	Reroute lines to provide proper pitch
	Refrigerant velocity too low in risers	Resize risers or install oil return traps

REFRIGERATION SYSTEMS (cont.)

Problem	Analysis	Correction
Compressor loses oil (cont.)	Shortage of refrigerant	Repair leak and recharge
	Liquid refrigerant flooding back to compressor	Adjust expansion valve; alter refrigerant charge on capillary tube system
	Gas–oil ratio low	Add 1 pt. of oil for each 10 lb. of refrigerant added to factory charge
	Plugged expansion valve or strainer	Clean or replace
	Superheat too high at compressor suction	Change location of TXV bulb or adjust superheat to return wet refrigerant to the compressor
Compressor noisy	Lack of compressor oil	Add oil to correct level
	Tubing rattle	Reroute tubing
	Mounting loose	Repair mounting
	Oil slugging	Adjust oil level or refrigerant charge
	Refrigerant flooding compressor	Check expansion valve for leak or oversized orifice
	Dry or scored shaft seal	Check oil level
	Internal parts of compressor broken or worn	Overhaul compressor
	Compressor drive coupling loose	Tighten coupling and check alignment
Unit low on capacity	Ice or dirt on evaporator	Clean coil or defrost
	Expansion valve struck or dirty	Clean or replace expansion valve
	Improper TXV superheat adjustment	Adjust expansion valve
	Wrong size expansion valve	Replace valve
	Excessive pressure drop in evaporator	Adjust expansion valve
	Clogged strainer	Clean or replace strainer
	Liquid flashing in liquid line	Subcool liquid or add refrigerant

REFRIGERATION SYSTEMS (cont.)

Problem	Analysis	Correction
Space temperature too high	Control setting too high	Adjust control
	Expansion valve too small	Replace valve
	Evaporator too small	Replace coil
	Insufficient air circulation	Correct circulation
	Shortage of refrigerant	Repair leak and recharge
	Expansion valve plugged	Clean or replace
	Inefficient compressor	Check efficiency
	Restricted or undersized refrigerant lines	Clear restriction or resize lines
	Evaporator iced or dirty	Clean and defrost evaporator
Suction line frosted or sweating	Superheat setting too low	Adjust superheat setting
	Expansion valve stuck open	Clean or replace valve
	Evaporator fan not running	Replace fan
	Overcharge of refrigerant	Drain charge
Liquid line frosted or sweating	Restricted drier or strainer	Replace drier or strainer
	Liquid line shut-off valve partially closed	Open valve
Hot liquid line	Expansion valve open too wide	Adjust expansion valve
	Refrigerant shortage	Repair leak and recharge
Top of condenser coils cool when unit is operating	Refrigerant shortage	Repair leak and recharge
	Refrigerant overcharge	Remove part of charge
	Inefficient compressor	Check efficiency and correct
Unit in vacuum–frost on expansion valve only	Ice plugging expansion valve orifice	Apply heat to expansion valve body; an increase in suction pressure indicates moisture; install new drier
	Expansion valve strainer plugged	Clean strainer or replace valve
High head pressure	Overcharge of refrigerant	Remove overcharge
	Air in system	Remove air
	Dirty condenser	Clean condenser
	Unit in too hot location	Relocate unit
	Water-cooled condenser plugged	Clean or replace condenser

REFRIGERATION SYSTEMS (cont.)

Problem	Analysis	Correction
High head pressure (cont.)	Condenser water too warm	Provide sufficient cool water; adjust water regulating valve
	Cooling water shut off	Turn on water
Low head pressure	Shortage of refrigerant	Repair leak and recharge
	Cold unit location	Provide warm condenser air
	Cold condenser water	Adjust water regulating valve or provide less water
	Inefficient compressor valves	Replace leaky valves
	Leaky oil return valve in oil separator	Repair or replace
High suction pressure	Expansion valve stuck open	Repair or replace valve
	Expansion valve too large	Replace valve
	Leaking compressor suction valves	Replace suction valves or compressor
	Evaporator too large	Resize evaporator
Low suction pressure	Shortage of refrigerant	Repair leak and recharge
	Evaporator underloaded	Clean or defrost evaporator
	Liquid line strainer clogged	Clean or replace strainer
	Plugged expansion valve	Clean or replace valve
	Lost charge on TXV power assembly	Replace power assembly
	Space temperature too low	Adjust or replace thermostat
	Expansion valve too small	Replace valve
	Excessive pressure drop through evaporator	Check for plugged external equalizer
	Oversized compressor	Resize compressor
Loss of compressor oil pressure	Malfunctioning oil pump	Repair or replace oil pump
	Oil pump inlet screen plugged	Clean or replace screen
Starting relay burned out	Improper relay mounting	Mount relay properly
	Relay vibrating	Mount relay in rigid position
	Wrong relay	Replace with proper relay
	Wrong running capacitor	Replace with proper capacitor

REFRIGERATION SYSTEMS (cont.)

Problem	Analysis	Correction
Starting relay burned out (cont.)	Excessive line voltage	Reduce voltage to a maximum of 10% over motor rating
	Low line voltage	Increase voltage to not less than 10% below motor rating
Starting relay contacts stuck	Bad bleed resistor	Replace resistor or capacitor
Starting capacitors burned out	Prolonged operation on starting winding	Reduce starting load; increase low voltage
	Sticking relay contacts	Replace relay
	Wrong capacitor	Replace with proper capacitor
Running capacitors burned out	Excessive line voltage	Reduce line voltage to not more than 10% over motor rating
	Wrong capacitor	Replace with proper capacitor
	Light compressor load	Check voltage on capacitor and replace capacitor
Evaporator freezes but defrosts while unit is running	Moisture in system	Remove refrigerant evacuate system; install new drier; recharge system
Evaporator coil iced over	Automatic defrost control erratic or inoperative	Replace control
	Automatic defrost control improperly wired	Rewire control
	Defective defrost control thermal element	Replace control
	Improperly installed control thermal element	Relocate element
	Defrost control termination point too low	Replace or adjust control
	Defrost valve solenoid burned out	Replace solenoid
	Stuck closed defrost valve	Repair or replace valve
	Restricted hot gas bypass line	Replace line

REFRIGERATION SYSTEMS (cont.)

Problem	Analysis	Correction
Evaporator coil iced over (cont.)	Inoperative freezer compartment door switch	Replace switch
	Inoperative freezer compartment fan	Clear fan or replace motor
	Freezer defrost element burned out	Replace element
	Freezer compartment drain trough or drain pan heater burned out	Replace heater
	Freezer compartment drain line plugged	Clean drain line
Refrigerator remains in the defrost cycle	Defrost control incorrectly wired	Rewire defrost control
	Automatic defrost control inoperative	Replace defrost control
	Defrost control termination point too high	Replace or adjust control
	Defrost solenoid valve stuck open	Clean or replace valve
	Room temperature too low (below 55°F)	Relocate unit or provide heat
Water collects in bottom of refrigerator	Drain tube plugged	Clean tube
	Drain tube frozen	Check drain heater element and repair or replace
	Split drain trough	Replace trough
	Water leakage between trough and cabinet liner	Seal with suitable sealer
	Fresh food compartment liner warped	Replace liner or seal with suitable sealer
	Evaporator baffle not properly installed	Install baffle properly
	Humidiplate not adjusted properly	Adjust humidiplate
	Door gasket not sealing properly	Adjust door or replace gasket
Condensation on outside of cabinet	Door gaskets leaking	Adjust door or replace gasket
	Mullion heater burned out	Replace mullion heater

Problem	Analysis	Correction
Condensation on outside of cabinet *(cont.)*	Wire loose to mullion heater	Reconnect wire
	Abnormally high humidity	None (explain reason to owner)
Water or ice collects in bottom of freezer compartment	Drain tube frozen	Defrost and repair heater
	Drain tube plugged	Clean tube
	Drain trough heater burned out	Replace heater
	Evaporator cover plate mislocated	Install properly
Fresh food compartment too warm	Poor air distribution to food compartment	Adjust controls for better air distribution
	Thermostat setting too high	Adjust thermostat
	Thermostat control bulb not making good contact	Provide good contact
	Bad thermostat	Replace thermostat
Freezer compartment warm	Thermostat set warm	Adjust thermostat
	Bad thermostat	Replace thermostat
	Freezer fan motor not running	Free blade or replace motor
	Light stays on	Replace switch or rewire
	Freezer door gaskets not sealing	Adjust door or replace gasket
	Freezer compartment door switch erratic	Replace switch
	Defective automatic defrost control	Replace defrost control
	Defrost valve solenoid burned out	Replace solenoid
	Restricted hot gas bypass line	Replace hot gas bypass line
	Loose wire at automatic defrost control or solenoid valve	Reconnect wire
	Excessive freezer compartment load	Advise customer

REFRIGERATION SYSTEMS (cont.)

Problem	Analysis	Correction
Freezer compartment warm *(cont.)*	Drain trough heater burned out	Replace heater
	Abnormally low room temperature	Relocate cabinet or provide heat
	Packages blocking air distribution	Advise customer
Food compartment too warm	System short of refrigerant	Repair leak and recharge
	Inefficient compressor	Replace compressor
	Thermostat set too high	Adjust thermostat
	Dirty condenser	Clear obstruction
	Inoperative condenser fan	Replace fan motor
	Freezer compartment fan inoperative	Free blade or replace motor
	Fresh food compartment fan inoperative	Free blade or replace motor
	Freezer compartment door switch inoperative	Replace switch
	Door gasket not sealing	Adjust door or replace gasket
	Evaporator baffle not installed properly	Install properly
	Shelves covered, restricting air flows	Remove covering and advise customer
	Excessive food compartment load	Advise customer
	Restricted strainer, filter drier, or capillary tube	Replace and charge unit
Inoperative defrost circuit	Defrost timer motor inoperative	Replace timer
	Inoperative defrost heater	Replace heater
	Faulty defrost limiter	Replace defrost limiter
Unit will not run	Blown fuse	Replace fuse
	Low voltage	Check outlet with voltmeter, should check 120V plus or minus 10%
		If circuit overloaded, reduce load or install separate circuit

REFRIGERATORS/FREEZERS

Problem	Analysis	Correction
Unit will not run *(cont.)*	Low voltage *(cont.)*	If unable to remedy, install auto-transformer
	Broken motor or temperature control	Jumper across terminals of control. If unit runs and connections are all tight, replace control
	Broken relay	Check relay, replace if necessary
	Broken overload	Check overload, replace if necessary
	Broken compressor	Check compressor, replace if necessary
	Defective service cord	If no circuit and current is indicated at outlet, replace or repair
	Broken lead to compressors, timer, or cold control	Repair or replace broken leads
	Broken timer	Check with test light and replace if necessary
Refrigerator section too warm	Repeated door openings	Explain to owner
	Overloading of shelves, blocking normal air circulation in cabinet	Explain to owner
	Warm or hot foods placed in cabinet	Instruct user to allow foods to cool to room temperature before placing in cabinet
	Poor door seal	Level cabinet, adjust door seal
	Interior light stays on	Check light switch; if faulty, replace
	Refrigerator section airflow control	Turn control knob to colder position. Check airflow heater.
	Damper control	Check if damper is opening by removing grille. With door open, damper should open. If control inoperative, replace control.

Problem	Analysis	Correction
Refrigerator section too warm *(cont.)*	Cold control knob set at too warm a position, not allowing unit to operate often enough	Turn knob to colder position
	Freezer section grille not properly positioned	Reposition grille
	Freezer fan not running properly	Replace fan, fan switch, or defective wiring
	Defective intake valve	Replace motor compressor
	Air duct seal not properly sealed or positioned	Check and reseal or put in correct position
Refrigerator section too cold	Refrigerator section airflow control knob turned to coldest position	Turn control knob to warmer position
	Airflow control remains open	Remove obstruction
	Broken airflow control	Replace control
	Broken airflow heater	Replace heater
Freezer section and refrigerator section too warm	Fan motor not running	Check and replace fan motor if necessary
	Cold control set too warm or broken	Check and replace if necessary
	Finned evaporator blocked with ice	Check defrost heater thermostat or timer
	Shortage of refrigerant	Check for leak, repair, evacuate and recharge system. Recover/recycle refrigerant.
	Not enough air circulation around cabinet	Relocate cabinet or provide clearances to allow sufficient circulation
	Dirty condenser or obstructed condenser ducts	Clean the condenser and the ducts
	Poor door seal	Level cabinet, adjust door seal
	Too many door openings	Explain to owner

REFRIGERATORS/FREEZERS (cont.)

Problem	Analysis	Correction
Freezer section too cold	Cold control knob improperly set	Turn knob to warner position
	Cold control capillary not properly clamped to evaporator	Tighten clamp or reposition
	Broken cold control	Check control. Replace if necessary.
Unit runs all the time	Not enough air circulation around cabinet or air circulation is restricted	Relocate cabinet or remove restriction
	Poor door seal	Check and make necessary adjustments
	Freezing large quantities of ice cubes, or heavy loading after shopping	Advise against heavy loading
	Refrigerant charge	Undercharge or overcharge—check, evacuate, and recharge with proper charge
	Room temperature too warm	Ventilate room
	Cold control	Check control; replace
	Defective light switch	Replace switch
	Excessive door openings	Advise owner
Noisy operation	Loose flooring or floor not firm	Tighten flooring or brace floor
	Tubing contacting cabinet or other tubing	Move tubing
	Cabinet not level	Level cabinet
	Drip tray vibrating	Move tray—place on pad if necessary
	Fan hitting liner or mechanically grounding	Move fan
	Compressor mechanically grounded	Replace compressor mounts
Unit cycles on overload	Broken relay	Replace relay
	Weak overload protector	Replace overload protector
	Low voltage	Check outlet with voltmeter. Underload voltage should be 120 V plus or minus 10%.

REFRIGERATORS/FREEZERS *(cont.)*

Problem	Analysis	Correction
Unit cycles on overload *(cont.)*	Low voltage *(cont.)*	Check for several appliances on same circuit or long or undersized extension cord being used.
	Poor compressor	Check with test cord and for ground before replacing
Stuck motor compressor	Broken valve	Replace motor compressor
	Insufficient oil	Add oil; replace motor compressor
	Overheated compressor	Replace compressor
Frost or ice on finned evaporator	Broken timer	Check with test light; replace
	Defective defrost heater	Replace
	Defective thermostat	Replace
Ice in drip catcher	Defective drip catcher heater	Replace
Unit runs all the time, temperature normal	Ice builds up on the evaporator	Check door gaskets; replace
	Control bulb on thermostat not in contact with evaporator surface	Place control bulb in contact with the evaporator surface
Freezer runs all the time. Temperature too cold.	Faulty thermostat	Check thermostat—test and replace if necessary
Freezer runs all the time. Temperature too warm.	Ice buildup in insulation	Remove breaker strips, stop unit, melt ice and dry insulation, seal outer shell leaks and joints, and then assemble.
Rapid ice buildup on the evaporator	Leaky door gasket	Adjust door hinges. Replace door gasket if cracked, brittle, or worn.
Door on freezer compartment freezes shut	Faulty electric gasket heater	Use alternate gasket heater or install new one
	Faulty gasket seal	Inspect and check gasket. If worn, cracked or hardened, replace.
Freezer works then warms up	Moisture in refrigerator	Install drier in liquid line
Gradual reduction in freezing capacity	Wax buildup in capillary tube	Use capillary tube cleaning tool or replace capillary tube

ICE MACHINES—CUBER

Problem	Analysis	Correction
Unit will not run	Defective operation of control switch	Replace switch
	Blown fuse	Replace fuse and correct cause
	Storage bin switch off	Adjust bin switch
Unit running but produces little or no ice	Low refrigerant charge	Repair leak and recharge
	Dirty condenser (aircooled)	Clean condenser
	Expansion valve plugged	Clean or replace valves
	Hot gas check valves leaking	Replace
	Hot gas solenoid leaking	Replace
	Defective compressor	Replace
	No water over freeze plates	Repair or replace water supply float valve. Repair or replace water pump. Clean water distributor. Clear restricted or pinched water hose. Raise water level in reservoir.
	Unit running in defrost cycle	Replace ice thickness switch. Replace freeze relay. Replace defrost control relay. Replace or repair mercury switch.
	Too thick ice slab	Adjust ice thickness control to about ½" to ¾"
	Low head pressure	Relocate unit; repair or replace defective water valve
Condensing unit cycles with bin switch closed	High discharge pressure	Remove air or noncondensables from system. Clean condenser.
	Low suction pressure	Repair leak and recharge system
	Defective pressure control	Replace

ICE MACHINES—CUBER *(cont.)*

Problem	Analysis	Correction
Condensing unit cycles with bin switch closed *(cont.)*	Defective condenser water valve	Repair or replace
	Defective condenser fan	Replace fan motor or free fan blade
Ice slab stuck to evaporator plate	Lime on evaporator plate	Clean plate
	Warped evaporator plate	Replace
	Low refrigerant charge	Repair leak and/or recharge
	Inoperative hot gas solenoid	Repair or replace
	Low discharge pressure	Relocate unit or condenser water valve
	Ice thickness switch imbedded into slab	Adjust switch to provide proper thickness
Ice slab; uneven slab hollow in center	Low refrigerant charge	Repair leak and recharge system
	Inoperative expansion valve	Repair or replace valve
Ice slab uneven; slab has hookup back edge with hang-over front edge	Leaking hot gas solenoid valve	Repair or replace solenoid valve
	Leaking gas check valve	Repair or replace check valve
Ice slab uneven; insufficient water over freeze plate	Water pump not running	Replace water pump or repair loose electrical connection
	Plugged water line	Clear water line
	Low water level in sump	Adjust or repair float valve
	Plugged water distributor	Clean distributor
Cloudy ice cubes	High mineral content in water	Install water softener
	Water not siphoning from reservoir at end of freezing cycle	Clear siphoning tube; adjust for proper operation

ICE MACHINES—CUBER *(cont.)*

Problem	Analysis	Correction
Cloudy ice cubes *(cont.)*	Insufficient water supply	Provide water supply in excess of 20 psi
	Restricted water distributor	Clean distributor
	Water float valve set too low	Adjust float
Slabs stack up on cutting grid	Low voltage to cutting grids	Supply proper voltage
	Grid plug, wires, or pins not making good contact	Provide good contact
	Blown grid fuse	Correct condition and replace fuse
	Grid fuse loose in receptacle	Replace receptacle
	High mineral content in water	Install water softener
	Lime buildup on grid wires	Clean grid wires
	Wrong grid wires	Replace with proper wires
Slow harvest	Low ambient air	Move; provide 50°F air
	High mineral content in water	Install water softener
	Water valve passing too much water	Adjust valve to 125 psig head pressure
	Water valve leaking	Repair or replace

ICE MACHINES—FLAKER

Problem	Analysis	Correction
Unit will not run	Blown fuse	Repair cause; replace fuse
	Ice storage bin thermostat keeps unit off too long	Adjust bin thermostat
	Power relay contacts stuck in lockout position	Replace power relay
	Ice storage bin thermostat inoperative	Repair or replace
	Loose electrical connection	Repair connection
Compressor cycling	Dirty condenser	Clean condenser
	Condenser air circulation restricted	Remove restriction
	Defective start relay	Replace
	Defective overload protector or lockout relay	Replace overload protector or relay
	Defective start capacitor	Replace
	Inoperative condenser fan	Free fan or replace motor
	Loose electrical connection	Repair loose connection
No ice, but unit running	Low refrigerant charge	Repair leak and recharge system
	O-ring at evaporator shell leaking	Replace O-ring seal
	No water	Provide water as required
Water leaking from unit	Float assembly stuck	Free or replace
	Bad float ball	Replace float ball
	Storage bin drain plugged	Clear drain
	Evaporator drain plugged	Clear drain
	Inlet water connector leaking	Repair connections
	O-ring at evaporator shell leaking	Replace O-ring

ICE MACHINES—FLAKER *(cont.)*

Problem	Analysis	Correction
Soft or wet Ice	Shortage of refrigerant	Repair leak and recharge unit
	High discharge pressure	Clean condenser or provide cooler condenser air
	Water in float reservoir too high	Adjust water float or repair float valve
Excessive or chattering noise	Intermittent water supply	Provide constant water supply
	Loose hold-down on auger gear motor	Tighten or replace
	Scale deposits on inside of evaporator shell	Remove evaporator shell and clean in proper solution
	Reservoir water level too low	Adjust water level
	End play in auger gear motor	Repair gear motor
	Worn auger gear bearings	Replace bearings
	Air lock in gravity water line to evaporator shell	Eliminate air lock
Unit operating with ice storage bin full of ice	Ice storage bin thermostat set too cold	Adjust thermostat
	Defective ice storage bin thermostat	Replace thermostat
Wet ice	High water level	Lower water level
	Undercharge, bubbles going through sight glass	Check for leaks, add R-12
	Misadjusted TXV	Adjust TXV
Ice too hard	Low water level	Raise water level
	TXV closed too far	See service manual for proper suction pressure and suction line temperature
	Moisture in system and TXV partially frozen shut	Dehydrate and recharge system

ICE MACHINES—FLAKER *(cont.)*

Problem	Analysis	Correction
No ice with gearmotor, compressor and condenser fan operating. (Red light not on, no power to defrost valve)	Very low refrigerant	Repair leak; restore to proper charge
	Stuck defrost valve, defrost line warm, suction pressure above 20 lbs.	Repair or replace defrost valve
Flaker does not turn off	Misadjusted bin thermostat	Adjust bin thermostat (ccw), check with ice on coil
	Bin thermostat will not open when set warmest with ice on thermostat cap tube	Replace bin thermostat
	Mislocated bin thermostat cap tube	Check location of ¼" stainless cap tube holder parallel and below ice path coming from ice tube. Check thermostat cap tube located in the ¼" stainless tube.
Flaker cycles off and on	Ice falling on bin thermostat capillary tube	Relocate tube
	Ice tube not on outlet spout	Remount ice tube and check for restriction
Low production	High head pressure	Clean condenser. Improve water supply to water-cooled condenser. Improve ventilation. Replace head pressure control valve on remote condenser models.
	Inadequate water supply	Check and clean filters. Raise water level
	TXV misadjusted	Adjust TXV
	High ambient	Decrease ambient to 90°F max
	Low head pressure	Adjust water regulating valve. Add R-12 to remote condenser models. Refer to Remote Condenser section of Service Manual.

ICE MACHINES—FLAKER *(cont.)*

Problem	Analysis	Correction
Flaker spouts coming off	IO of discharge tubing too small	Change to braided nylon tubing
	Roll pins breaking on deflector	Readjust defrost thermostat
	Hose kinking	Change to braided nylon tubing reroute presently used tubing
Flaker will not operate. No lights on in control box.	Line fuse blown	Check circuits for short or ground. Replace fuse.
	Loose connection in control box or in power supply line	Check for power supply at controls in control box. Check connections to bin thermostat.
Flaker will not operate. Yellow light on, but no ice on bin thermostat.	Bin control set too warm in a cold room between 45° and 55°F.	Set bin thermostat colder (cw) but recheck with ice to be sure it will shut off
	Room below 45°F. Bin control has lost charge.	Add heat to the room. Replace bin control.
Condensing fan and gearmotor operate (green light on) but not the compressor	Inoperative capacitors or relay	Replace capacitors or relay
	Overload switch defective	Replace overload switch
	Loose connections or defective compressor	Check for power at compressor C-R terminals, C-S terminals. With power off, remove C connection; check ohms between C and R, also C and S.
Water-cooled flaker: gearmotor operates (green light on) but not the condensing fan or compressor	High-pressure cut-out open; inadequate water supply	Check water supply and condenser water valve
	Water supply okay; high-pressure cut-out won't close with condenser cool.	Replace defective high-pressure cut-out.
Compressor operating but fan off	Circuit not complete	Check circuit
	Fan motor burned out	Replace motor

ICE MACHINES—FLAKER (cont.)

Problem	Analysis	Correction
Condenser fan operating but compressor unit operating intermittently	Dirty condenser coil	Clean coil
	High or low voltage	Correct to proper voltage within 10% of nameplate
	Excessive refrigerant	Remove some refrigerant, check sight glass
Intermittent defrost, red light cycling on and off. Water level normal in float tank.	Water line elbow in bottom bearing in too far	Back out elbow 1 turn
	Defrost thermostat misadjusted, very cold supply water	Adjust defrost thermostat setting
	TXV too far open	Close TXV ¼ turn; verify proper setting
	Deflector partially closed	Check proper deflector setting
	Gearmotor not running	Check power to gearmotor receptacle in bottom of control box
	Gearmotor stalled with power on; bottom evaporator mounting screw too tight	Back off evaporator mounting screw 1 turn to see if gearmotor will operate after it cools down sufficiently for overload to cut back in
	Gearmotor stalled	Remove gearmotor and auger assembly and check operation at workbench. Remove bottom bearing, check for condition, and check fit on bottom of auger.

BOILERS	
Problem	Cause
Boiler does not deliver enough heat	Poor draft
	Heating surfaces are covered with soot
	Boiler is too small for the heating system
	Poor fuel
	Improper arrangement of boiler sections in cast-iron boilers
Too much time is required to get up steam in a steam boiler	Poor fuel or fuel firing
	Boiler defective
	Too little or badly arranged heating surface
	Poor draft
	Boiler too small
	Heating passages too short
	Heating surfaces covered with soot
Boiler is slow to respond to the operation of the dampers	Poor fuel or fuel firing
	Clinkers on grate or ashpit full of ashes (coal-fired boilers)
	Boiler too small
	Air leakage into the chimney or stack
Water line is unsteady	Varying pressure differnces on the system
	Dirt or grease in the water
	Excessive boiler output

BOILERS *(cont.)*	
Problem	**Cause**
Water is carried over into the steam main	Water line is too high
	Boiler output excessive
	Outlet connections from boiler too small
	Priming or foaming
	Steam-liberating surface too small
Water disappears from the gauge glass	Pressure drop too great in return line
	Foaming
	Valve closed in the return line
	Improper water gauge connection
Flues require cleaning too frequently	Poor draft
	Excess air in firebox
	Combustion rate too slow
	Smoky combustion
Low carbon dioxide	Improper conversion job
	Air leakage between cast-iron sections
	Problem with burners
Smoke from boiler fire door	Incorrect setting of dampers
	Incorrect reduction in the breeching size
	Poor or defective draft in the chimney
	Dirty or clogged flues

BOILER FLAME PROBLEMS

Problem	Analysis	Correction
Yellow flame	Lack of primary air	Reset primary air and check for blockage
Lifting flame	Gas velocity faster than speed gas can burn	Reduce input gas or primary air
Popping	Flashback during shutoff; burning continued	Increase gas pressure reduce primary air, reduce orifice, check gas valve and burner

MAXIMUM ALLOWABLE IMPURITIES IN BOILER WATER

Chemical Name	Chemical Symbol	ppm
Sodium sulphite	Na_2SO_3	1.0
Sodium chloride	$NaCl$	10.0
Sodium phosphate	Na_3PO_4	25.0
Sodium sulphate	Na_2SO_4	25.0
Silica oxide	SiO_2	0.20
Total dissolved solids	—	50.0

FUEL PUMPS AND FUEL UNITS

Problem	Analysis	Correction
No oil blow at nozzle	Frozen pump shaft	Remove unit and return to factory for repair
	Clogged strainer or filter	Remove and clean strainer; repack filter element
	Air binding in two-pipe system	Check and insert bypass plug
Oil leak	Loose plugs or fittings	Dope with good-quality thread sealer
	Blown seal	Replace fuel unit
	Leak at pressure-adjusting end cap nut	Fiber washer may have been left out after adjustment of valve spring; replace washer
	Seal leaking	Replace fuel unit
Noisy operation	Bad coupling alignment	Loosen mounting screws and shift fuel unit to a position where noise is eliminated. Retighten mounting screws.
	Air inlet line	Tighten all connections and fittings in the intake line and unused intake port plugs
	Pump noise	Work in gears by continued running or replace

FUEL PUMPS AND FUEL UNITS (cont.)

Problem	Analysis	Correction
Pulsating pressure	Air leak in intake line	Tighten all fittings and valve packing in intake line
	Partially clogged strainer	Remove and clean strainer
	Air leaking around strainer cover	Tighten strainer cover screws
	Partially clogged filter	Replace filter element
Low oil pressure	Nozzle capacity is greater than fuel-unit capacity	Replace fuel unit with unit of correct capacity
	Defective gauge	Check against another and replace if necessary
Improper nozzle cutoff	Filter leaks	Check face of filter cover and gasket for damage
	Strainer cover loose	Tighten screws
	Partially clogged nozzle strainer	Clean strainer or change nozzle
	Air leak in intake line	Tighten intake fittings and packing nut on shut-off valve. Tighten unused intake port plug.
	Air pockets between cutoff valve and nozzle	Start and stop burner until smoke and after-fire disappears

STEAM SYSTEMS

Problem	Analysis	Correction
Trap Blows Live Steam—No prime (bucket traps)	Trap not primed when originally installed	Prime the trap
	Open or leaking by-pass valve	Remove or repair by-pass valve
	Trap not primed after cleanout	Prime the trap
	Sudden pressure drops	Install check valve ahead of trap
Valve mechanism does not close	Scale or dirt lodged in orifice	Clean out the trap
	Worn or defective valve or disc mechanism	Repair or replace defective parts
Back pressure too high (thermodynamic trap)	Trap stuck open	Clean out the trap
	Worn or defective	Repair or replace
	parts	defective parts
	Condensate return line or pig tank undersized	Increase line or pig tank size
Ruptured bellows (thermostatic traps)	—	Replace bellows
Blowing flash steam (normal condition)	Forms when condensate released to lower or atmospheric pressure	No corrective action necessary

STEAM SYSTEMS *(cont.)*		
Problem	**Analysis**	**Correction**
Trap Does Not Discharge— Pressure too high	Trap pressure rating too low	Install correct trap
	Pressure reducing valve set too high or broken	Readjust or replace pressure reducing valve
	Orifice enlarged by normal wear	Replace worn orifice
	System pressure raised	Install correct pressure change assembly
Condensate not reaching trap	Strainer clogged	Blow out screen or replace
	By-pass opening or leaking	Remove or repair by-pass valve
	Obstruction in line to trap inlet	Remove obstruction
	Steam supply shut off	Open steam supply valve
High vacuum in condensate return line	—	Install correct pressure change assembly
Trap clogged with foreign matter	—	Clean out and install strainer
Trap held closed by defective mechanism	—	Repair or replace mechanism
No pressure differential across trap	Blocked or restricted condensate return line	Remove restriction
	Incorrect pressure change assembly	Install correct pressure change assembly
Continuous Discharge from Trap—Trap too small	Capacity undersized	Install properly sized larger trap
	Pressure rating of trap too high	Install correct pressure change assembly

STEAM SYSTEMS *(cont.)*

Problem	Analysis	Correction
Trap clogged with foreign matter	Dirt or foreign matter in trap internals	Clean out and install strainer
	Strainer plugged	Clean out strainer
Bellows overstressed (thermostatic traps)	—	Replace bellows
Loss of prime	—	Install check valve on inlet side
Failure of valve to seat	Worn valve and seat	Replace worn parts
	Scale or dirt under valve and in orifice	Clean out the trap
	Worn guide pins and lever	Replace worn parts
Sluggish or Uneven Heating—Trap has no capacity margin for heavy starting loads	—	Install properly sized larger trap
Inadequate steam supply	Steam supply pressure valve has changed	Restore normal steam pressure
	Pressure reducing valve setting off	Readjust or replace reducing valve
Insufficient air handling capacity (bucket traps)	—	Use thermic buckets or increase vent size
Short circuiting (group traps)	—	Trap each unit individually
Back Pressure Troubles—Condensate return line too small	—	Install larger condensate return line
Obstruction in condensate return line	—	Remove obstruction
Other traps blowing steam into header	—	Locate and repair other faulty traps
Pig tank vent line plugged	—	Clean out pig tank vent line
Excess vacuum in condensate return line	—	Install correct pressure change assembly

STEAM AND HOT WATER BOILERS

Problem	Analysis	Correction
Low water	Feedwater pump not functioning	Replace
	Feedwater stop valve closed	Open stop valve
Burner shut down	Low water	Reset safety circuit
	Burner on too long	Shorten burner time
	Operating control pressure set too high	Set operating control lower than high-pressure limit
	Safety circuit open	Reset combustion air proving switch
No low fire on startup	Modulating control switch set too high	Set modulating switch significantly below the operating setpoint
Frequent burner on/off cycling	Burner uncalibrated	Set burner to actual load demands
Burner will not ignite	No electricity	Check disconnect, fuses or breakers, and overloads
	Low gas pressure switch tripped	Replace gas pressure regulator
	High gas pressure switch tripped	Set switch to slightly above operating pressure
	Low oil pressure switch tripped	Set switch to slightly below operating pressure
	Pilot flame failure	Turn burner switch on
No ignition	Electrode grounded or cracked	Replace electrode
	Ignition transformer failure	Replace ignition transformer
Gauge glass leaking	Loose valve nuts	Tighten packing nuts
	Bad gasket	Replace gasket
Try cock leaking	Stem leaking	Replace try cock
Temperature sensing controls not working	Sensing bulb bottomed out	Apply heat transfer compound and reinsert
	Kinked connecting tube	Replace connecting tube
	Unlevel mercury bulb	Level mercury bulb
	Dirty mechanism	Clean or replace

STEAM AND HOT WATER BOILERS (cont.)

Problem	Analysis	Correction
Building space too cold	Thermostat out of calibration	Calibrate thermostat
	No steam	Check; turn steam on
	Valve stuck closed	Check; rebuild valve
	Trap failed closed	Isolate and rebuild valve
	Thermostat has wrong setpoint	Change setpoint dial
Building space too hot	Thermostat out of calibration	Calibrate thermostat
	Hole in diaphragm	Disassemble and replace
	Valve wiredrawn	Rebuild valve
	Thermostat has wrong setpoint	Change setpoint dial

HUMIDISTATS

Problem	Analysis	Correction
Humidity too low	Humidistat has wrong setpoint	Change setpoint dial
	Humidifier out of calibration	Calibrate humidistat
	Valve stuck closed	Free up or lube
	No steam or water for humidifier	Check supply pipes
	Hole in diaphragm	Disassemble and replace
Humidity too high	Humidifier setpoint wrong	Change dial
	Humidifier out of calibration	Calibrate humidifier
	Valve stuck open	Free up or lube

THERMOCOUPLES

Problem	Analysis	Correction
Pilot flame lit but safety control fails to function	Thermocouple not hot enough to generate current	Wait at least 1 min. for thermocouple to become hot enough
	Loose or dirty electrical connections	Disconnect, clean, reconnect, and tighten
	Drafts deflecting flame away from thermocouple	Eliminate source of draft
	Pilot flame too small or yellow in color due to restricted pilot line or dirt in primary air opening or burner head	Disconnect, clean thoroughly, and reconnect. Change orifice if necessary.
	Thermocouple tip too low in pilot flame	Check installation to make sure thermocouple is properly mounted in bracket
Safety control operates but fails when main burner has been on a short time	Restriction in pilot or main gas tubing	Eliminate restriction. Provide normal pressure.
	Draft-deflecting flame couple	Eliminate draft or baffle

REVERSING VALVES

Problem	Analysis	Correction
Valve will not shift from heat to cool	Defective coil	Replace coil
	Pilot valve operating correctly Dirt in one bleeder hole	Deenergize solenoid, raise head pressure, and reenergize solenoid to break dirt loose. If unsuccessful, remove valve and wash out. Check on air before installing. If no movement, replace valve, add strainer to discharge tube, and mount valve horizontally.
	Low refrigeration charge	Repair leak and recharge system
	No voltage to coil	Repair electrical circuit
	Clogged pilot tubes	Raise head pressure, and operate solenoid to free. If still no shift, replace valve.
	Pressure differential too high	Recheck system
	Both ports of pilot open Back seat port did not close	Raise head pressure and operate solenoid to free partially clogged port
	Piston cup leak	Stop unit. After pressures equalize, restart with solenoid energized. If valve shifts, reattempt with compressor running. If still no shift, replace valve.
Apparent leak in heating position	Pilot needle and piston needle leaking	Operate valve several times; then recheck. If excessive leak, replace valve.
	Piston needle on end of slide leaking	Operate valve several times; then recheck. If excessive leak, replace valve.
Valve starts to shift but does not complete reversal	Body damage	Replace valve.
	Valve hung up at midstroke. Pumping volume of compressor not sufficient to maintain reversal.	Raise head pressure, and operate solenoid. If no shift, use valve with smaller ports.

REVERSING VALVES *(cont.)*

Problem	Analysis	Correction
Valve starts to shift but does not complete reversal *(cont.)*	Not enough pressure differential at start of stroke or not enough flow to maintain pressure differential	Check unit for correct operating pressures and charge. Raise head pressure. If no shift, use valve with smaller ports.
	Both ports of pilot open	Raise head pressure, and operate solenoid. If no shift, use valve with smaller ports.
Valve will not shift from heat to cool	Dirt in bleeder hole	Raise head pressure; operate solenoid. Remove valve and wash out. Check on air before reinstalling. If no movement, replace valve. Add strainer to discharge tube. Mount valve horizontally.
	Defective pilot	Replace valve
	Clogged pilot tube	Raise head pressure. Operate solenoid to free dirt. If still no shift, replace valve.
	Piston-cup leak	Stop unit. After pressures equalize, restart with solenoid deenergized. If valve shifts, reattempt with compressor running. If it still will not reverse while running, replace valve.
	Pressure differential too high	Stop unit; will reverse during equalization period. Recheck system.

DAMPER MOTORS

Problem	Analysis	Correction
Motor not operating	Motor damaged	Replace actuator
	Power off	Check switches and fuse
	Loose wiring	Check connections
Motor operates but output shaft does not move	Linkage binding	Check for free movement
	Actuator damaged	Replace actuator
Motor stalls at maximum travel position	Limit switch damage	Inspect and replace if necessary

ELECTRICAL COMPONENTS AND CONTROLS

Problem	Analysis	Correction
Compressor will not start (no hum)	No voltage	Reset breaker or replace blown fuses.
	Bad contactor	Check voltage (24 V). If voltage is present and contactor does not pull in, coil is bad in contactor; replace coil or contactor.
	Wire burnt off	Repair wiring
	Internal or external overload open	If infinite ohms, overload is bad.
		If there is power and compressor is at ambient temperature, replace compressor. If compressor is hot, find reason for compressor overloading.
	Contactor not pulling in	Replace contactor coil or contactor. Continue to check circuit.
	Bad thermostat	Replace thermostat
Compressor will not start but hums	Low voltage	Correct problem. If power is in distribution, call power company.
	Bad start assist	Either replace start assist or install start relay and capacitor.
	Wire loose on start terminal	Make repairs
	Compressor stuck	Replace compressor
Outdoor fan motor will not run	No voltage	If there is voltage, move on to next check. If no voltage, the relay is bad if the unit is out of defrost. If the motor is not trying to start and cool to touch, repair or replace motor. If too hot, repair or replace.
	Bad capacitor	Replace motor. Replace capacitor. If motor still does not start, replace motor.

ELECTRICAL COMPONENTS AND CONTROLS (cont.)

Problem	Analysis	Correction
Outdoor fan motor runs in high speed	Contacts stuck closed	Replace controls
	During defrost, outdoor fan runs all the time	If there is voltage, replace fan
Outdoor fan motor runs in low speed constantly	Contacts stuck	Replace contacts
	Bad capacitor	Check capacitor and replace if needed. Temporarily substitute a capacitor to verify your findings.
Indoor fan motor will not come on (outdoor unit operating)	No voltage	Replace fuses and determine why the fuses blew
		Replace fan relay
	Bad thermostat	Replace thermostat
Indoor fan motor runs continuously	Fan relay stuck	Replace relay. Replace sequencer.
	Thermostat set in on position	Turn fan switch to auto position
System will not go into reverse cycle	Stuck reversing valve	Check system charge before replacing reversing valve. There may not be enough pressure differential across valve to make it operate. Also bad valves in the compressor can cause the same problem.
System will not defrost	Bad defrost timer	Replace timer
	Bad defrost thermostat	Replace the defrost thermostat
	Bad defrost relay	Replace defrost relay
	Outdoor fan runs all the time	Replace relay if contacts are stuck
Auxiliary heat will not come on during normal operation	Bad thermostat	If no power at contacts, replace thermostat
	Blown fuses	Replace fuses
	Bad sequencer	Replace sequencers
	Open limit switch	Replace limit
	Blown heater fuse	Replace fuse
	Burnt out element	Replace element

ELECTRICAL COMPONENTS AND CONTROLS (cont.)

Problem	Analysis	Correction
Nothing will come on when thermostat is switched to different positions	No power to indoor unit	Check for tripped breaker or blown fuse at distribution panel
	Bad transformer	Replace transformer
Compressor hums but will not start	Low voltage	Check voltage at main panel. If low, call power company.
	Blown fuse	Replace blown fuse
	Bad start capacitor	Replace capacitor
	Start relay contacts open	Replace relay
	Shorted run capacitor	Replace run capacitor
	Compressor stuck	Replace compressor
Compressor starts but will not get out of start winding	Low voltage Relay contacts stuck closed	Change relay; do not file or sandpaper contacts
	Bad run capacitor	Replace capacitor
	Bad compressor (open winding)	Replace compressor
	Bad compressor (grounded compressor)	Replace compressor
	Bad compressor (shorted winding)	Replace compressor
	High head pressure—will not allow compressor to reach speed	If air cooled, replace condenser fan motor. Clean condenser if dirty. Correct charge if overcharged. If water cooled, correct water problem across condenser.
Unit blows fuses	Short	If found, correct wiring
	Bad compressor	Replace compressor
	Shorted pump motor	Replace pump
	Condenser fan motor shorted	Replace condenser fan motor
	Water splashing on controls due to a water curtain or other part out of place	Repair problem with the water

ELECTRICAL COMPONENTS AND CONTROLS (cont.)

Problem	Analysis	Correction
System will not go into defrost	Hot gas solenoid will not open	Replace solenoid coil
	Bridge control will not close	Correct setting. If setting is OK, replace control.
	Harvest control open with head pressure sufficiently high	Replace harvest control
	Bad relay	Replace relay
Compressor, fan motor and water pump will not come on	Relay contact open	Replace relay
	Contactor not making contact	Replace contactor
System goes into defrost, but immediately goes out of defrost	Relay contacts not closing because of dirty contact or coil in relay burnt out	Replace relay
System will not come out of harvest cycle	Delay timer contacts stuck closed	Replace timer
	Safety thermodisc will not open	Replace the thermodisc
Nothing will come on	No power	Replace blown fuses or check for tripped breakers
	Open circuit in toggle switch	Replace toggle switch
	Bad contactor coil	Replace contactor or coil
Pump motor and condenser fan motor will come on, but compressor will not run	If unit is water cooled, high-pressure control open (no condenser fan motor if water cooled)	Replace control
	One or both bin controls open	Adjust or replace switches
	Contactor contact open	If contacts are bad, replace contactor

ELECTRICAL COMPONENTS AND CONTROLS *(cont.)*

Problem	Analysis	Correction
Pump motor and condenser fan motor will come on, but compressor will not run *(cont.)*	Compressor overload open	Determine why overload is open. If the compressor is at ambient temperature, the overload is bad and compressor will have to be replaced (internal overload). If hot, check the compressor and starting devices.

ELECTRICAL—COMMON PROBLEMS

Problem	Correction
Blown fuses	Needless to say, the reason for the overload must be eliminated, and the fuse replaced with the proper type.
Loose connections	There could be dozens of connections on any given machine. Each of these spots may be a source of trouble. A loose connection in a power circuit can generate local heat. This spreads to other parts of the same component, other components, or conductors. An example of where direct trouble can arise is thermally sensitive elements. These can be overload relays or thermally operated circuit breakers.
Faulty contacts	This applies to such components as motor starters, contactors, relays, push buttons, and switches.
Incorrect wiremarkers	This problem usually appears on the builder's assembly floor or in reassembly. The error can be difficult to locate, as a cable may have many conductors running some distance to various parts of the machine.
Combination problems	The following are typical types of combination problems: 1. Electrical–mechanical 2. Electrical–pressure (fluid power or pneumatic) 3. Electrical–temperature

ELECTRICAL—COMMON PROBLEMS *(cont.)*

Problem	Correction
Combination problems *(cont.)*	The greatest problem is that the observed or reported trouble is not always indicative of which aspect is at fault. It may be both. It is usually faster to check the electrical circuit first. However, both systems must be checked as both may contribute to the problem. For example, very few solenoid coils burn out due to a defect in the coil. Most solenoid trouble on valves develops from a faulty mechanical or pressure condition that prevents the solenoid plunger from seating properly, thus drawing excessive current. The result is an overload or a burned-out solenoid coil.
Low voltage	If no immediate indication of trouble is apparent check the line and control voltage. Due to inadequate power supply or conductor size, low voltage can be a problem. This problem generally shows up when starting or energizing a component, such as a motor starter or solenoid. However, it can cause trouble at other spots in the cycle.

Heat is one result of low voltage that may not be noticed in the functioning of a machine. As the voltage drops, the current in motor windings increases. This produces heat in the coils of the components (motor starters, relays, solenoids), which not only shortens the life of the components but may cause malfunctioning. For example, where there are relatively close moving metal parts, heat can cause these parts to expand to a point of sticking. |
| **Grounds** | There are many locations on a machine where a grounded condition can occur. However, there are a few spots in which grounds occur most often.

1. Connection points in solenoid valves, limit switches, and pressure switches
2. Raceway openings |

ELECTRIC FURNACES

Problem	Analysis	Correction
Fan operates with low or no heat	Defective time-delay sequencer	Replace sequencer
	Blown or faulty heater element fuse	Replace fuse
Unit fails to operate	Defective thermostat	Replace thermostat
	Defective furnace transformer	Replace
	Blown or faulty furnace fuse	Replace fuse
	Blown or defective transformer fuse	Replace fuse
	Open fused disconnect switch	Correct problem and close furnace circuit
	Improperly set room	Change thermostat to proper setting
Heats without fan operation	Defective fan motor	Repair or replace
	Defective fan motor run capacitor	Repair or replace
	Faulty fan control relay	Replace
	Faulty fan motor wiring or loose connections	Repair or replace wiring
Fan operates on heating, not on cooling	Improperly connected or faulty room thermostat	Make proper connections to thermostat
	Defective cooling-cycle control relay	Replace
	Defective or improper fan motor connections	Repair or replace
Individual heater fails to operate	Defective high-limit control	Replace
	Blown or faulty heater circuit fuse	Replace fuse
	Faulty heater element	Replace
	Defective time-delay sequencer	Replace

TROUBLESHOOTING NOTES

CHAPTER 8
Electrical and Motors

OHM'S LAW/POWER FORMULAS

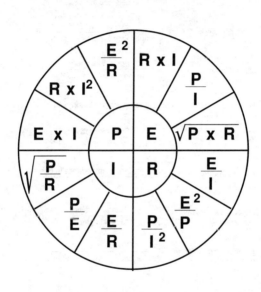

P = Power = Watts

R = Resistance = Ohms

I = Current = Amperes

E = Force = Volts

OHM'S LAW FOR ALTERNATING CURRENT

For the following Ohm's law formulas for AC current, θ is the phase angle in degrees where current lags voltage (in inductive circuit) or by which current leads voltage (in a capacitive circuit). In a resonant circuit (such as 120 VAC) the phase angle is 0° and impedance = resistance

$$\text{Current in amps} = \frac{\text{Voltage in volts}}{\text{Impedance in ohms}}$$

$$\text{Current in amps} = \sqrt{\frac{\text{Power in watts}}{\text{Impedance in ohms} \times \cos\theta}}$$

$$\text{Current in amps} = \frac{\text{Power in watts}}{\text{Voltage in volts} \times \cos\theta}$$

$$\text{Voltage in volts} = \text{Current in amps} \times \text{impedance in ohms}$$

$$\text{Voltage in volts} = \frac{\text{Power in watts}}{\text{Current in amps} \times \cos\theta}$$

$$\text{Voltage in volts} = \sqrt{\frac{\text{Power in watts} \times \text{impedance in ohms}}{\cos\theta}}$$

$$\begin{array}{l}\text{Impedance in ohms} = \\ \text{(Resistance)}\end{array} \quad \frac{\text{Voltage in volts}}{\text{Current in amps}}$$

$$\begin{array}{l}\text{Impedance in ohms} = \\ \text{(Resistance)}\end{array} \quad \frac{\text{Power in watts}}{(\text{Current in amps}^2 \times \cos\theta)}$$

$$\begin{array}{l}\text{Impedance in ohms} = \\ \text{(Resistance)}\end{array} \quad \frac{(\text{Voltage in volts}^2 \times \cos\theta)}{\text{Power in watts}}$$

$$\text{Power in watts} = \text{Current in amps}^2 \times \text{impedance in ohms} \times \cos\theta$$

$$\text{Power in watts} = \text{Current in amps} \times \text{voltage in volts} \times \cos\theta$$

$$\text{Power in watts} = \frac{(\text{Voltage in volts}^2 \times \cos\theta)}{\text{Impedance in ohms}}$$

OHM'S LAW FOR DIRECT CURRENT

Current in amps $= \dfrac{\text{Voltage in volts}}{\text{Resistance in ohms}} = \dfrac{\text{Power in watts}}{\text{Voltage in volts}}$

Current in amps $= \sqrt{\dfrac{\text{Power in watts}}{\text{Resistance in ohms}}}$

Voltage in volts = Current in amps × resistance in ohms

Voltage in volts $= \dfrac{\text{Power in watts}}{\text{Current in amps}}$

Voltage in volts $= \sqrt{\text{Power in watts} \times \text{resistance in ohms}}$

Power in watts = (Current in amps)^2 × resistance in ohms

Power in watts = Voltage in volts × current in amps

Power in watts $= \dfrac{(\text{Voltage in volts})^2}{\text{Resistance in ohms}}$

Resistance in ohms $= \dfrac{\text{Voltage in volts}}{\text{Current in amps}}$

Resistance in ohms $= \dfrac{\text{Power in watts}}{(\text{Current in amps})^2}$

POWER FACTOR

An AC electrical system carries two types of power:
(1) true power, watts, that pulls the load (note: mechanical load reflects back into an AC system as resistance) and (2) reactive power, vars, that generates magnetism within inductive equipment. The vector sum of these two will give actual volt-amperes flowing in the circuit (see diagram right). Power factor is the cosine of the angle between true power and volt-amperes.

Volt-amperes

Power-factor angle

Reactive power / volts

True power, watts

SINGLE-PHASE POWER

Power of a single-phase AC circuit equals voltage times current times power factor:

$P\text{watts} = E\text{volts} \times I\text{amps} \times PF.$

To figure reactive power, vars squared equals volt-amperes squared minus power squared, or

$\text{Vars} = \sqrt{(VA)^2 - (P)^2}$

8-3

AC/DC POWER FORMULAS

To Find	For Direct Current	1φ, 115 or 120 V	1φ, 208, 230, or 240 V	3φ — All Voltages
		For Alternating Current		
Amperes when horsepower is known	$\dfrac{HP \times 746}{E \times E_{FF}}$	$\dfrac{HP \times 746}{E \times E_{FF} \times PF}$	$\dfrac{HP \times 746}{E \times E_{FF} \times PF}$	$\dfrac{HP \times 746}{1.73 \times E \times E_{FF} \times PF}$
Amperes when kilowatts is known	$\dfrac{kW \times 1000}{E}$	$\dfrac{kW \times 1000}{E \times PF}$	$\dfrac{kW \times 1000}{E \times PF}$	$\dfrac{kW \times 1000}{1.73 \times E \times PF}$
Amperes when kVA is known		$\dfrac{kVA \times 1000}{E}$	$\dfrac{kVA \times 1000}{E}$	$\dfrac{kVA \times 1000}{1.73 \times E}$
Kilowatts	$\dfrac{I \times E}{1000}$	$\dfrac{I \times E \times PF}{1000}$	$\dfrac{I \times E \times PF}{1000}$	$\dfrac{I \times E \times 1.73 \times PF}{1000}$
Kilovolt-amps		$\dfrac{I \times E}{1000}$	$\dfrac{I \times E}{1000}$	$\dfrac{I \times E \times 1.73}{1000}$
Horsepower (output)	$\dfrac{I \times E \times E_{FF}}{746}$	$\dfrac{I \times E \times E_{FF} \times PF}{746}$	$\dfrac{I \times E \times E_{FF} \times PF}{746}$	$\dfrac{I \times E \times 1.73 \times E_{FF} \times PF}{746}$

8-4

HORSEPOWER FORMULAS

Current and Voltage Known

$$HP = \frac{E \times I \times E_{ff}}{746}$$

where
HP = horsepower
I = current (amps)
E = voltage (volts)
E_{ff} = efficiency
746 = constant

Speed and Torque Known

$$HP = \frac{rpm \times T}{5252}$$

where
HP = horsepower
rpm = revolutions per minute
T = torque (lb.-ft.)
5252 = constant

EFFICIENCY FORMULAS

Input and Output Power Known

$$E_{ff} = \frac{P_{out}}{P_{in}}$$

where
E_{ff} = efficiency (%)
P_{out} = output power (W)
P_{in} = input power (W)

Horsepower and Power Loss Known

$$E_{ff} = \frac{746 \times HP}{(746 \times HP) + W_l}$$

where
E_{ff} = efficiency (%)
746 = constant
HP = horsepower
W_l = watts lost

VOLTAGE UNBALANCE

$$V_u = \frac{V_d}{V_a} \times 100$$

where
V_u = voltage unbalance (%)
V_d = voltage deviation (V)
V_a = voltage average (V)
100 = constant

TEMPERATURE CONVERSIONS

Convert °C to °F

$$°F = (1.8 \times °C) + 32$$

Convert °F to °C

$$°C = \frac{(°F - 32)}{1.8}$$

SUMMARY OF SERIES, PARALLEL, AND COMBINATION CIRCUITS

To Find	Series Circuits	Parallel Circuits	Series/Parallel
Resistance (R) Ohm Ω	$R_T = R_1 + R_2 + R_3$ Sum of individual resistances	$\frac{1}{R_T} = \frac{1}{R_1} + \frac{1}{R_2} + \frac{1}{R_3}$	Total resistance equals resistance of parallel portion and sum of series resistors
Current (I) Ampere A	$I_t = I_1 = I_2 = I_3$ The same throughout entire circuit	$I_t = I_1 + I_2 + I_3$ Sum of individual currents	Series rules apply to series portion of circuit. Parallel rules apply to parallel part of circuit.
Voltage (E) Volt V, E	$E_T = E_1 + E_2 + E_3$ Sum of individual voltages	$E_T = E_1 = E_2 = E_3$ Total voltage and branch voltage are the same	Total voltage is sum of voltage drops across each series resistor and each of the branches of parallel portion
Power (P) Watt W	$P_T = P_1 + P_2 + P_3$ Sum of individual wattages	$P_T = P_1 + P_2 + P_3$ Sum of individual wattages	$P_T = P_1 + P_2 + P_3$ Sum of individual wattages

8-6

TYPICAL POWER WIRING COLOR CODE

120/240 Volt		277/480 Volt	
Black	Phase 1	Brown	Phase 1
Red	Phase 2	Orange	Phase 2
Blue	Phase 3	Yellow	Phase 3
White or with 3 white stripes	Neutral	Gray or with 3 white stripes	Neutral
Green	Ground	Green w/yellow stripe	Ground

POWER TRANSFORMER COLOR CODES

Wire Color (solid)	Circuit Type
Black	If a transformer does not have a tapped primary, both leads are black.
Black	If a transformer does have a tapped primary, the black is the common lead.
Black & yellow . .	Tap for a tapped primary.
Black & red	End for a tapped primary.

HAZARDOUS LOCATIONS

Class	Group	Material
I	A	Acetylene
	B	Hydrogen, butadiene, ethylene oxide, propylene oxide
	C	Carbon monoxide, ether, ethylene, hydrogen sulfide, morpholine, cyclopropane
	D	Gasoline, benzene, butane, propane, alcohol, acetone, ammonia, vinyl chloride
II	E	Metal dusts
	F	Carbon black, coke dust, coal
	G	Grain dust, flour, starch, sugar, plastics
III	No groups	Wood chips, cotton, flax, and nylon

ENCLOSURE TYPES

Type	Use	Service Conditions	UL Tests	Comments
1	Indoor	None	Rod entry, rust resistance	—
3	Outdoor	Windblown dust, rain, sleet, and ice on enclosure	Rain, external icing, dust, and rust resistance	Do not provide protection against internal condensation, or internal icing
3R	Outdoor	Falling rain and ice on enclosure	Rod entry, rain, external icing, and rust resistance	Do not provide protection against dust, internal condensation, or internal icing
4	Indoor/outdoor	Windblown dust and rain, splashing water, hose-directed water, and ice on enclosure	Hosedown, external icing, and rust resistance	Do not provide protection against internal condensation or internal icing
4X	Indoor/outdoor	Corrosion, windblown dust and rain, splashing water, hose-directed water, and ice on enclosure	Hosedown, external icing, and corrosion resistance	Do not provide protection against internal condensation or internal icing
6	Indoor/outdoor	Occasional temporary submersion at a limited depth	—	—

ENCLOSURE TYPES (cont.)

Type	Use	Service Conditions	UL Tests	Comments
6P	Indoor/outdoor	Prolonged submersion at a limited depth	—	—
7	Indoor locations classified as Class I, or Groups A, B, C, or D, as defined in the NEC®	Withstand and contain an internal explosion of specified gases, contain an explosion sufficiently so an explosive gas-air mixture in the atmosphere is not ignited	Explosion, hydrostatic, and temperature	Enclosed heat-generating devices shall not cause external surfaces to reach temperatures capable of igniting explosive gas-air mixtures in the atmosphere
9	Indoor locations classified as Class II, Groups E or G, as defined in the NEC®	Dust	Dust penetration, temperature, and gasket aging	Enclosed heat-generating devices shall not cause external surfaces to reach temperatures capable of igniting explosive gas-air mixtures in the atmosphere
12	Indoor	Dust, falling dirt, and dripping noncorrosive liquids	Drip, dust, and rust resistance	Do not provide protection against internal condensation
13	Indoor	Dust, spraying water oil and noncorrosive coolant	Oil explosion and rust resistance	Do not provide protection against internal condensation

JUNCTION BOX CALCULATIONS

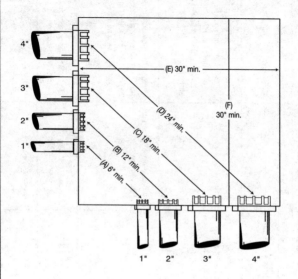

Distance (A) is 6 × 1" = 6" minimum

Distance (B) is 6 × 2" = 12" minimum

Distance (C) is 6 × 3" = 18" minimum

Distance (D) is 6 × 4" = 24" minimum

Distance (E) is (6 × 4") + 3" + 2" + 1" = 30" minimum

Distance (F) is (6 × 4") + 3" + 2" + 1" = 30" minimum

BOX FILL

Max. Number of Conductors in Outlet, Device, and Junction Boxes

Box Dimension (in.) Trade Size or Type	Min. Capacity (in.³)	Maximum Number of Conductors						
		18	16	14	12	10	8	6
4 × 1¼ round or octagonal	12.5	8	7	6	5	5	5	2
4 × 1½ round or octagonal	15.5	10	8	7	6	6	5	3
4 × 2⅛ round or octagonal	21.5	14	12	10	9	8	7	4
4 × 1¼ square	18.0	12	10	9	8	7	6	3
4 × 1½ square	21.0	14	12	10	9	8	7	4
4 × 2⅛ square	30.3	20	17	15	13	12	10	6
4¹¹⁄₁₆ × 1¼ square	25.5	17	14	12	11	10	8	5
4¹¹⁄₁₆ × 1½ square	29.5	19	16	14	13	11	9	5
4¹¹⁄₁₆ × 2⅛ square	42.0	28	24	21	18	16	14	8
3 × 2 × 1½ device	7.5	5	4	3	3	3	2	1
3 × 2 × 2 device	10.0	6	5	5	4	4	3	2
3 × 2 × 2¼ device	10.5	7	6	5	4	4	3	2
3 × 2 × 2½ device	12.5	8	7	6	5	5	4	2
3 × 2 × 2¾ device	14.0	9	8	7	6	5	4	2
3 × 2 × 3½ device	18.0	12	10	9	8	7	6	3
4 × 2⅛ × 1½ device	10.3	6	5	5	4	4	3	2
4 × 2⅛ × 1⅞ device	13.0	8	7	6	5	5	4	2
4 × 2⅛ × 2⅛ device	14.5	9	8	7	6	5	4	2
3¾ × 2 × 2½ masonry box/gang	14.0	9	8	7	6	5	4	2
3¾ × 2 × 3½ masonry box/gang	21.0	14	12	10	9	8	7	2

If one or more cable clamp is in a box, it is counted the same as the largest conductor. A loop of conductor 12" or more counts as two conductors.

CONDUCTOR VOLUME ALLOWANCE

Wire Size (AWG)	Volume Each (in.³)	Formula
18	1.50	$V = L \times W \times D$
16	1.75	
14	2.00	Volume =
12	2.25	Length times width
10	2.50	times depth
8	3.00	(in cu. in.)
6	5.00	

To find box size needed, add up total volume for all wires to be used. Then use the volume formula. Example: If total volume of all wires is 420 cu. in. use an 8" × 10" × 6" box = 480 cu. in.

SIZES OF PANELBOARDS

Single-Phase — 3-Wire Systems

40 A	100 A	150 A	225 A	400 A
70 A	125 A	200 A	300 A	600 A

Three-Phase — 4-Wire Systems

60 A	150 A	225 A	400 A
125 A	200 A	300 A	600 A

SIZES OF GUTTERS AND WIREWAYS

2½" × 2½"	6" × 6"	10" × 10"
4" × 4"	8" × 8"	

These sizes are available in 12", 24", 36", 48", and 60" lengths.

SIZES OF DISCONNECTS

30 A	200 A	800 A	1600 A
60 A	400 A	1200 A	1800 A
100 A	600 A	1400 A	

SIZES OF PULL BOXES AND JUNCTION BOXES

4" × 4" × 4"	10" × 8" × 4"	12" × 12" × 6"
6" × 4" × 4"	10" × 8" × 6"	12" × 12" × 8"
6" × 6" × 4"	10" × 10" × 4"	15" × 12" × 4"
6" × 6" × 6"	10" × 10" × 6"	15" × 12" × 6"
8" × 6" × 4"	10" × 10" × 8"	18" × 12" × 4"
8" × 6" × 6"	12" × 8" × 4"	18" × 12" × 6"
8" × 6" × 8"	12" × 8" × 6"	18" × 18" × 4"
8" × 8" × 4"	12" × 10" × 4"	18" × 18" × 6"
8" × 8" × 6"	12" × 10" × 6"	24" × 18" × 6"
8" × 8" × 8"	12" × 12" × 4"	24" × 24" × 6"
		24" × 24" × 8"

BUSWAY OR BUSDUCT

1 φ	3 φ
225 A	225 A
400 A	400 A
600 A	600 A
800 A	800 A
1,000 A	1,000 A
1,200 A	1,200 A
1,350 A	1,350 A
1,600 A	1,600 A
2,000 A	2,000 A
2,500 A	2,500 A
3,000 A	3,000 A
4,000 A	4,000 A
5,000 A	5,000 A

CBs AND FUSES

15	70	225	800
20	80	250	1,000
25	90	300	1,200
30	100	350	1,600
35	110	400	2,000
40	125	450	2,500
45	150	500	3,000
50	175	600	4,000
60	200	700	5,000
			6,000

For fuses only, additional standard sizes are 1, 3, 6, and 10. All sizes are rated in amps.

SWITCHBOARDS OR SWITCHGEARS

1 φ	3 φ
200 A	400 A
400 A	600 A
600 A	800 A
800 A	1,200 A
1,200 A	1,600 A
1,600 A	2,000 A
2,000 A	2,500 A
2,500 A	3,000 A
3,000 A	4,000 A
4,000 A	

BENDING STUB-UPS

Conduit or EMT Size (in.)	Deduction from Mark or Arrow (in.)
½	5"
¾	6"
1	8"

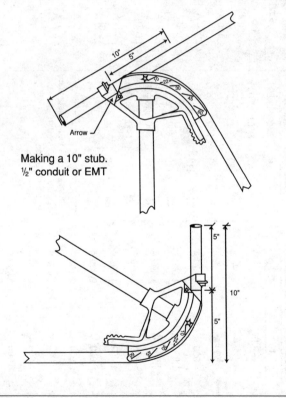

Making a 10" stub.
½" conduit or EMT

BACK-TO-BACK BENDING

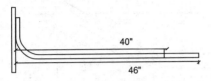

40"

46"

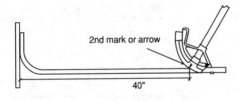

2nd mark or arrow

40"

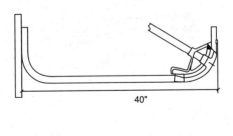

40"

8-15

OFFSET AND SADDLE BENDS

2-Bend Offset

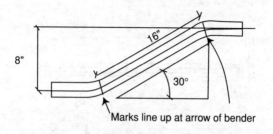

Marks line up at arrow of bender

Angle of Bends	Inches of Run per in. of Offset	Loss of Conduit Length per in. of Offset
10°	5.76"	1/16"
22.5°	2.6"	3/16"
30°	2.0"	1/4"
45°	1.414"	3/8"
60°	1.15"	1/2"

3-Bend Saddle

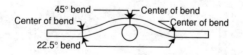

45° bend — Center of bend
Center of bend — Center of bend
22.5° bend

ELECTRICAL CABLE CLASS RATINGS

Electrical cable is rated according to the following parameters: the number of wires, the wire size, the type of insulation and the moisture condition of the environment of the wire. Therefore, an electrical cable designated $^{10}\!/_{2}$ with ground—type UF—600 V-(UL) meets the specifications below:

- The "10" relates to wire size—10 gauge wire.
- 2 defines an electrical cable with two wires.
- The word "ground" indicates the cable has a third wire to be connected to ground.
- The term "type UF" means the insulation type has an acceptable moisture rating.
- "600 V" defines the cable as being rated at 600 V maximum.
- "(UL)" means the cable has certification from Underwriters Laboratory.

CABLE INSULATION MOISTURE RATINGS

Dry	Indoor aboveground level; moisture usually not encountered.
Moist	Indoor belowground level (basement); locations are partially protected; moisture level is moderate.
Wet	Locations affected by weather (outside); concrete slabs, underground, etc.; water saturation likely.

CONDUCTOR PREFIX CODES

B	Outer braid		**O**	Neoprene jacket
F	Fixture wire		**R**	Rubber covering
FEP	Fluorinated ethylene propylene. Use in dry locations only, hotter than 90° C		**S**	Appliance cord
			SP	Lamp cord, rubber
			SPT	Lamp cord, plastic
H	Load temp up to 75° C		**T**	Load temp up to 60° C
HH	Load temp up to 90° C		**W**	Wet use only
L	Seamless lead jacket		**X**	Moisture and heat resistant
M	Machine tool wire			
N	Resistant to oil and gas			

VOLTAGE DROP FORMULAS

The NEC® recommends a maximum 3% voltage drop for either the branch circuit or the feeder.

Single-phase:

$$VD = \frac{2 \times R \times I \times L}{CM}$$

Three-phase:

$$VD = \frac{1.732 \times R \times I \times L}{CM}$$

VD = Volts (voltage drop of the circuit)

R = 12.9 ohms/copper or 21.2 ohms/aluminum (resistance constants for a 1,000 circular mils conductor that is 1,000 ft. long, at an operating temperature of 75°C).

I = Amps (load at 100 %)

L = Ft. (length of circuit from load to power supply)

CM = Circular-mils (conductor wire size)

2 = Single-phase constant

1.732 = Three-phase constant

CONDUCTOR LENGTH/VOLTAGE DROP

Voltage drop can be reduced by limiting the length of the conductors.

Single-phase:

$$L = \frac{CM \times VD}{2 \times R \times I}$$

Three-phase:

$$L = \frac{CM \times VD}{1.732 \times R \times I}$$

CONDUCTOR SIZE/VOLTAGE DROP

Increase the size of the conductor to decrease the voltage drop of circuit (reduce its resistance).

Single-phase:

$$CM = \frac{2 \times R \times I \times L}{VD}$$

Three-Phase:

$$CM = \frac{1.732 \times R \times I \times L}{VD}$$

COPPER WIRE SPECIFICATIONS

Wire Size (AWG)	Area (Circular Mils)	Diameter (Mils, 1000th in.)	Diameter (mm)	Weight (lbs. per 1000 ft.)
40	9.9	3.1	.080	.0200
38	15.7	4	.101	.0476
36	25	5	.127	.0757
34	39.8	6.3	.160	.120
32	63.2	8	.202	.191
30	101	10	.255	.304
28	160	12.6	.321	.484
26	254	15.9	.405	.769
24	404	20.1	.511	1.22
22	642	25.3	.644	1.94
20	1,020	32	.812	3.09
18	1,620	40	1.024	4.92
16	2,580	51	1.291	7.82
14	4,110	64	1.628	12.4
12	6,530	81	2.053	19.8
10	10,400	102	2.688	31.4
8	16,500	128	3.264	50
6	26,300	162	4.115	79.5
4	41,700	204	5.189	126
3	52,600	229	5.827	159
2	66,400	258	6.544	201
1	83,700	289	7.348	253
0	106,000	325	8.255	319
00	133,000	365	9.271	403
000	168,000	410	10.414	508
0000	212,000	460	11.684	641

COPPER WIRE RESISTANCE

Wire Size (AWG)	77°F (ft. per ohm)	149°F (ft. per ohm)	77°F (ohms per 1,000 ft.)	149°F (ohms per 1,000 ft.)
40	.93	.81	1,070	1,230
38	1.5	1.3	673	776
36	2.4	2.0	423	488
34	3.8	3.3	266	307
32	6.0	5.2	167	193
30	9.5	8.3	105	121
28	15.1	13.1	66.2	76.4
26	24.0	20.8	41.6	48.0
24	38.2	33.1	26.2	30.2
22	60.6	52.6	16.5	19.0
20	96.2	84.0	10.4	11.9
18	153.6	133.2	6.51	7.51
16	244.5	211.4	4.09	4.73
14	387.6	336.7	2.58	2.97
12	617.3	534.8	1.62	1.87
10	980.4	847.5	1.02	1.18
8	1,560	1,353	.641	.739
6	2,481	2,151	.403	.465
4	3,953	3,425	.253	.292
3	4,975	4,310	.201	.232
2	6,289	5,435	.159	.184
1	7,936	6,849	.126	.146
0	10,000	8,621	.100	.116
00	12,658	10,870	.079	.092
000	15,873	13,699	.063	.073
0000	20,000	17,544	.050	.057

VOLTAGE DROP AMPERE-FEET

Copper Conductors, 70°C Copper Temp., 600 V Class Single Conductor Cables in Steel Conduit

Conductor Size (AWG or MCM)	DC Circuits 1% Drop on 120 V	60 Cycle AC Circuits				
		1% Drop, 1.00 PF		3% Drop, 0.85 PF		
		120 V 1-Phase	208 V 3-Phase	115 V 1-Phase	208 V 3-Phase	220 V 3-Phase
14	191	191	382	623	1,300	1,380
12	305	305	612	998	2,080	2,200
10	484	484	968	1,580	3,280	3,470
8	770	770	1,540	2,450	5,110	5,410
6	1,200	1,190	2,380	3,800	7,850	8,310
4	1,900	1,890	3,780	5,750	12,000	12,700
2	3,030	2,970	5,950	8,620	18,000	19,100
1	3,820	3,710	7,430	10,400	21,800	23,100
1/0	4,820	4,560	9,120	12,500	26,000	27,500
2/0	6,060	5,610	11,200	14,800	30,800	32,700
3/0	7,650	6,940	13,900	16,900	35,200	37,100
4/0	9,760	8,520	17,100	20,200	42,100	44,600
250	11,500	9,930	19,800	22,300	46,500	49,300
300	13,800	11,500	23,100	24,800	52,000	55,000
350	16,100	13,200	26,300	27,000	56,200	59,500
400	18,400	14,000	28,100	28,700	60,000	63,500
500	22,700	16,400	32,800	31,800	66,400	70,300
750	34,600	20,400	40,800	36,700	76,600	81,200

Maximum circuit ampere-ft. without exceeding specified percentage voltage drop, various circuit voltages, and power factors.

Note: Length to be used is the distance from point of supply to load.

WIRE LENGTH VS WIRE SIZE (MAX. VOLTAGE DROP)

Max. Wire Ft. @ 240 V, Single-Phase, 2% Max. Voltage Drop

3/0	2/0	1/0	#2	#4	Watts	Amps
180	—	—	—	—	48,000	200
240	190	185	—	—	36,000	150
360	280	230	—	—	24,000	100
440	365	290	180	—	19,200	80
520	415	330	205	130	16,800	70
600	485	385	240	150	14,400	60
720	580	460	290	180	12,000	50
880	725	575	360	230	5,600	40
1,200	970	770	485	300	7,200	30
1,440	1,100	920	580	365	6,000	25

#6	#8	#10	#12	#14	Watts	Amps
105	—	—	—	—	12,000	50
130	90	—	—	—	5,600	40
175	120	75	—	—	7,200	30
210	144	90	—	—	6,000	25
265	180	110	70	—	4,800	20
350	240	150	95	60	3,600	15
525	360	225	140	90	2,400	10
1,020	720	455	285	180	1,200	5

Max. Wire Ft. @ 120 V, Single-Phase, 2% Max. Voltage Drop

3/0	2/0	1/0	#2	#4	Watts	Amps
230	180	144	90	—	9,600	80
260	205	165	105	65	8,400	70
305	240	190	120	76	7,200	60
360	290	230	145	90	6,000	50
440	360	290	175	115	4,800	40
600	490	385	240	150	3,600	30
720	580	460	290	180	3,000	25
900	725	575	365	230	2,400	20
1,200	965	770	485	305	1,800	15

#6	#8	#10	#12	#14	Watts	Amps
57	—	—	—	—	6,000	50
72	45	—	—	—	4,800	40
95	60	38	—	—	3,600	30
115	72	45	—	—	3,000	25
140	90	57	36	—	2,400	20
190	120	75	47	30	1,800	15
285	180	115	70	45	1,200	10
575	360	225	140	90	600	5

VOLTAGE DROP TABLE

Conductor Size	DC	Volts Drop per 1000 Ampere-Ft.						
		AC System						
		Load Power Factor (%)						
		100	95	90	85	80	75	70

For DC circuit or single phase, 60-cycle, 2-wire system or 3-wire system with balanced load. Copper conductors, 70°C copper temperature. 600 V class single-conductor cables in steel conduit.

Conductor Size	DC	100	95	90	85	80	75	70
14	6.29	6.29	6.06	5.78	5.54	5.26	4.97	4.74
12	3.93	3.93	3.81	3.64	3.46	3.29	3.13	2.95
10	2.48	2.48	2.44	2.31	2.19	2.08	1.96	1.85
8	1.56	1.56	1.51	1.47	1.41	1.34	1.27	1.20
6	0.999	1.011	0.987	0.953	0.918	0.872	0.826	0.774
4	0.631	0.635	0.641	0.624	0.600	0.578	0.554	0.528
2	0.396	0.404	0.418	0.413	0.400	0.386	0.372	0.358
1	0.314	0.323	0.356	0.337	0.330	0.322	0.311	0.300
1/0	0.249	0.263	0.280	0.282	0.277	0.269	0.263	0.255
2/0	0.198	0.214	0.233	0.236	0.233	0.230	0.226	0.222
3/0	0.157	0.173	0.196	0.206	0.204	0.200	0.194	0.188
4/0	0.123	0.141	0.163	0.170	0.171	0.170	0.169	0.166
250 MCM	0.1041	0.121	0.146	0.152	0.155	0.155	0.155	0.154
300 MCM	0.0870	0.1040	0.128	0.135	0.139	0.140	0.141	0.141
350 MCM	0.0746	0.0912	0.117	0.125	0.128	0.131	0.131	0.131
400 MCM	0.0652	0.0855	0.1086	0.117	0.120	0.122	0.124	0.125
500 MCM	0.0528	0.0733	0.0959	0.1040	0.1086	0.111	0.113	0.114
750 MCM	0.0347	0.0589	0.0808	0.0884	0.0940	0.0976	0.0999	0.1020

For 3-phase, 60-cycle, 3-wire or 4-wire balanced system. Copper conductors, 70°C copper temperature, 600 V class single-conductor cables in steel conduit.

Conductor Size	DC	100	95	90	85	80	75	70
14		5.45	5.25	5.00	4.80	4.55	4.30	4.10
12		3.40	3.30	3.15	3.00	2.85	2.70	2.55
10		2.15	2.10	2.00	1.90	1.80	1.70	1.60
8		1.35	1.31	1.27	1.22	1.18	1.10	1.04
6		0.875	0.855	0.825	0.795	0.755	0.715	0.670
4		0.550	0.555	0.540	0.520	0.500	0.480	0.457
2		0.350	0.362	0.358	0.346	0.334	0.322	0.310
1		0.280	0.308	0.292	0.286	0.279	0.269	0.260
1/0		0.228	0.242	0.244	0.240	0.233	0.228	0.221
2/0		0.185	0.202	0.204	0.202	0.199	0.196	0.192
3/0		0.150	0.170	0.178	0.177	0.173	0.168	0.163
4/0		0.122	0.141	0.147	0.148	0.147	0.146	0.144
250 MCM		0.105	0.126	0.132	0.134	0.134	0.134	0.133
300 MCM		0.0900	0.111	0.117	0.120	0.121	0.122	0.122
350 MCM		0.0790	0.101	0.108	0.111	0.113	0.114	0.114
400 MCM		0.0740	0.0940	0.101	0.104	0.106	0.107	0.108
500 MCM		0.0635	0.0830	0.0900	0.0940	0.0964	0.0974	0.0988
750 MCM		0.0510	0.0700	0.0765	0.0814	0.0845	0.0865	0.0883

Note: Length to be used is the distance from point of supply to load, not amount of wire in circuit.

AMPACITIES OF COPPER CONDUCTORS (3)

Ampacities of Not More Than 3 Insulated Conductors Rated 0–2000 V in Cable or Raceway

Wire Size AWG (kcmil)	In Cable, Raceway, or Earth, Ambient Temperature 30°C (86°F)			In Cable or Raceway, Ambient Temperature 40°C (104°F)			Wire Size AWG (kcmil)
	60°C (140°F)	75°C (167°F)	90°C (194°F)	150°C (302°F)	200°C (392°F)	250°C (482°F)	
	Types TW UF	Types RHW, THHW, THW, THWN, XHHW, USE, ZW	Types TBS, SA, SIS, FEP, MI FEPB, ZW-2, RHH, RHW-2, THHN, THHW, THW-2, XHH, USE-2, THWN-2 XHHW, XHHW-2.	Type Z	Types FEP, FEPB, PFA	Types PFAH, TFE Nickel or nickel-coated copper	
14*	20	20	25	34	36	39	14
12*	25	25	30	43	45	54	12
10*	30	35	40	55	60	73	10
8	40	50	55	76	83	93	8
6	55	65	75	96	110	117	6
4	70	85	95	120	125	148	4
3	85	100	110	143	152	166	3
2	95	115	130	160	171	191	2
1	110	130	150	186	197	215	1
1/0	125	150	170	215	229	244	1/0
2/0	145	175	195	251	260	273	2/0
3/0	165	200	225	288	297	308	3/0
4/0	195	230	260	332	346	361	4/0
250	215	255	290				250

Temperature Correction Factors

Ambient Temp.°C	For Other Than 30°C			For Other Than 40°C			Ambient Temp.°C
	Multiply the Ampacities above by the Factors Below						
21–25	1.08	1.05	1.04	.95	.97	.98	41–50
26–30	1.00	1.00	1.00	.90	.94	.95	51–60
31–35	.91	.94	.96	.85	.90	.93	61–70
36–40	.82	.88	.91	.80	.87	.90	71–80
41–45	.71	.82	.87	.74	.83	.87	81–90
46–50	.58	.75	.82	.67	.79	.85	91–100
51–55	.41	.67	.76	.52	.71	.79	101–120
56–60		.58	.71	.30	.61	.72	121–140
61–70		.33	.58		.50	.65	141–160
71–80			.41		.35	.58	161–180
						.49	181–200
						.35	201–225

*Unless specifically permitted by the NEC®, overcurrent protection for copper conductors shall not exceed 15 amps for no. 14 AWG, 20 amps for no. 12 AWG, and 30 amps for no. 10 AWG.

AMPACITIES OF COPPER CONDUCTORS (3) *(cont.)*

Ampacities of Not More Than 3 Insulated Conductors Rated 0–2000 V in Cable or Raceway

Wire Size	In Cable, Raceway, or Earth, Ambient Temperature 30°C (86°F)			In Cable or Raceway, Ambient Temperature 40°C (104°F)			Wire Size
	60°C (140°F)	75°C (167°F)	90°C (194°F)	150°C (302°F)	200°C (392°F)	250°C (482°F)	
	Types	Types	Types	Type	Types	Types	
(kcmil)	TW, UF	RHW, THHW, THW, THWN, XHHW, USE, ZW	TBS, SA, SIS, FEP, MI FEPB, ZW-2, RHH, RHW-2, THHN, THHW, THW-2, XHH, USE-2, THWN-2 XHHW, XHHW-2.	Z	FEP, FEPB, PFA	PFAH, TFE Nickel or nickel-coated copper	(kcmil)
300	240	285	320	—	—	—	300
350	260	310	350	—	—	—	350
400	280	335	380	—	—	—	400
500	320	380	430	—	—	—	500
600	355	420	475	—	—	—	600
700	385	460	520	—	—	—	700
750	400	475	535	—	—	—	750
800	410	490	555	—	—	—	800
900	435	520	585	—	—	—	900
1,000	455	545	615	—	—	—	1,000
1,250	495	590	665	—	—	—	1,250
1,500	520	625	705	—	—	—	1,500
1,750	545	650	735	—	—	—	1,750
2,000	560	665	750	—	—	—	2,000

	Temperature Correction Factors		
Ambient Temp.°C	For Other Than 30°C Multiply the Ampacities above by the Factors Below		
21–25	1.08	1.05	1.04
26–30	1.00	1.00	1.00
31–35	.91	.94	.96
36–40	.82	.88	.91
41–45	.71	.82	.87
46–50	.58	.75	.82
51–55	.41	.67	.76
56–60	—	.58	.71
61–70	—	.33	.58
71–80	—	—	.41

AMPACITIES OF COPPER CONDUCTORS (1)

Ampacities of Single Insulated Conductors Rated 0–2000 V in Free Air

Wire Size	Ambient Temperature 30°C (86°F)			Ambient Temperature 40°C (104°F)			Bare Conductors with Max. Temp. 80°C (176°F)
	60°C (140°F)	75°C (167°F)	90°C (194°F)	150°C (302°F)	200°C (392°F)	250°C (482°F)	
	Types	Types	Types	Type	Types	Types	
AWG (kcmil)	TW UF	RHW, THHW, THW, THWN, XHHW, ZW	TBS, SA, SIS, FEP, MI, FEPB, ZW-2, RHH, RHW-2, THHN, THHW, THW-2, XHH, USE-2, XHHW, XHHW-2.	Z	FEP, FEPB, PFA	PFAH, TFE Nickel or nickel-coated copper	
14*	25	30	35	46	54	59	—
12*	30	35	40	60	68	78	—
10*	40	50	55	80	90	107	—
8	60	70	80	106	124	142	98
6	80	95	105	155	165	205	124
4	105	125	140	190	220	278	155
3	120	145	165	214	252	327	—
2	140	170	190	255	293	381	209
1	165	195	220	293	344	440	—
1/0	195	230	260	339	399	532	282
2/0	225	265	300	390	467	591	329
3/0	260	310	350	451	546	708	382
4/0	300	360	405	529	629	830	444
250	340	405	455				494

Temperature Correction Factors

Ambient Temp.°C	For Other Than 30°C			For Other Than 40°C			Ambient Temp.°C
	Multiply the Ampacities above by the Factors Below						
21–25	1.08	1.05	1.04	.95	.97	.98	41–50
26–30	1.00	1.00	1.00	.90	.94	.95	51–60
31–35	.91	.94	.96	.85	.90	.93	61–70
36–40	.82	.88	.91	.80	.87	.90	71–80
41–45	.71	.82	.87	.74	.83	.87	81–90
46–50	.58	.75	.82	.67	.79	.85	91–100
51–55	.41	.67	.76	.52	.71	.79	101–120
56–60	—	.58	.71	.30	.61	.72	121–140
61–70	—	.33	.58	—	.50	.65	141–160
71–80	—	—	.41	—	.35	.58	161–180
—	—	—	—	—	—	.49	181–200
—	—	—	—	—	—	.35	201–225

*Unless specifically permitted by the NEC®, overcurrent protection for copper conductors shall not exceed 15 amps for no. 14 AWG, 20 amps for no. 12 AWG, and 30 amps for no. 10 AWG.

AMPACITIES OF COPPER CONDUCTORS (1) *(cont.)*

Ampacities of Single Insulated Conductors Rated 0–2000 V in Free Air

Wire Size	Ambient Temperature 30°C (86°F)			Ambient Temperature 40°C (104°F)			Bare Conductors with Max. Temp. 80°C (176°F)
	60°C (140°F)	75°C (167°F)	90°C (194°F)	150°C (302°F)	200°C (392°F)	250°C (482°F)	
	Types	Types	Types	Type	Types	Types	
	TW, UF	RHW, THHW, THW, THWN, XHHW, ZW	TBS, SA, SIS, FEP, MI, FEPB, ZW-2, RHH, RHW-2, THHN, THHW, THW-2, XHH, USE-2, THWN-2, XHHW, XHHW-2.	Z	FEP, FEPB, PFA	PFAH, TFE Nickel or nickel-coated copper	
(kcmil)							
300	375	445	505	—	—	—	556
350	420	505	570	—	—	—	
400	455	545	615	—	—	—	
500	515	620	700	—	—	—	773
600	575	690	780	—	—	—	
700	630	755	855	—	—	—	
750	655	785	885	—	—	—	1,000
800	680	815	920	—	—	—	
900	730	870	985	—	—	—	
1,000	780	935	1,055	—	—	—	1,193
1,250	890	1,065	1,200	—	—	—	
1,500	980	1,175	1,325	—	—	—	
1,750	1,070	1,280	1,445	—	—	—	
2,000	1,155	1,385	1,560	—	—	—	

Temperature Correction Factors

Ambient Temp.°C	For Other Than 30°C Multiply the Ampacities above by the Factors Below		
21–25	1.08	1.05	1.04
26–30	1.00	1.00	1.00
31–35	.91	.94	.96
36–40	.82	.88	.91
41–45	.71	.82	.87
46–50	.58	.75	.82
51–55	.41	.67	.76
56–60	—	.58	.71
61–70	—	.33	.58
71–80	—	—	.41

AMPACITIES OF ALUMINUM AND COPPER-CLAD ALUMINUM CONDUCTORS (3)

Ampacities of Not More Than 3 Single Insulated Conductors Rated 0–2000 V in Cable or Raceway

Wire Size	In Cable, Raceway, or Earth, Ambient Temperature 30°C (86°F)			In Cable or Raceway Ambient Temp. 40°C (104°F)	Wire Size
	60°C (140°F)	75°C (167°F)	90°C (194°F)	150°C (302°F)	
AWG (kcmil)	Types TW, UF	Types THWN, THHW, XHHW, USE, THW, RHW	Types TBS, SA, SIS, THHN, THHW, THW-2, THWN-2, RHH, RHW-2, USE-2, XHH, XHHW, XHHW-2, ZW-2	Type Z	**AWG (kcmil)**
12*	20	20	25	30	12
10*	25	30	35	44	10
8	30	40	45	57	8
6	40	50	60	75	6
4	55	65	75	94	4
3	65	75	85	109	3
2	75	90	100	124	2
1	85	100	115	145	1
1/0	100	120	135	169	1/0
2/0	115	135	150	198	2/0
3/0	130	155	175	227	3/0
4/0	150	180	205	260	4/0
250	170	205	230	—	250

Temperature Correction Factors					
Ambient Temp.°C	For Other Than 30°C			40°C	Ambient Temp.°C
	Multiply the Ampacities above by the Factors Below				
21–25	1.08	1.05	1.04	.95	41–50
26–30	1.00	1.00	1.00	.90	51–60
31–35	.91	.94	.96	.85	61–70
36–40	.82	.88	.91	.80	71–80
41–45	.71	.82	.87	.74	81–90
46–50	.58	.75	.82	.67	91–100
51–55	.41	.67	.76	.52	101–120
56–60	—	.58	.71	.30	121–140
61–70	—	.33	.58	—	—
71–80	—	—	.41	—	—

*Unless specifically permitted by the NEC®, overcurrent protection for aluminum and copper-clad aluminum conductors shall not exceed 15 amps for no. 12 AWG and 25 amps for no. 10 AWG.

AMPACITIES OF ALUMINUM AND COPPER-CLAD ALUMINUM CONDUCTORS (3) *(cont.)*

Ampacities of Not More Than 3 Single Insulated Conductors Rated 0–2000 V in Cable or Raceway

Wire Size	In Cable, Raceway, or Earth, Ambient Temperature 30°C (86°F)			In Cable or Raceway Ambient Temp. 40°C (104°F)	Wire Size
	60°C (140°F)	75°C (167°F)	90°C (194°F)	150°C (302°F)	
	Types	Types	Types	Type	
(kcmil)	TW, UF	THWN, THHW, XHHW, USE, THW, RHW	TBS, SA, SIS, THHN, THHW, THW-2, THWN-2, RHH, RHW-2, USE-2, XHH, XHHW, XHHW-2, ZW-2	Z	(kcmil)
300	190	230	255	—	300
350	210	250	280	—	350
400	225	270	305	—	400
500	260	310	350	—	500
600	285	340	385	—	600
700	310	375	420	—	700
750	320	385	435	—	750
800	330	395	450	—	800
900	355	425	480	—	900
1,000	375	445	500	—	1,000
1,250	405	485	545	—	1,250
1,500	435	520	585	—	1,500
1,750	455	545	615	—	1,750
2,000	470	560	630	—	2,000

Temperature Correction Factors					
Ambient Temp.°C	For Other Than 30°C Multiply the Ampacities above by the Factors Below				
21–25	1.08	1.05	1.04		
26–30	1.00	1.00	1.00		
31–35	.91	.94	.96		
36–40	.82	.88	.91		
41–45	.71	.82	.87		
46–50	.58	.75	.82		
51–55	.41	.67	.76		
56–60	—	.58	.71		
61–70	—	.33	.58		
71–80	—	—	.41		

AMPACITIES OF ALUMINUM OR COPPER-CLAD ALUMINUM CONDUCTORS (1)

Ampacities of Single Insulated Conductors Rated 0–2000 V in Free Air

Wire Size AWG (kcmil)	Ambient Temperature 30°C (86°F)			Ambient Temp. 40°C (104°F)	Wire Size AWG (kcmil)
	60°C (140°F)	75°C (167°F)	90°C (194°F)	150°C (302°F)	
	Types TW, UF	Types THWN, THHW, XHHW, RHW, THW	Types TBS, XHH, RHH, RHW-2, XHHW, USE-2, THHN, SA, THHW, THW-2, XHHW-2, ZW-2, THWN-2, SIS	Type Z	
12*	25	30	35	47	12
10*	35	40	40	63	10
8	45	55	60	83	8
6	60	75	80	112	6
4	80	100	110	148	4
3	95	115	130	170	3
2	110	135	150	198	2
1	130	155	175	228	1
1/0	150	180	205	263	1/0
2/0	175	210	235	305	2/0
3/0	200	240	275	351	3/0
4/0	235	280	315	411	4/0
250	265	315	355	—	250

Temperature Correction Factors

Ambient Temp.°C	For Other Than 30°C			40°C	Ambient Temp.°C
	Multiply the Ampacities above by the Factors Below				
21–25	1.08	1.05	1.04	.95	41–50
26–30	1.00	1.00	1.00	.90	51–60
31–35	.91	.94	.96	.85	61–70
36–40	.82	.88	.91	.80	71–80
41–45	.71	.82	.87	.74	81–90
46–50	.58	.75	.82	.67	91–100
51–55	.41	.67	.76	.52	101–120
56–60	—	.58	.71	.30	121–140
61–70	—	.33	.58	—	—
71–80	—	—	.41	—	—

*Unless specifically permitted by the NEC®, overcurrent protection for aluminum and copper-clad aluminum conductors shall not exceed 15 amps for no. 12 AWG and 25 amps for no. 10 AWG.

AMPACITIES OF ALUMINUM OR
COPPER-CLAD ALUMINUM CONDUCTORS (1) *(cont.)*

Ampacities of Single Insulated Conductors Rated 0–2000 V in Free Air

Wire Size	Ambient Temperature 30°C (86°F)			Ambient Temp. 40°C (104°F)	Wire Size
	60°C (140°F)	75°C (167°F)	90°C (194°F)	150°C (302°F)	
	Types	Types	Types	Type	
(kcmil)	TW, UF	THWN, THHW, XHHW, RHW, THW	TBS, XHH RHH, RHW-2, XHHW, USE-2, THHN, SA, THHW, THW-2, XHHW-2, ZW-2, THWN-2, SIS	Z	(kcmil)
300	290	350	395	—	300
350	330	395	445	—	350
400	355	425	480	—	400
500	405	485	545	—	500
600	455	540	615	—	600
700	500	595	675	—	700
750	515	620	700	—	750
800	535	645	725	—	800
900	580	700	785	—	900
1,000	625	750	845	—	1,000
1,250	710	855	960	—	1,250
1,500	795	950	1,075	—	1,500
1,750	875	1,050	1,185	—	1,750
2,000	960	1,150	1,335	—	2,000

Temperature Correction Factors

Ambient Temp.°C	For Other Than 30°C Multiply the Ampacities above by the Factors Below			
21–25	1.08	1.05	1.04	
26–30	1.00	1.00	1.00	
31–35	.91	.94	.96	
36–40	.82	.88	.91	
41–45	.71	.82	.87	
46–50	.58	.75	.82	
51–55	.41	.67	.76	
56–60	—	.58	.71	
61–70	—	.33	.58	
71–80	—	—	.41	

AMPACITY ADJUSTMENTS FOR 4 OR MORE CONDUCTORS IN A CABLE OR RACEWAY

Number of Current-Carrying Conductors	Percent of Values in Ampacity Charts/Adjust for Ambient Temperature (if Necessary)
4 to 6	80%
7 to 9	70%
10 to 20	50%
21 to 30	45%
31 to 40	40%
41 and more	35%

Note: For use with ampacity charts on pages 8–24 through 8–31.

AMPERAGE RATINGS FOR SINGLE-PHASE SERVICE OR FEEDER CONDUCTORS IN NORMAL DWELLING UNITS

Copper	Aluminum or Copper-Clad Aluminum	Service or Feeder Rating in Amps
4 AWG	2 AWG	100
3	1	110
2	1/0	125
1	2/0	150
1/0	3/0	175
2/0	4/0	200
3/0	250 kcmil	225
4/0	300	250
250 kcmil	350	300
350	500	350
400	600	400

MINIMUM SIZE CONDUCTORS FOR GROUNDING RACEWAY AND EQUIPMENT

Amperage Rating or Setting of Automatic Overcurrent Device Not to Exceed	Conductor Size	
	Copper	Aluminum or Copper-Clad Aluminum
15	14 AWG	12 AWG
20	12	10
30	10	8
40	10	8
60	10	8
100	8	6
200	6	4
300	4	2
400	3	1
500	2	1/0
600	1	2/0
800	1/0	3/0
1,000	2/0	4/0
1,200	3/0	250 kcmil
1,600	4/0	350
2,000	250 kcmil	400
2,500	350	600
3,000	400	600
4,000	500	800
5,000	700	1,200
6,000	800	1,200

The equipment grounding conductor shall be sized larger than this table per NEC® installation restrictions in Article 250.

Type	Size	RMC Trade Size (in.)											
		½	¾	1	1¼	1½	2	2½	3	3½	4	5	6
RHH, RHW, RHW-2	14	4	7	12	21	28	46	66	102	136	176	276	398
	12	3	6	10	17	23	38	55	85	113	146	229	330
	10	3	5	8	14	19	31	44	68	91	118	185	267
	8	1	2	4	7	10	16	23	36	48	61	97	139
	6	1	1	3	6	8	13	18	29	38	49	77	112
	4	1	1	2	4	6	10	14	22	30	38	60	87
	3	1	1	2	4	5	9	12	19	26	34	53	76
	2	1	1	1	3	4	7	11	17	23	29	46	66
	1	0	1	1	1	3	5	7	11	15	19	30	44
	1/0	0	1	1	1	2	4	6	10	13	17	26	38
	2/0	0	1	1	1	2	4	5	8	11	14	23	33
	3/0	0	1	1	1	1	3	4	7	10	12	20	28
	4/0	0	0	1	1	1	3	4	6	8	11	17	24
	250	0	0	0	1	1	1	3	4	6	8	13	18
	300	0	0	0	1	1	1	2	4	5	7	11	16
	350	0	0	0	1	1	1	2	4	5	6	10	15
	400	0	0	0	1	1	1	1	3	4	6	9	13
	500	0	0	0	1	1	1	1	3	4	5	8	11
	600	0	0	0	0	1	1	1	2	3	4	6	9
	700	0	0	0	0	1	1	1	1	3	3	6	8
	750	0	0	0	0	1	1	1	1	3	3	5	8
	800	0	0	0	0	0	1	1	1	2	3	5	7
	1,000	0	0	0	0	0	1	1	1	1	3	4	6
TW, THHW, THW, THW-2	14	9	15	25	44	59	98	140	216	288	370	581	839
	12	7	12	19	33	45	75	107	165	221	284	446	644
	10	5	9	14	25	34	56	80	123	164	212	332	480
	8	3	5	8	14	19	31	44	68	91	118	185	267
RHH*, RHW*, RHW-2*	14	6	10	17	29	39	65	93	143	191	246	387	558
	12	5	8	13	23	32	52	75	115	154	198	311	448
	10	3	6	10	18	25	41	58	90	120	154	242	350
	8	1	4	6	11	15	24	35	54	72	92	145	209
RHH*, RHW*, RHW-2*, TW, THHW, THW-2	6	1	3	5	8	11	18	27	41	55	71	111	160
	4	1	1	3	6	8	14	20	31	41	53	83	120
	3	1	1	3	5	7	12	17	26	35	45	71	103
	2	1	1	2	4	6	10	14	22	30	38	60	87
	1	1	1	1	3	4	7	10	15	21	27	42	61
	1/0	0	1	1	2	3	6	8	13	18	23	36	52
	2/0	0	1	1	2	3	5	7	11	15	19	31	44
NOTE: (*) Denotes Types Minus Outer Covering	3/0	0	1	1	1	2	4	6	9	13	16	26	37
	4/0	0	0	1	1	2	3	5	8	10	14	21	31
	250	0	0	1	1	1	3	4	6	8	11	17	25
	300	0	0	1	1	1	2	3	5	7	9	15	22
	350	0	0	0	1	1	1	3	5	6	8	13	19
	400	0	0	0	1	1	1	3	4	6	7	12	17
	500	0	0	0	1	1	1	2	3	5	6	10	14
	600	0	0	0	1	1	1	1	3	4	5	8	12
	700	0	0	0	0	1	1	1	2	3	4	7	10
	750	0	0	0	0	1	1	1	2	3	4	7	10
	800	0	0	0	0	1	1	1	2	3	4	6	9
	1,000	0	0	0	0	0	1	1	1	2	3	5	8
THHN, THWN, THWN-2	14	13	22	36	63	85	140	200	309	412	531	833	1,202
	12	9	16	26	46	62	102	146	225	301	387	608	877
	10	6	10	17	29	39	64	92	142	189	244	383	552
	8	3	6	9	16	22	37	53	82	109	140	221	318
	6	2	4	7	12	16	27	38	59	79	101	159	230
	4	1	2	4	7	10	16	23	36	48	62	98	141
	3	1	1	3	6	8	14	20	31	41	53	83	120
	2	1	1	3	5	7	11	17	26	34	44	70	100
	1	1	1	1	4	5	8	12	19	25	33	51	74

RIGID METALLIC CONDUIT—MAXIMUM NUMBER OF CONDUCTORS (cont.)

Type	Size	RMC Trade Size (in.)											
		½	¾	1	1¼	1½	2	2½	3	3½	4	5	6
THHN,	1/0	1	1	1	3	4	7	10	16	21	27	43	63
THWN,	2/0	0	1	1	2	3	6	8	13	18	23	36	52
THWN-2	3/0	0	1	1	1	3	5	7	11	15	19	30	43
(cont.)	4/0	0	1	1	1	2	4	6	9	12	16	25	36
	250	0	0	1	1	1	3	5	7	10	13	20	29
	300	0	0	1	1	1	3	4	6	8	11	17	25
	350	0	0	1	1	1	2	3	5	7	10	15	22
	400	0	0	1	1	1	2	3	5	7	8	13	20
	500	0	0	0	1	1	1	2	4	5	7	11	16
	600	0	0	0	1	1	1	1	3	4	6	9	13
	700	0	0	0	1	1	1	1	3	4	5	8	11
	750	0	0	0	0	1	1	1	3	4	5	7	11
	800	0	0	0	0	1	1	1	2	3	4	7	10
	1,000	0	0	0	0	0	1	1	1	3	4	6	8
FEP, FEPB,	14	12	22	35	61	83	136	194	300	400	515	808	1,166
PFA, PFAH,	12	9	16	26	44	60	99	142	219	292	376	590	851
TFE	10	6	11	18	32	43	71	102	157	209	269	423	610
	8	3	6	10	18	25	41	58	90	120	154	242	350
	6	2	4	7	13	17	29	41	64	85	110	172	249
	4	1	3	5	9	12	20	29	44	59	77	120	174
	3	1	2	4	7	10	17	24	37	50	64	100	145
	2	1	1	3	6	8	14	20	31	41	53	83	120
PFA, PFAH, TFE	1	1	1	2	4	6	9	14	21	28	37	57	83
PFA, PFAH,	1/0	1	1	1	3	5	8	11	18	24	30	48	69
TFE, Z	2/0	1	1	1	3	4	6	9	14	19	25	40	57
	3/0	0	1	1	2	3	5	8	12	16	21	33	47
	4/0	0	1	1	1	2	4	6	10	13	17	27	39
Z	14	15	26	42	73	100	164	234	361	482	621	974	1,405
	12	10	18	30	52	71	116	166	256	342	440	691	997
	10	6	11	18	32	43	71	102	157	209	269	423	610
	8	4	7	11	20	27	45	64	99	132	170	267	386
	6	3	5	8	14	19	31	45	69	93	120	188	271
	4	1	3	5	9	13	22	31	48	64	82	129	186
	3	1	2	4	7	9	16	22	35	47	60	94	136
	2	1	1	3	6	8	13	19	29	39	50	78	113
	1	1	1	2	5	6	10	15	23	31	40	63	92
XHH,	14	9	15	25	44	59	98	140	216	288	370	581	839
XHHW,	12	7	12	19	33	45	75	107	165	221	284	446	644
XHHW-2,	10	5	9	14	25	34	56	80	123	164	212	332	480
ZW	8	3	5	8	14	19	31	44	68	91	118	185	267
	6	1	3	6	10	14	23	33	51	68	87	137	197
	4	1	2	4	7	10	16	24	37	49	63	99	143
	3	1	1	3	6	8	14	20	31	41	53	84	121
	2	1	1	3	5	7	12	17	26	35	45	70	101
XHH,	1	1	1	1	4	5	9	12	19	26	33	52	76
XHHW,	1/0	1	1	1	3	4	7	10	16	22	28	44	64
XHHW-2	2/0	0	1	1	2	3	6	9	13	18	23	37	53
	3/0	0	1	1	1	3	5	7	11	15	19	30	44
	4/0	0	1	1	1	2	4	6	9	12	16	25	36
	250	0	0	1	1	1	3	5	7	10	13	20	30
	300	0	0	1	1	1	3	4	6	9	11	18	25
	350	0	0	1	1	1	2	3	6	7	10	15	22
	400	0	0	1	1	1	2	3	5	7	9	14	20
	500	0	0	0	1	1	1	2	4	5	7	11	16
	600	0	0	0	1	1	1	1	3	4	6	9	13
	700	0	0	0	1	1	1	1	3	4	5	8	11
	750	0	0	0	0	1	1	1	3	4	5	7	11
	800	0	0	0	0	1	1	1	2	3	4	7	10

LIQUIDTIGHT FLEXIBLE METALLIC CONDUIT – MAXIMUM NUMBER OF CONDUCTORS

Type	Size	LFMC Trade Size (in.)									
		½	¾	1	1¼	1½	2	2½	3	3½	4
RHH, RHW, RHW-2	14	4	7	12	21	27	44	66	102	133	173
	12	3	6	10	17	22	36	55	84	110	144
	10	3	5	8	14	18	29	44	68	89	116
	8	1	2	4	7	9	15	23	36	46	61
	6	1	1	3	6	7	12	18	28	37	48
	4	1	1	2	4	6	9	14	22	29	38
	3	1	1	1	4	5	8	13	19	25	33
	2	1	1	1	3	4	7	11	17	22	29
	1	0	1	1	1	3	5	7	11	14	19
	1/0	0	1	1	1	2	4	6	10	13	16
	2/0	0	1	1	1	1	3	5	8	11	14
	3/0	0	0	1	1	1	3	4	7	9	12
	4/0	0	0	1	1	1	2	4	6	8	10
	250	0	0	0	1	1	1	3	4	6	8
	300	0	0	0	1	1	1	2	4	5	7
	350	0	0	0	1	1	1	2	3	5	6
	400	0	0	0	1	1	1	1	3	4	6
	500	0	0	0	1	1	1	1	3	4	5
	600	0	0	0	0	1	1	1	2	3	4
	700	0	0	0	0	1	1	1	1	3	3
	750	0	0	0	0	1	1	1	1	2	3
	800	0	0	0	0	1	1	1	1	2	3
	1,000	0	0	0	0	0	1	1	1	1	3
TW, THHW, THW, THW-2	14	9	15	25	44	57	93	140	215	280	365
	12	7	12	19	33	43	71	108	165	215	280
	10	5	9	14	25	32	53	80	123	160	209
	8	3	5	8	14	18	29	44	68	89	116
RHH*, RHW*, RHW-2*	14	6	10	16	29	38	62	93	143	186	243
	12	5	8	13	23	30	50	75	115	149	195
	10	3	6	10	18	23	39	58	90	117	152
	8	1	4	6	11	14	23	35	53	70	91
RHH*, RHW*, RHW-2*, TW, THHW, THW-2	6	1	3	5	8	11	18	27	41	53	70
	4	1	1	3	6	8	13	20	30	40	52
	3	1	1	3	5	7	11	17	26	34	44
	2	1	1	2	4	6	9	14	22	29	38
	1	1	1	1	3	4	7	10	15	20	26
	1/0	0	1	1	2	3	6	8	13	17	23
	2/0	0	1	1	2	3	5	7	11	15	19
Note:	3/0	0	1	1	1	2	4	6	9	12	16
(٠) Denotes	4/0	0	0	1	1	1	3	5	8	10	13
types minus	250	0	0	1	1	1	3	4	6	8	11
outer	300	0	0	1	1	1	2	3	5	7	9
covering	350	0	0	0	1	1	1	3	5	6	8
	400	0	0	0	1	1	1	3	4	6	7
	500	0	0	0	1	1	1	2	3	5	6
	600	0	0	0	1	1	1	1	3	4	5
	700	0	0	0	1	1	1	1	2	3	4
	750	0	0	0	0	1	1	1	2	3	4
	800	0	0	0	0	1	1	1	2	3	4
	1,000	0	0	0	0	0	1	1	1	2	3
THHN, THWN, THWN-2	14	13	22	36	63	81	133	201	308	401	523
	12	9	16	26	46	59	97	146	225	292	381
	10	6	10	16	29	37	61	92	141	184	240
	8	3	6	9	16	21	35	53	81	106	138
	6	2	4	7	12	15	25	38	59	76	100
	4	1	2	4	7	9	15	23	36	47	61
	3	1	1	3	6	8	13	20	30	40	52
	2	1	1	3	5	7	11	17	26	33	44
	1	1	1	1	4	5	8	12	19	25	32

Type	Size	½	¾	1	1¼	1½	2	2½	3	3½	4
THHN, THWN, THWN-2 (cont.)	1/0	1	1	1	3	4	7	10	16	21	27
	2/0	0	1	1	2	3	6	8	13	17	23
	3/0	0	1	1	1	3	5	7	11	14	19
	4/0	0	0	1	1	2	4	6	9	12	15
	250	0	0	1	1	1	3	5	7	10	12
	300	0	0	1	1	1	3	4	6	8	11
	350	0	0	1	1	1	2	3	5	7	9
	400	0	0	0	1	1	1	3	5	6	8
	500	0	0	0	1	1	1	2	4	5	7
	600	0	0	0	1	1	1	1	3	4	6
	700	0	0	0	1	1	1	1	3	4	5
	750	0	0	0	1	1	1	1	3	3	5
	800	0	0	0	0	1	1	1	2	3	4
	1,000	0	0	0	0	0	1	1	1	3	3
FEP, FEPB, PFA, PFAH, TFE	14	12	21	35	61	79	129	195	299	389	507
	12	9	15	25	44	57	94	142	218	284	370
	10	6	11	18	32	41	68	102	156	203	266
	8	3	6	10	18	23	39	58	89	117	152
	6	2	4	7	13	17	27	41	64	83	108
	4	1	3	5	9	12	19	29	44	58	75
	3	1	2	4	7	10	16	24	37	48	63
	2	1	1	3	6	8	13	20	30	40	52
PFA, PFAH, TFE	1	1	1	2	4	5	9	14	21	28	36
PFA, PFAH, TFE, Z	1/0	1	1	1	3	4	7	11	18	23	30
	2/0	1	1	1	3	4	6	9	14	19	25
	3/0	1	1	2	2	3	5	8	12	16	20
	4/0	0	1	1	1	2	4	6	10	13	17
Z	14	20	26	42	73	95	156	235	360	469	611
	12	14	18	30	52	67	111	167	255	332	434
	10	8	11	18	32	41	68	102	156	203	266
	8	5	7	11	20	26	43	64	99	129	168
	6	4	5	8	14	18	30	45	69	90	118
	4	2	3	5	9	12	20	31	48	62	81
	3	2	2	4	7	9	15	23	35	45	59
	2	1	1	3	6	7	12	19	29	38	49
	1	1	1	2	5	6	10	15	23	30	40
XHH, XHHW, XHHW-2, ZW	14	9	15	25	44	57	93	140	215	280	366
	12	7	12	19	33	43	71	108	165	215	280
	10	5	9	14	25	32	53	80	123	160	209
	8	3	5	8	14	18	29	44	68	89	116
	6	1	3	6	10	13	22	33	50	66	86
	4	1	2	4	7	9	16	24	36	48	62
	3	1	1	3	6	8	13	20	31	40	52
	2	1	1	3	5	7	11	17	26	34	44
XHH, XHHW, XHHW-2	1	1	1	1	4	5	8	12	19	25	33
	1/0	1	1	1	3	4	7	10	16	21	28
	2/0	0	1	1	2	3	6	9	13	17	23
	3/0	0	1	1	1	3	5	7	11	14	19
	4/0	0	1	1	1	2	4	6	9	12	16
	250	0	0	1	1	1	3	5	7	10	13
	300	0	0	1	1	1	3	4	6	8	11
	350	0	0	1	1	1	2	3	5	7	10
	400	0	0	0	1	1	1	3	5	6	8
	500	0	0	0	1	1	1	2	4	5	7
	600	0	0	0	1	1	1	1	3	4	6
	700	0	0	0	1	1	1	1	3	4	5
	750	0	0	0	0	1	1	1	3	3	5
	800	0	0	0	0	1	1	1	2	3	4

NONMETALLIC TUBING–MAXIMUM NUMBER OF CONDUCTORS

Type	Size	ENT Trade Size (in.)					
		½	¾	1	1¼	1½	2
RHH, RHW, RHW-2	14	3	6	10	19	26	43
	12	2	5	9	16	22	36
	10	1	4	7	13	17	29
	8	1	1	3	6	9	15
	6	1	1	3	5	7	12
	4	1	1	2	4	6	9
	3	1	1	1	3	5	8
	2	0	1	1	3	4	7
	1	0	1	1	1	3	5
	1/0	0	0	1	1	2	4
	2/0	0	0	1	1	1	3
	3/0	0	0	1	1	1	3
	4/0	0	0	1	1	1	2
	250	0	0	0	1	1	1
	300	0	0	0	1	1	1
	350	0	0	0	1	1	1
	400	0	0	0	1	1	1
	500	0	0	0	1	1	1
	600	0	0	0	0	1	1
	700	0	0	0	0	1	1
	750	0	0	0	0	0	1
	800	0	0	0	0	0	1
	1,000	0	0	0	0	0	1
TW, THHW, THW,THW-2	14	7	13	22	40	55	92
	12	5	10	17	31	42	71
	10	4	7	13	23	32	52
	8	1	4	7	13	17	29
RHH*, RHW*, RHW-2*	14	4	8	15	27	37	61
	12	3	7	12	21	29	49
	10	3	5	9	17	23	38
	8	1	3	5	10	14	23
RHH*, RHW*, RHW-2*, TW, THW, THHW, THW-2	6	1	2	4	7	10	17
	4	1	1	3	5	8	13
	3	1	1	2	5	7	11
	2	1	1	2	4	6	9
	1	0	1	1	3	4	6
	1/0	0	1	1	2	3	5
	2/0	0	1	1	1	3	5
NOTE:	3/0	0	0	1	1	2	4
(•) Denotes	4/0	0	0	1	1	1	3
types minus	250	0	0	1	1	1	2
outer	300	0	0	0	1	1	2
covering	350	0	0	0	1	1	1
	400	0	0	0	1	1	1
	500	0	0	0	1	1	1
	600	0	0	0	0	1	1
	700	0	0	0	0	1	1
	750	0	0	0	0	1	1
	800	0	0	0	0	1	1
	1,000	0	0	0	0	0	1
THHN, THWN, THWN-2	14	10	18	32	58	80	132
	12	7	13	23	42	58	96
	10	4	8	15	26	36	60
	8	2	5	8	15	21	35
	6	1	3	6	11	15	25
	4	1	1	4	7	9	15
	3	1	1	3	5	8	13
	2	1	1	2	5	6	11
	1	1	1	1	3	5	8

NONMETALLIC TUBING—MAXIMUM NUMBER OF CONDUCTORS *(cont.)*

Type	Size	½	¾	1	1¼	1½	2
				ENT Trade Size (in.)			
THHN,	1/0	0	1	1	3	4	7
THWN,	2/0	0	1	1	2	3	5
THWN-2	3/0	0	1	1	1	3	4
(cont.)	4/0	0	0	1	1	2	4
	250	0	0	1	1	1	3
	300	0	0	1	1	1	2
	350	0	0	0	1	1	2
	400	0	0	0	1	1	1
	500	0	0	0	1	1	1
	600	0	0	0	1	1	1
	700	0	0	0	0	1	1
	750	0	0	0	0	1	1
	800	0	0	0	0	1	1
	1,000	0	0	0	0	0	1
FEP, FEPB,	14	10	18	31	56	77	128
PFA, PFAH,	12	7	13	23	41	56	93
TFE	10	5	9	16	29	40	67
	8	3	5	9	17	23	38
	6	1	4	6	12	16	27
	4	1	2	4	8	11	19
	3	1	1	4	7	9	16
	2	1	1	3	5	8	13
PFA, PFAH, TFE	1	1	1	1	4	5	9
PFA, PFAH,	1/0	0	1	1	3	4	7
TFE, Z	2/0	0	1	1	2	4	6
	3/0	0	1	1	1	3	5
	4/0	0	1	1	1	2	4
Z	14	12	22	38	68	93	154
	12	8	15	27	48	66	109
	10	5	9	16	29	40	67
	8	3	6	10	18	25	42
	6	1	4	7	13	18	30
	4	1	3	5	9	12	20
	3	1	1	3	6	9	15
	2	1	1	3	5	7	12
	1	1	1	2	4	6	10
XHH,	14	7	13	22	40	55	92
XHHW,	12	5	10	17	31	42	71
XHHW-2,	10	4	7	13	23	32	52
ZW	8	1	4	7	13	17	29
	6	1	3	5	9	13	21
	4	1	1	4	7	9	15
	3	1	1	3	6	8	13
	2	1	1	2	5	6	11
XHH,	1	1	1	1	3	5	8
XHHW,	1/0	0	1	1	3	4	7
XHHW-2	2/0	0	1	1	2	3	6
	3/0	0	1	1	1	3	5
	4/0	0	0	1	1	2	4
	250	0	0	1	1	1	3
	300	0	0	1	1	1	3
	350	0	0	1	1	1	2
	400	0	0	0	1	1	1
	500	0	0	0	1	1	1
	600	0	0	0	1	1	1
	700	0	0	0	0	1	1
	750	0	0	0	0	1	1
	800	0	0	0	0	1	1

ELECTRICAL METALLIC TUBING—MAXIMUM NUMBER OF CONDUCTORS

Type	Size	EMT Trade Size (in.)									
		½	¾	1	1¼	1½	2	2½	3	3½	4
RHH, RHW, RHW-2	14	4	7	11	20	27	46	80	120	157	201
	12	3	6	9	17	23	38	66	100	131	167
	10	2	5	8	13	18	30	53	81	105	135
	8	1	2	4	7	9	16	28	42	55	70
	6	1	1	3	5	8	13	22	34	44	56
	4	1	1	2	4	6	10	17	26	34	44
	3	1	1	1	4	5	9	15	23	30	38
	2	1	1	1	3	4	7	13	20	26	33
	1	0	1	1	1	3	5	9	13	17	22
	1/0	0	1	1	1	2	4	7	11	15	19
	2/0	0	1	1	1	2	4	6	10	13	17
	3/0	0	0	1	1	1	3	5	8	11	14
	4/0	0	0	1	1	1	3	5	7	9	12
	250	0	0	0	1	1	1	3	5	7	9
	300	0	0	0	1	1	1	3	5	6	8
	350	0	0	0	1	1	1	3	4	6	7
	400	0	0	0	1	1	1	2	4	5	7
	500	0	0	0	0	1	1	2	3	5	6
	600	0	0	0	0	1	1	1	3	4	5
	700	0	0	0	0	0	1	1	2	3	4
	750	0	0	0	0	0	1	1	2	3	4
	800	0	0	0	0	0	1	1	2	3	4
	1,000	0	0	0	0	0	1	1	1	2	3
TW, THHW, THW, THW-2	14	8	15	25	43	58	96	168	254	332	424
	12	6	11	19	33	45	74	129	195	255	326
	10	5	8	14	24	33	55	96	145	190	243
	8	2	5	8	13	18	30	53	81	105	135
RHH*, RHW*, RHW-2*	14	6	10	16	28	39	64	112	169	221	282
	12	4	8	13	23	31	51	90	136	177	227
	10	3	6	10	18	24	40	70	106	138	177
	8	1	4	6	10	14	24	42	63	83	106
RHH*, RHW*, RHW-2*, TW, THHW, THW-2	6	1	3	4	8	11	18	32	48	63	81
	4	1	1	3	6	8	13	24	36	47	60
	3	1	1	3	5	7	12	20	31	40	52
	2	1	1	2	4	6	10	17	26	34	44
	1	1	1	1	3	4	7	12	18	24	31
	1/0	0	1	1	2	3	6	10	16	20	26
	2/0	0	1	1	1	3	5	9	13	17	22
	3/0	0	1	1	1	2	4	7	11	15	19
	4/0	0	0	1	1	1	3	6	9	12	16
	250	0	0	1	1	1	3	5	7	10	13
	300	0	0	1	1	1	2	4	6	8	11
	350	0	0	0	1	1	1	4	6	7	10
	400	0	0	0	1	1	1	3	5	7	9
	500	0	0	0	1	1	1	3	4	6	7
	600	0	0	0	1	1	1	2	3	4	6
	700	0	0	0	0	1	1	1	3	4	5
	750	0	0	0	0	1	1	1	3	4	5
	800	0	0	0	0	1	1	1	3	3	5
	1,000	0	0	0	0	0	1	1	2	3	4
THHN, THWN, THWN-2	14	12	22	35	61	84	138	241	364	476	608
	12	9	16	26	45	61	101	176	266	347	443
	10	5	10	16	28	38	63	111	167	219	279
	8	3	6	9	16	22	36	64	96	126	161
	6	2	4	7	12	16	26	46	69	91	116
	4	1	2	4	7	10	16	28	42	56	71
	3	1	1	3	6	8	13	24	36	47	60
	2	1	1	3	5	7	11	20	30	40	51
	1	1	1	1	4	5	8	15	22	29	37

NOTE:
(·) Denotes types minus outer covering

ELECTRICAL METALLIC TUBING - MAXIMUM NUMBER OF CONDUCTORS (cont.)

Type	Size	EMT Trade Size (in.)									
		½	¾	1	1¼	1½	2	2½	3	3½	4
THHN, THWN, THWN-2 (cont.)	1/0	1	1	1	3	4	7	12	19	25	32
	2/0	0	1	1	2	3	6	10	16	20	26
	3/0	0	1	1	1	3	5	8	13	17	22
	4/0	0	1	1	1	2	4	7	11	14	18
	250	0	0	1	1	1	3	6	9	11	15
	300	0	0	1	1	1	3	5	7	10	13
	350	0	0	1	1	1	2	4	6	9	11
	400	0	0	0	1	1	1	4	6	8	10
	500	0	0	0	1	1	1	3	5	6	8
	600	0	0	0	1	1	1	2	4	5	7
	700	0	0	0	1	1	1	2	3	4	6
	750	0	0	0	0	1	1	1	3	4	5
	800	0	0	0	0	1	1	1	3	4	5
	1,000	0	0	0	0	1	1	1	2	3	4
FEP, FEPB, PFA, PFAH, TFE	14	12	21	34	60	81	134	234	354	462	590
	12	9	15	25	43	59	98	171	258	337	430
	10	6	11	18	31	42	70	122	185	241	309
	8	3	6	10	18	24	40	70	106	138	177
	6	2	4	7	12	17	28	50	75	98	126
	4	1	3	5	9	12	20	35	53	69	88
	3	1	2	4	7	10	16	29	44	57	73
	2	1	1	3	6	8	13	24	36	47	60
PFA, PFAH, TFE	1	1	1	2	4	6	9	16	25	33	42
PFA, PFAH, TFE, Z	1/0	1	1	1	3	5	8	14	21	27	35
	2/0	0	1	1	3	4	6	11	17	22	29
	3/0	0	1	1	2	3	5	9	14	18	24
	4/0	0	1	1	1	2	4	8	11	15	19
Z	14	14	25	41	72	98	161	282	426	556	711
	12	10	18	29	51	69	114	200	302	394	504
	10	6	11	18	31	42	70	122	185	241	309
	8	4	7	11	20	27	44	77	117	153	195
	6	3	5	8	14	19	31	54	82	107	137
	4	1	3	5	9	13	21	37	56	74	94
	3	1	2	4	7	9	15	27	41	54	69
	2	1	1	3	6	8	13	22	34	45	57
	1	1	1	2	4	6	10	18	28	36	46
XHH, XHHW, XHHW-2, ZW	14	8	15	25	43	58	96	168	254	332	424
	12	6	11	19	33	45	74	129	195	255	326
	10	5	8	14	24	33	55	96	145	190	243
	8	2	5	8	13	18	30	53	81	105	135
	6	1	3	6	10	14	22	39	60	78	100
	4	1	2	4	7	10	16	28	43	56	72
	3	1	1	3	6	8	14	24	36	48	61
	2	1	1	3	5	7	11	20	31	40	51
XHH, XHHW, XHHW-2	1	1	1	1	4	5	8	15	23	30	38
	1/0	1	1	1	3	4	7	13	19	25	32
	2/0	0	1	1	2	3	6	10	16	21	27
	3/0	0	1	1	1	3	5	9	13	17	22
	4/0	0	1	1	1	2	4	7	11	14	18
	250	0	0	1	1	1	3	6	9	12	15
	300	0	0	1	1	1	3	5	8	10	13
	350	0	0	1	1	1	2	4	7	9	11
	400	0	0	0	1	1	1	4	6	8	10
	500	0	0	0	1	1	1	3	5	6	8
	600	0	0	0	1	1	1	2	4	5	6
	700	0	0	0	0	1	1	2	3	4	6
	750	0	0	0	0	1	1	1	3	4	5
	800	0	0	0	0	1	1	1	3	4	5

FLEXIBLE METALLIC CONDUIT–MAXIMUM NUMBER OF CONDUCTORS

Type	Size	½	¾	1	1¼	1½	2	2½	3	3½	4
RHH, RHW, RHW-2	14	4	7	11	17	25	44	67	96	131	171
	12	3	6	9	14	21	37	55	80	109	142
	10	3	5	7	11	17	30	45	64	88	115
	8	1	2	4	6	9	15	23	34	46	60
	6	1	1	3	5	7	12	19	27	37	48
	4	1	1	2	4	5	10	14	21	29	37
	3	1	1	1	3	5	8	13	18	25	33
	2	1	1	1	3	4	7	11	16	22	28
	1	0	1	1	1	2	5	7	10	14	19
	1/0	0	1	1	1	2	4	6	9	12	16
	2/0	0	1	1	1	1	3	5	8	11	14
	3/0	0	0	1	1	1	3	5	7	9	12
	4/0	0	0	1	1	1	2	4	6	8	10
	250	0	0	0	1	1	1	3	4	6	8
	300	0	0	0	1	1	1	2	4	5	7
	350	0	0	0	1	1	1	2	3	5	6
	400	0	0	0	1	1	1	1	3	4	6
	500	0	0	0	0	1	1	1	3	4	5
	600	0	0	0	0	1	1	1	2	3	4
	700	0	0	0	0	0	1	1	1	3	3
	750	0	0	0	0	0	1	1	1	2	3
	800	0	0	0	0	0	1	1	1	2	3
	1,000	0	0	0	0	0	1	1	1	2	3
TW, THHW, THW, THW-2	14	9	15	23	36	53	94	141	203	277	361
	12	7	11	18	28	41	72	108	156	212	277
	10	5	8	13	21	30	54	81	116	158	207
	8	3	5	7	11	17	30	45	64	88	115
RHH*, RHW*, RHW-2*	14	6	10	15	24	35	62	94	135	184	240
	12	5	8	12	19	28	50	75	108	148	193
	10	4	6	10	15	22	39	59	85	115	151
	8	1	4	6	9	13	23	35	51	69	90
RHH*, RHW*, RHW-2*, TW, THHW, THW, THW-2	6	1	3	4	7	10	18	27	39	53	69
	4	1	1	3	5	7	13	20	29	39	51
	3	1	1	3	4	6	11	17	25	34	44
	2	1	1	2	4	5	10	14	21	29	37
	1	1	1	1	2	4	7	10	15	20	26
	1/0	0	1	1	1	3	6	9	12	17	22
	2/0	0	1	1	1	3	5	7	10	14	19
Note:	3/0	0	1	1	1	2	4	6	9	12	16
(*) Denotes	4/0	0	0	1	1	1	3	5	7	10	13
types minus	250	0	0	1	1	1	3	4	6	8	11
outer	300	0	0	1	1	1	2	3	5	7	9
covering	350	0	0	0	1	1	1	3	4	6	8
	400	0	0	0	1	1	1	3	4	6	7
	500	0	0	0	1	1	1	2	3	5	6
	600	0	0	0	0	1	1	1	3	4	5
	700	0	0	0	0	1	1	1	2	3	4
	750	0	0	0	0	1	1	1	2	3	4
	800	0	0	0	0	1	1	1	1	3	4
	1,000	0	0	0	0	0	1	1	1	2	3
THHN, THWN, THWN-2	14	13	22	33	52	76	134	202	291	396	518
	12	9	16	24	38	56	98	147	212	289	378
	10	6	10	15	24	35	62	93	134	182	238
	8	3	6	9	14	20	35	53	77	105	137
	6	2	4	6	10	14	25	38	55	76	99
	4	1	2	4	6	9	16	24	34	46	61
	3	1	1	3	5	7	13	20	29	39	51
	2	1	1	3	4	6	11	17	24	33	43
	1	1	1	1	3	4	8	12	18	24	32

FLEXIBLE METALLIC CONDUIT — MAXIMUM NUMBER OF CONDUCTORS (cont.)

Type	AWG/kcmil	½	¾	1	1¼	1½	2	2½	3	3½	4
THHN, THWN, THWN-2 (cont.)	1/0	1	1	1	2	4	7	10	15	20	27
	2/0	0	1	1	1	3	6	9	12	17	22
	3/0	0	1	1	1	2	5	7	10	14	18
	4/0	0	1	1	1	1	4	6	8	12	15
	250	0	0	1	1	1	3	5	7	9	12
	300	0	0	1	1	1	3	4	6	8	11
	350	0	0	1	1	1	2	3	5	7	9
	400	0	0	0	1	1	1	3	5	6	8
	500	0	0	0	1	1	1	2	4	5	7
	600	0	0	0	0	1	1	1	3	4	5
	700	0	0	0	0	1	1	1	3	4	5
	750	0	0	0	0	1	1	1	2	3	4
	800	0	0	0	0	1	1	1	2	3	4
	1,000	0	0	0	0	0	1	1	1	3	3
FEP, FEPB, PFA, PFAH, TFE	14	12	21	32	51	74	130	196	282	385	502
	12	9	15	24	37	54	95	143	206	281	367
	10	6	11	17	26	39	68	103	148	201	263
	8	4	6	10	15	22	39	59	85	115	151
	6	2	4	7	11	16	28	42	60	82	107
	4	1	3	5	7	11	19	29	42	57	75
	3	1	2	4	6	9	16	24	35	48	62
	2	1	1	3	5	7	13	20	29	39	51
PFA, PFAH, TFE	1	1	1	2	3	5	9	14	20	27	36
PFA, PFAH, TFE, Z	1/0	1	1	1	3	4	8	11	17	23	30
	2/0	1	1	1	2	3	6	9	14	19	24
	3/0	0	1	1	1	3	5	8	11	15	20
	4/0	0	1	1	1	2	4	6	9	13	16
Z	14	15	25	39	61	89	157	236	340	463	605
	12	11	18	28	43	63	111	168	241	329	429
	10	6	11	17	26	39	68	103	148	201	263
	8	4	7	11	17	24	43	65	93	127	166
	6	3	5	7	12	17	30	45	65	89	117
	4	1	3	5	8	12	21	31	45	61	80
	3	1	2	4	6	8	15	23	33	45	58
	2	1	1	3	5	7	12	19	27	37	49
	1	1	1	2	4	6	10	15	22	30	39
XHH, XHHW, XHHW-2, ZW	14	9	15	23	36	53	94	141	203	277	361
	12	7	11	18	28	41	72	108	156	212	277
	10	5	8	13	21	30	54	81	116	158	207
	8	3	5	7	11	17	30	45	64	88	115
	6	1	3	5	8	12	22	33	48	65	85
	4	1	2	4	6	9	16	24	34	47	61
	3	1	1	3	5	7	13	20	29	40	52
	2	1	1	2	4	6	11	17	24	33	44
XHH, XHHW, XHHW-2	1	1	1	1	3	5	8	13	18	25	32
	1/0	1	1	1	2	4	7	10	15	21	27
	2/0	0	1	1	1	3	6	9	13	17	23
	3/0	0	1	1	1	3	5	7	10	14	19
	4/0	0	1	1	1	2	4	6	9	12	15
	250	0	0	1	1	1	3	5	7	10	13
	300	0	0	1	1	1	3	4	6	8	11
	350	0	0	1	1	1	2	4	5	7	9
	400	0	0	0	1	1	1	3	5	6	8
	500	0	0	0	1	1	1	3	4	5	7
	600	0	0	0	0	1	1	1	3	4	5
	700	0	0	0	0	1	1	1	3	4	5
	750	0	0	0	0	1	1	1	2	3	4
	800	0	0	0	0	1	1	1	2	3	4

RIGID PVC SCHEDULE 40 CONDUIT – MAXIMUM NUMBER OF CONDUCTORS

Type	Size	½	¾	1	1¼	1½	2	2½	3	3½	4	5	6
							PVC 40 Trade Size (in.)						
RHH, RHW, RHW-2	14	4	7	11	20	27	45	64	99	133	171	269	390
	12	3	5	9	16	22	37	53	82	110	142	224	323
	10	2	4	7	13	18	30	43	66	89	115	181	261
	8	1	2	4	7	9	15	22	35	46	60	94	137
	6	1	1	3	5	7	12	18	28	37	48	76	109
	4	1	1	2	4	6	10	14	22	29	37	59	85
	3	1	1	1	4	5	8	12	19	25	33	52	75
	2	1	1	1	3	4	7	10	16	22	28	45	65
	1	0	1	1	1	3	5	7	11	14	19	29	43
	1/0	0	1	1	1	2	4	6	9	13	16	26	37
	2/0	0	0	1	1	1	3	5	8	11	14	22	32
	3/0	0	0	1	1	1	3	4	7	9	12	19	28
	4/0	0	0	1	1	1	2	4	6	8	10	16	24
	250	0	0	0	1	1	1	3	4	6	8	12	18
	300	0	0	0	1	1	1	2	4	5	7	11	16
	350	0	0	0	1	1	1	2	3	5	6	10	14
	400	0	0	0	1	1	1	1	3	4	6	9	13
	500	0	0	0	0	1	1	1	3	4	5	8	11
	600	0	0	0	0	1	1	1	2	3	4	6	9
	700	0	0	0	0	0	1	1	1	3	3	6	8
	750	0	0	0	0	0	1	1	1	2	3	5	8
	800	0	0	0	0	0	1	1	1	2	3	5	7
	1,000	0	0	0	0	0	1	1	1	1	2	4	6
TW, THHW, THW, THW-2	14	8	14	24	42	57	94	135	209	280	361	568	822
	12	6	11	18	32	44	72	103	160	215	277	436	631
	10	4	8	13	24	32	54	77	119	160	206	325	470
	8	2	4	7	13	18	30	43	66	89	115	181	261
RHH*, RHW*, RHW-2*	14	5	9	16	28	38	63	90	139	186	240	378	546
	12	4	8	12	22	30	50	72	112	150	193	304	439
	10	3	6	10	17	24	39	56	87	117	150	237	343
	8	1	3	6	10	14	23	33	52	70	90	142	205
RHH*, RHW*, RHW-2*, TW, THW, THHW, THW-2	6	1	2	4	8	11	18	26	40	53	69	109	157
	4	1	1	3	6	8	13	19	30	40	51	81	117
	3	1	1	3	5	7	11	16	25	34	44	69	100
	2	1	1	2	4	6	10	14	22	29	37	59	85
	1	0	1	1	3	4	7	10	15	20	26	41	60
	1/0	0	1	1	2	3	6	8	13	17	22	35	51
	2/0	0	1	1	1	3	5	7	11	15	19	30	43
	3/0	0	1	1	1	2	4	6	9	12	16	25	36
	4/0	0	0	1	1	1	3	5	8	10	13	21	30
	250	0	0	1	1	1	3	4	6	8	11	17	25
	300	0	0	1	1	1	2	3	5	7	9	15	21
	350	0	0	1	1	1	1	3	5	6	8	13	19
	400	0	0	0	1	1	1	3	4	6	7	12	17
	500	0	0	0	1	1	1	2	3	5	6	10	14
	600	0	0	0	1	1	1	1	3	4	5	8	11
	700	0	0	0	0	1	1	1	3	3	4	7	10
	750	0	0	0	0	1	1	1	2	3	4	6	10
	800	0	0	0	0	1	1	1	2	3	4	6	9
	1,000	0	0	0	0	0	1	1	1	2	3	4	7
THHN, THWN, THWN-2	14	11	21	34	60	82	135	193	299	401	517	815	1,178
	12	8	15	25	43	59	99	141	218	293	377	594	859
	10	5	9	15	27	37	62	89	137	184	238	374	541
	8	3	5	9	16	21	36	51	79	106	137	216	312
	6	1	4	6	11	15	26	37	57	77	99	156	225
	4	1	2	4	7	9	16	22	35	47	61	96	138
	3	1	1	3	6	8	13	19	30	40	51	81	117
	2	1	1	3	5	7	11	16	25	33	43	68	98
	1	1	1	1	3	5	8	12	18	25	32	50	73

Note: (-) Denotes types minus outer covering

RIGID PVC SCHEDULE 40 CONDUIT – MAXIMUM NUMBER OF CONDUCTORS *(cont.)*

Type	Size	½	¾	1	1¼	1½	2	2½	3	3½	4	5	6
						PVC 40 Trade Size (in.)							
THHN, THWN, THWN-2 *(cont.)*	1/0	1	1	1	3	4	7	10	15	21	27	42	61
	2/0	0	1	1	2	3	6	8	13	17	22	35	51
	3/0	0	1	1	1	3	5	7	11	14	18	29	42
	4/0	0	1	1	1	2	4	6	9	12	15	24	35
	250	0	0	1	1	1	3	4	7	10	12	20	28
	300	0	0	1	1	1	3	4	6	8	11	17	24
	350	0	0	1	1	1	2	3	5	7	9	15	21
	400	0	0	0	1	1	1	3	5	6	8	13	19
	500	0	0	0	1	1	1	2	4	5	7	11	16
	600	0	0	0	1	1	1	1	3	4	5	9	13
	700	0	0	0	0	1	1	1	3	4	5	8	11
	750	0	0	0	0	1	1	1	2	3	4	7	11
	800	0	0	0	0	1	1	1	2	3	4	7	10
	1,000	0	0	0	0	0	1	1	1	3	3	6	8
FEP, FEPB, PFA, PFAH, TFE	14	11	20	33	58	79	131	188	290	389	502	790	1,142
	12	8	15	24	42	58	96	137	212	284	366	577	834
	10	6	10	17	30	41	69	98	152	204	263	414	598
	8	3	6	10	17	24	39	56	87	117	150	237	343
	6	2	4	7	12	17	28	40	62	83	107	169	244
	4	1	3	5	8	12	19	28	43	58	75	118	170
	3	1	2	4	7	10	16	23	36	48	62	98	142
	2	1	1	3	6	8	13	19	30	40	51	81	117
PFA, PFAH, TFE	1	1	1	2	4	5	9	13	20	28	36	56	81
PFA, PFAH, TFE, Z	1/0	1	1	1	3	4	8	11	17	23	30	47	68
	2/0	0	1	1	3	4	6	9	14	19	24	39	56
	3/0	0	1	1	2	3	5	7	12	16	20	32	46
	4/0	0	1	1	1	2	4	6	9	13	16	26	38
Z	14	13	24	40	70	95	158	226	350	469	605	952	1,376
	12	9	17	28	49	68	112	160	248	333	429	675	976
	10	6	10	17	30	41	69	98	152	204	263	414	598
	8	3	6	11	19	26	43	62	96	129	166	261	378
	6	2	4	7	13	18	30	43	67	90	116	184	265
	4	1	3	5	9	12	21	30	46	62	80	126	183
	3	1	2	4	6	9	15	22	34	45	58	92	133
	2	1	1	3	5	7	12	18	28	38	49	77	111
	1	1	1	2	4	6	10	14	23	30	40	62	90
XHH, XHHW, XHHW-2, ZW	14	8	14	24	42	57	94	135	209	280	361	568	822
	12	6	11	18	32	44	72	103	160	215	277	436	631
	10	4	8	13	24	32	54	77	119	160	206	325	470
	8	2	4	7	13	18	30	43	66	89	115	181	261
	6	1	3	5	10	13	22	32	49	66	85	134	193
	4	1	2	4	7	9	16	23	35	48	61	97	140
	3	1	1	3	6	8	13	19	30	40	52	82	118
	2	1	1	3	5	7	11	16	25	34	44	69	99
XHH, XHHW, XHHW-2	1	1	1	1	3	5	8	12	19	25	32	51	74
	1/0	1	1	1	3	4	7	10	16	21	27	43	62
	2/0	0	1	1	2	3	6	8	13	17	23	36	52
	3/0	0	1	1	1	3	5	7	11	14	19	30	43
	4/0	0	1	1	1	2	4	6	9	12	15	24	35
	250	0	0	1	1	1	3	5	7	10	13	20	29
	300	0	0	1	1	1	3	4	6	8	11	17	25
	350	0	0	1	1	1	2	3	5	7	9	15	22
	400	0	0	0	1	1	1	3	5	6	8	13	19
	500	0	0	0	1	1	1	2	4	5	7	11	16
	600	0	0	0	1	1	1	1	3	4	5	9	13
	700	0	0	0	0	1	1	1	3	4	5	8	11
	750	0	0	0	0	1	1	1	2	3	4	7	11
	800	0	0	0	0	1	1	1	2	3	4	7	10

RIGID PVC SCHEDULE 80 CONDUIT – MAXIMUM NUMBER OF CONDUCTORS

Type	Size	½	¾	1	1¼	1½	2	2½	3	3½	4	5	6
						PVC 80 Trade Size (in.)							
RHH, RHW, RHW-2	14	3	5	9	17	23	39	56	88	118	153	243	349
	12	2	4	7	14	19	32	46	73	98	127	202	290
	10	1	3	6	11	15	26	37	59	79	103	163	234
	8	1	1	3	6	8	13	19	31	41	54	85	122
	6	1	1	2	4	6	11	16	24	33	43	68	98
	4	1	1	1	3	5	8	12	19	26	33	53	77
	3	0	1	1	3	4	7	11	17	23	29	47	67
	2	0	1	1	3	4	6	9	14	20	25	41	58
	1	0	1	1	1	2	4	6	9	13	17	27	38
	1/0	0	0	1	1	1	3	5	8	11	15	23	33
	2/0	0	0	1	1	1	3	4	7	10	13	20	29
	3/0	0	0	1	1	1	3	4	6	8	11	17	25
	4/0	0	0	0	1	1	2	3	5	7	9	15	21
	250	0	0	0	1	1	1	2	4	5	7	11	16
	300	0	0	0	1	1	1	2	3	5	6	10	14
	350	0	0	0	1	1	1	1	3	4	5	9	13
	400	0	0	0	0	1	1	1	3	4	5	8	12
	500	0	0	0	0	1	1	1	2	3	4	7	10
	600	0	0	0	0	0	1	1	1	3	3	6	8
	700	0	0	0	0	0	1	1	1	2	3	5	7
	750	0	0	0	0	0	1	1	1	2	3	5	7
	800	0	0	0	0	0	1	1	1	2	3	4	7
	1,000	0	0	0	0	0	1	1	1	1	2	4	5
TW, THHW, THW, THW-2	14	6	11	20	35	49	82	118	185	250	324	514	736
	12	5	9	15	27	38	63	91	142	192	248	394	565
	10	3	6	11	20	28	47	67	106	143	185	294	421
	8	1	3	6	11	15	26	37	59	79	103	163	234
RHH*, RHW*, RHW-2*	14	4	8	13	23	32	55	79	123	166	215	341	490
	12	3	6	10	19	26	44	63	99	133	173	274	394
	10	2	5	8	15	20	34	49	77	104	135	214	307
	8	1	3	5	9	12	20	29	46	62	81	128	184
RHH*, RHW*, RHW-2*, TW, THW, THHW, THW-2	6	1	1	1	3	7	9	16	22	35	48	62	98	141
	4	1	1	3	5	7	12	17	26	35	46	73	105
	3	1	1	2	4	6	10	14	22	30	39	63	90
	2	1	1	1	3	5	8	12	19	26	33	53	77
	1	0	1	1	2	3	6	8	13	18	23	37	54
	1/0	0	1	1	1	3	5	7	11	15	20	32	46
	2/0	0	1	1	1	2	4	6	10	13	17	27	39
NOTE: (·) Denotes types minus outer covering	3/0	0	0	1	1	1	3	5	8	11	14	23	33
	4/0	0	0	1	1	1	3	4	7	9	12	19	27
	250	0	0	0	1	1	2	3	5	7	9	15	22
	300	0	0	0	1	1	1	3	5	6	8	13	19
	350	0	0	0	1	1	1	2	4	6	7	12	17
	400	0	0	0	1	1	1	2	4	5	7	10	15
	500	0	0	0	1	1	1	1	3	4	5	9	13
	600	0	0	0	0	1	1	1	2	3	4	7	10
	700	0	0	0	0	1	1	1	2	3	4	6	9
	750	0	0	0	0	0	1	1	1	3	3	6	8
	800	0	0	0	0	0	1	1	1	3	3	6	8
	1,000	0	0	0	0	0	1	1	1	2	3	5	7
THHN, THWN, THWN-2	14	9	17	28	51	70	118	170	265	358	464	736	1,055
	12	6	12	20	37	51	86	124	193	261	338	537	770
	10	4	7	13	23	32	54	78	122	164	213	338	485
	8	2	4	7	13	18	31	45	70	95	123	195	279
	6	1	3	5	9	13	22	32	51	68	89	141	202
	4	1	1	3	6	8	14	20	31	42	54	86	124
	3	1	1	3	5	7	12	17	26	35	46	73	105
	2	1	1	2	4	6	10	14	22	30	39	61	88
	1	0	1	1	3	4	7	10	16	22	29	45	65

RIGID PVC SCHEDULE 80 CONDUIT – MAXIMUM NUMBER OF CONDUCTORS *(cont.)*												
		PVC 80 Trade Size (in.)										
Type	Size	½	¾	1	1¼	1½	2	2½	3	3½	4	5	6
THHN, THWN, THWN-2 *(cont.)*	1/0	0	1	1	2	3	6	9	14	18	24	38	55
	2/0	0	1	1	1	3	5	7	11	15	20	32	46
	3/0	0	1	1	1	2	4	6	9	13	17	26	38
	4/0	0	0	1	1	1	3	5	8	10	14	22	31
	250	0	0	1	1	1	3	4	6	8	11	18	25
	300	0	0	0	1	1	2	3	5	7	9	15	22
	350	0	0	0	1	1	1	3	5	6	8	13	19
	400	0	0	0	1	1	1	3	4	6	7	12	17
	500	0	0	0	1	1	1	2	3	5	6	10	14
	600	0	0	0	0	1	1	1	3	4	5	8	12
	700	0	0	0	0	1	1	1	2	3	4	7	10
	750	0	0	0	0	1	1	1	2	3	4	7	9
	800	0	0	0	0	1	1	1	2	3	4	6	9
	1,000	0	0	0	0	0	1	1	1	2	3	5	7
FEP, FEPB, PFA, PFAH, TFE	14	8	16	27	49	68	115	164	257	347	450	714	1,024
	12	6	12	20	36	50	84	120	188	253	328	521	747
	10	4	8	14	26	36	60	86	135	182	235	374	536
	8	2	5	8	15	20	34	49	77	104	135	214	307
	6	1	3	6	10	14	24	35	55	74	96	152	218
	4	1	2	4	7	10	17	24	38	52	67	106	153
	3	1	1	3	6	8	14	20	32	43	56	89	127
	2	1	1	3	5	7	12	17	26	35	46	73	105
PFA, PFAH, TFE	1	1	1	1	3	5	8	11	18	25	32	51	73
PFA, PFAH, TFE, Z	1/0	0	1	1	3	4	7	10	15	20	27	42	61
	2/0	0	1	1	2	3	5	8	12	17	22	35	50
	3/0	0	1	1	1	2	4	6	10	14	18	29	41
	4/0	0	0	1	1	1	4	5	8	11	15	24	34
Z	14	10	19	33	59	82	138	198	310	418	542	860	1,233
	12	7	14	23	42	58	98	141	220	297	385	610	875
	10	4	8	14	26	36	60	86	135	182	235	374	536
	8	2	5	9	16	22	38	54	85	115	149	236	339
	6	2	4	6	11	16	26	38	60	81	104	166	238
	4	1	2	4	8	11	18	26	41	55	72	114	164
	3	1	2	3	5	8	13	19	30	40	52	83	119
	2	1	1	2	5	6	11	16	25	33	43	69	99
	1	0	1	2	4	5	9	13	20	27	35	56	80
XHH, XHHW, XHHW-2, ZW	14	6	11	20	35	49	82	118	185	250	324	514	730
	12	5	9	15	27	38	63	91	142	192	248	394	565
	10	3	6	11	20	28	47	67	106	143	185	294	421
	8	1	3	6	11	15	26	37	59	79	103	163	234
	6	1	2	4	8	11	19	28	43	59	76	121	173
	4	1	1	3	6	8	14	20	31	42	55	87	125
	3	1	1	3	5	7	12	17	26	36	47	74	106
	2	1	1	2	4	6	10	14	22	30	39	62	89
XHH, XHHW, XHHW-2	1	0	1	1	3	4	7	10	16	22	29	46	66
	1/0	0	1	1	2	3	6	9	14	19	24	39	56
	2/0	0	1	1	1	3	5	7	11	16	20	32	46
	3/0	0	1	1	1	2	4	6	9	13	17	27	38
	4/0	0	0	1	1	1	3	5	8	11	14	22	32
	250	0	0	1	1	1	3	4	6	9	11	18	26
	300	0	0	1	1	1	2	3	5	7	10	15	22
	350	0	0	1	1	1	1	3	5	6	8	14	20
	400	0	0	0	1	1	1	3	4	6	7	12	17
	500	0	0	0	1	1	1	2	3	5	6	10	14
	600	0	0	0	1	1	1	1	3	4	5	8	11
	700	0	0	0	0	1	1	1	2	3	4	7	10
	750	0	0	0	0	1	1	1	2	3	4	6	9
	800	0	0	0	0	1	1	1	1	3	4	6	9

DESIGNING MOTOR CIRCUITS

For one motor:
1. Determine full-load current of motor(s).

2. Multiply full-load current × 1.25 to determine minimum conductor ampacity.

3. Determine wire size.

4. Determine conduit size.

5. Determine minimum fuse or circuit breaker size.

6. Determine overload rating.

For more than one motor:
1. Perform steps 1 through 6 as shown above for each motor.

2. Add full-load current of all motors, plus 25% of the full-load current of the largest motor to determine minimum conductor ampacity.

3. Determine wire size.

4. Determine conduit size.

5. Add the fuse or circuit breaker size of the largest motor, plus the full-load currents of all other motors to determine the maximum fuse or circuit breaker size for the feeder.

MOTOR SELECTION REQUIREMENTS

Requirements of Driven Machine:
1. Hp needed
2. Torque range
3. Operating cycle—frequency of starts and stops
4. Speed
5. Operating position—horizontal, vertical, or tilted
6. Direction of rotation
7. Endplay and thrust
8. Ambient (room) temperature
9. Surrounding conditions water, gas, corrosion, dust, outdoor, etc.

Electrical Supply:
1. Voltage of power system
2. Number of phases
3. Frequency
4. Limitations on starting current
5. Effect of demand, energy on power rates

A WYE-WOUND MOTOR FOR USE ON 240/480 V

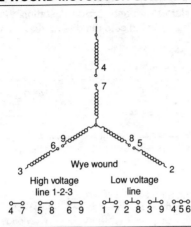

1

4

7

9

6

8 5

3

Wye wound

2

High voltage
line 1-2-3

Low voltage
line

4 7   5 8   6 9

1 7   2 8   3 9   4 5 6

NUMBERING OF DUAL-VOLTAGE WYE-WOUND MOTOR

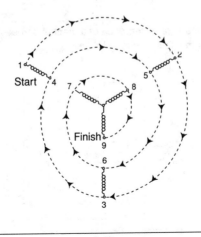

1

Start

4

7

8

Finish

9

6

5

2

3

MOTOR CONTROL CIRCUITS

Magnetic starter with one stop–start station and a pilot lamp that burns to indicate that the motor is running.

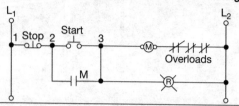

Magnetic starter with three stop-start stations.

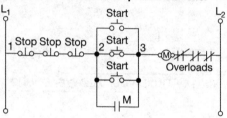

Diagrammatic representation of a magnetic three-phase starter with one start–stop station.

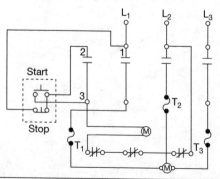

8-50

MOTOR CONTROL CIRCUITS (cont.)

Jogging using a selector push button.

Magnetic starter with a plugging switch.

MAXIMUM OCPD

$$OCPD = FLC \times R_M$$

where

R_M = maximum rating of OCPD

and

FLC = full load current (from motor nameplate or NEC® Table 430.150)

Motor Type	Code Letter	Motor Size	FLC %			
			TDF	NTDF	ITB	ITCB
AC*	—	—	175	300	150	700
AC*	A	—	150	150	150	700
AC*	B–E	—	175	250	200	700
AC*	F–V	—	175	300	250	700
DC	—	1/8 to 50 hp	150	150	150	250
DC	—	More than 50 hp	150	150	150	175

*Full-voltage and resistor starting

STANDARD SIZES OF FUSES AND CBs

NEC® 240.6 lists standard amperage ratings of fuses and fixed-trip circuit breakers as follows:

15	20	25	30	35	40
45	50	60	70	80	90
100	110	125	150	175	200
225	250	300	350	400	450
500	600	700	800	1000	1,200
1,600	2,000	2,500	3,000	4,000	5,000	6,000

LOCKED ROTOR CURRENT

Apparent, 1φ	Apparent, 3φ	True, 1φ	True, 3φ
$LRC = \dfrac{1000 \times HP \times kVA/HP}{V}$	$LRC = \dfrac{1000 \times HP \times kVA/HP}{V \times \sqrt{3}}$	$LRC = \dfrac{1000 \times HP \times kVA/HP}{V \times PF \times E_{ff}}$	$LRC = \dfrac{1000 \times HP \times kVA/HP}{V \times \sqrt{3} \times PF \times E_{ff}}$
where	where	where	where
LRC = Locked rotor current (in amps)	LRC = Locked rotor current (in amps)	LRC = Locked rotor current (in amps)	LRC = Locked rotor current (in amps)
1000 = Multiplier for kilo	1000 = Multiplier for kilo	1000 = Multiplier for kilo	1000 = Multiplier for kilo
HP = Horsepower	HP = Horsepower	HP = Horsepower	HP = Horsepower
kVA/HP = Kilovolt amps per horsepower	kVA/HP = Kilovolt amps per horsepower	kVA/HP = Kilovolt amps per horsepower	kVA/HP = Kilovolt amps per horsepower
V = Volts	V = Volts	V = Volts	V = Volts
	$\sqrt{3}$ = 1.732	PF = Power factor	$\sqrt{3}$ = 1.732
		E_{ff} = Motor efficiency	

MOTOR TORQUE FORMULAS

Torque	Starting Torque	Nominal Torque Rating
$T = \dfrac{HP \times 5252}{rpm}$	$T = \dfrac{HP \times 5252 \times \%}{rpm}$	$T = \dfrac{HP \times 63{,}000}{rpm}$
where	where	where
T = Torque	HP = Horsepower	T = Nominal torque rating (in lb.-in.)
HP = Horsepower	5252 = Constant $\left(\dfrac{33{,}000 \text{ lb.-ft.}}{\pi \times 2} = 5252\right)$	$63{,}000$ = Constant
5252 = Constant $\left(\dfrac{33{,}000 \text{ lb.-ft.}}{\pi \times 2} = 5252\right)$	rpm = Revolutions per min.	HP = Horsepower
rpm = Revolutions per min.	$\%$ = Motor class percentage	rpm = Revolutions per min.

GEAR REDUCER FORMULAS

Output Torque	Output Speed	Output Horsepower
$O_T = I_T \times R_R \times R_E$	$O_S = \dfrac{I_S}{R_R} \times R_E$	$O_{HP} = I_{HP} \times R_E$
where	where	where
O_T = Output torque (in lb.-ft.)	O_S = Output speed (in rpm)	O_{HP} = Output horsepower
I_T = Input torque (in lb.-ft.)	I_S = Input speed (in rpm)	I_{HP} = Input horsepower
R_R = Gear reducer ratio	R_R = Gear reducer ratio	R_E = Reducer efficiency %
R_E = Reducer efficiency %	R_E = Reducer efficiency %	

MOTOR POWER FORMULAS—COST SAVINGS

Power Consumed	Operating Cost	Annual Savings
$P = \dfrac{\text{HP} \times 746}{E_{ff}}$	$C_{/hr.} = \dfrac{P_{/hr.} \times C_{/kWh}}{1000}$	$S_{Ann} = C_{Ann\ Std} - C_{Ann\ Eff}$
where	where	where
P = Power consumed (W)	$C_{/hr.}$ = Operating cost per hour	S_{Ann} = Annual cost savings
HP = Horsepower	$P_{/hr.}$ = Power consumed per hour	$C_{Ann\ Std}$ = Annual operating cost for standard motor
746 = Constant	C_{kWh} = Cost per kilowatt hour	$C_{Ann\ Eff}$ = Annual operating cost for energy-efficient motor
E_{ff} = Efficiency (%)	1000 = Constant to remove kilo	

8-55

FULL-LOAD CURRENTS—DC MOTORS

Motor Rating (hp)	Current (Amps)	
	120 V	240 V
¼	3.1	1.6
⅓	4.1	2.0
½	5.4	2.7
¾	7.6	3.8
1	9.5	4.7
1½	13.2	6.6
2	17	8.5
3	25	12.2
5	40	20
7½	48	29
10	76	38

FULL-LOAD CURRENTS—1ϕ, AC MOTORS

Motor Rating (hp)	Current (Amps)	
	115 V	230 V
⅙	4.4	2.2
¼	5.8	2.9
⅓	7.2	3.6
½	9.8	4.9
¾	18.8	6.9
1	16	8
1½	20	10
2	24	12
3	34	17
5	56	28
7½	80	40
10	100	50

FULL-LOAD CURRENTS—3φ, AC INDUCTION MOTORS

Motor Rating (hp)	Current (Amps)			
	208 V	230 V	460 V	575 V
¼	1.11	.96	.48	.38
⅓	1.34	1.18	.59	.47
½	2.2	2.0	1.0	.8
¾	3.1	2.8	1.4	1.1
1	4.0	3.6	1.8	1.4
1½	5.7	5.2	2.6	2.1
2	7.5	6.8	3.4	2.7
3	10.6	9.6	4.8	3.9
5	16.7	15.2	7.6	6.1
7½	24.0	22.0	11.0	9.0
10	31.0	28.0	14.0	11.0
15	46.0	42.0	21.0	17.0
20	59	54	27	22
25	75	68	34	27
30	88	80	40	32
40	114	104	52	41
50	143	130	05	52
60	169	154	77	62
75	211	192	96	77
100	273	248	124	99
125	343	312	156	125
150	396	360	180	144
200	—	480	240	192
250	—	602	301	242
300	—	—	362	288
350	—	—	413	337
400	—	—	477	382
500	—	—	590	472

STARTING METHODS: SQUIRREL-CAGE INDUCTION MOTORS

Starter Type	% Full-Voltage Value		
	Voltage at Motor	Line Current	Motor Output Torque
Full voltage	100	100	100
Autotransformer			
80 pc tap	80	64*	64
65 pc tap	65	42*	42
50 pc tap	50	25*	25
Primary reactor			
80 pc tap	80	80	64
65 pc tap	65	65	42
50 pc tap	50	50	25
Primary resistor			
Typical rating	80	80	64
Part-winding			
Low-speed motors ($\frac{1}{2}$–$\frac{1}{2}$)	100	50	50
High-speed motors ($\frac{1}{2}$–$\frac{1}{2}$)	100	70	50
High-speed motors ($\frac{2}{3}$–$\frac{1}{3}$)	100	65	42
Wye start-delta run ($\frac{1}{3}$–$\frac{1}{3}$)	100	33	33

*Autotransformer magnetizing current not included.
Magnetizing current usually less than 25% motor full-load current.

NEMA RATINGS OF 60-HZ AC CONTACTORS IN AMPERES

Size	8 Hr. Open Rating (A)	3φ			1φ	
		200 V	230 V	230/460 V	115 V	230V
00	9	1½	1½	2	⅓	1
0	18	3	3	5	1	2
1	27	7½	7½	10	2	3
2	45	10	15	25	3	7½
3	90	25	30	50	—	—
4	135	40	50	100	—	—
5	270	75	100	200	—	—
6	540	150	200	400	—	—
7	810	—	300	600	—	—
8	1,215	—	450	900	—	—
9	2,250	—	800	1600	—	—

SETTING BRANCH CIRCUIT PROTECTIVE DEVICES

Type of Motor	Percent of Full Load Current				
	Nontime Delay Fuse	Dual-Element (Time-Delay) Fuse	Instant. Trip-Type Breaker	Time-Limit Breaker
All AC single-phase and polyphase squirrel-cage and synchronous motors with full-voltage, resistance or reactor starting:				
No code letter	300	175	700	250
Code letter F to V	300	175	700	250
Code letter B to E	250	175	700	200
Code letter A	150	150	700	150
All AC squirrel-cage and synchronous motors with autotransformer starting:				
Code letter F to V	250	175	700	200
Code letter B to E	200	175	700	200
Code letter A	150	150	700	150
Wound Rotor	150	150	700	150
Direct current				
Not more than 50 hp	150	150	250	150
More than 50 hp	150	150	175	150

1φ 115 V MOTORS AND CIRCUITS—120 V SYSTEM

Size of Motor		Motor Overload Protection Low-Peak or Fusetron®		Switch 115% Minimum or HP Rated or Fuse Holder Size	Minimum Size of Starter	Controller Termination Temperature Rating				Minimum Size of Copper Wire and Trade Conduit	
		Motor Less Than 40°C or Greater Than 1.15 SF (Max. Fuse 125%)	All Other Motors (Max. Fuse 115%)			60°C		75°C		Wire Size (AWG or kcmil)	Conduit (in.)
HP	Amp					TW	THW	TW	THW		
1/6	4.4	5	5	30	00	•	•	•	•	14	1/2
1/4	5.8	7	6 1/4	30	00	•	•	•	•	14	1/2
1/3	7.2	9	8	30	00	•	•	•	•	14	1/2
1/2	9.8	12	10	30	00	•	•	•	•	14	1/2
3/4	13.8	15	15	30	00	•	•	•	•	12	1/2
1	16	20	17 1/2	30	01	•	•	•	•	12	1/2
1 1/2	20	25	20	30	01	•	•	•	•	10	1/2
2	24	30	25	30	01	•	•	•	•	10	1/2

8-61

1φ 230 V MOTORS AND CIRCUITS – 240 V SYSTEM

Size of Motor		Motor Overload Protection Low-Peak or Fusetron®		Switch 115% Minimum or HP Rated or Fuse Holder Size	Minimum Size of Starter	Controller Termination Temperature Rating				Minimum Size of Copper Wire and Trade Conduit	
		Motor Less Than 40°C or Greater Than 1.15 SF (Max. Fuse 125%)	All Other Motors (Max. Fuse 115%)			60°C		75°C		Wire Size (AWG or kcmil)	Conduit (in.)
HP	Amp					TW	THW	TW	THW		
1/6	2.2	2½	2½	30	00	•	•	•	•	14	½
1/4	2.9	3½	3³/₁₀	30	00	•	•	•	•	14	½
1/3	3.6	4½	4	30	00	•	•	•	•	14	½
1/2	4.9	5⁶/₁₀	5⁶/₁₀	30	00	•	•	•	•	14	½
3/4	6.9	8	7½	30	00	•	•	•	•	14	½
1	8	10	9	30	00	•	•	•	•	14	½
1½	10	12	10	30	00	•	•	•	•	14	½
2	12	15	12	30	0	•	•	•	•	14	½
3	17	20	17½	30	1	•	•	•	•	12	½

1φ 230 V MOTORS AND CIRCUITS—240 V SYSTEM (cont.)

Size of Motor		Motor Overload Protection Low-Peak or Fusetron®		Switch 115% Minimum or HP Rated or Fuse Holder Size	Minimum Size of Starter	Controller Termination Temperature Rating				Minimum Size of Copper Wire and Trade Conduit	
		Motor Less Than 40°C or Greater Than 1.15 SF (Max. Fuse 125%)	All Other Motors (Max. Fuse 115%)			60°C		75°C		Wire Size (AWG or kcmil)	Conduit (in.)
HP	Amp					TW	THW	TW	THW		
5	28	35	30*	60	2		•			8	¾
						•				8	½
									•	10	½
7½	40	50	45	60	2	•	•			6	¾
								•		8	¾
10	50	60	50	60	3	•	•			4	1
									•	6	¾

*Fuse reducers required.

3φ 230 V MOTORS AND CIRCUITS—240 V SYSTEM

Size of Motor		Motor Overload Protection Low-Peak or Fusetron®		Switch 115% Minimum or HP Rated or Fuse Holder Size	Minimum Size of Starter	Controller Termination Temperature Rating				Minimum Size of Copper Wire and Trade Conduit	
		Motor Less Than 40°C or Greater Than 1.15 SF (Max. Fuse 125%)	All Other Motors (Max. Fuse 115%)			60°C		75°C			
HP	Amp					TW	THW	TW	THW	Wire Size (AWG or kcmil)	Conduit (in.)
½	2	2½	2¼	30	00	•	•	•	•	14	½
¾	2.8	3½	3²⁄₁₀	30	00	•	•	•	•	14	½
1	3.6	4½	4	30	00	•	•	•	•	14	½
1½	5.2	6¼	5⁶⁄₁₀	30	00	•	•	•	•	14	½
2	6.8	8	7½	30	0	•	•	•	•	14	½
3	9.6	12	10	30	0	•	•	•	•	14	½
5	15.2	17½	17½	30	1	•	•	•	•	14	½
7½	22	25	25	30	1	•	•	•	•	10	½
10	28	35	30*	60	2	•	•			8	¾
								•	•	10	½
15	42	50	45	60	2	•	•			6	1
								•	•	6	¾

8-64

3φ 230 V MOTORS AND CIRCUITS—240 V SYSTEM (cont.)

Size of Motor HP	Amp	Motor Overload Protection Low-Peak or Fusetron® — Motor Less Than 40°C or Greater Than 1.15 SF (Max. Fuse 125%)	All Other Motors (Max. Fuse 115%)	Switch 115% Minimum or HP Rated or Fuse Holder Size	Minimum Size of Starter	60°C TW	60°C THW	75°C TW	75°C THW	Wire Size (AWG or kcmil)	Conduit (in.)
20	54	60*	60*	100	3	•	•		•	4	1
25	68	80	75	100	3	•	•			3	1¼
									•	3	1
30	80	100	90	100	3	•	•		•	4	1
										1	1¼
40	104	125	110	200	4	•	•			2/0	1½
									•	1	1¼
50	130	150	150	200	4	•	•		•	3/0	2
										2/0	1½

*Fuse reducers required.

3φ 230 V MOTORS AND CIRCUITS—240 V SYSTEM (cont.)

Size of Motor		Motor Overload Protection Low-Peak or Fusetron®		Switch 115% Minimum or HP Rated or Fuse Holder Size	Minimum Size of Starter	Controller Termination Temperature Rating				Minimum Size of Copper Wire and Trade Conduit	
		Motor Less Than 40°C or Greater Than 1.15 SF (Max. Fuse 125%)	All Other Motors (Max. Fuse 115%)			60°C		75°C		Wire Size (AWG or kcmil)	Conduit (in.)
HP	Amp					TW	THW	TW	THW		
75	192	225	200*	400	5	•	•			300	2½
									•	250	2½
100	248	300	250	400	5	•	•			500	3
									•	350	2½
150	360	450	400*	600	6	•	•			300-2φ	2-2½
									•	4/0-2φ	2-2

*Fuse reducers required.

8-66

3ϕ 460 V MOTORS AND CIRCUITS—480 V SYSTEM

Size of Motor		Motor Overload Protection Low-Peak or Fusetron®		Switch 115% Minimum or HP Rated or Fuse Holder Size	Minimum Size of Starter	Controller Termination Temperature Rating				Minimum Size of Copper Wire and Trade Conduit	
						60°C		75°C			
HP	Amp	Motor Less Than 40°C or Greater Than 1.15 SF (Max. Fuse 125%)	All Other Motors (Max. Fuse 115%)			TW	THW	TW	THW	Wire Size (AWG or kcmil)	Conduit (in.)
½	1	1¼	1⅛	30	00	•	•	•	•	14	½
¾	1.4	1⁶⁄₁₀	1⁶⁄₁₀	30	00	•	•	•	•	14	½
1	1.8	2¼	2	30	00	•	•	•	•	14	½
1½	2.6	3²⁄₁₀	2⁶⁄₁₀	30	00	•	•	•	•	14	½
2	3.4	4	3½	30	00	•	•	•	•	14	½
3	4.8	5⁶⁄₁₀	5	30	0	•	•	•	•	14	½
5	7.6	9	8	30	0	•	•	•	•	14	½
7½	11	12	12	30	1	•	•	•	•	14	½
10	14	17½	15	30	1	•	•	•	•	14	½
15	21	25	20	30	2	•	•	•	•	10	½
20	27	30*	30*	60	2	•	•	•		8	¾
									•	10	½

*Fuse reducers required.

3φ 460 V MOTORS AND CIRCUITS—480 V SYSTEM (cont.)

Size of Motor		Motor Overload Protection Low-Peak or Fusetron®		Switch 115% Minimum or HP Rated or Fuse Holder Size	Minimum Size of Starter	Controller Termination Temperature Rating				Minimum Size of Copper Wire and Trade Conduit	
		Motor Less Than 40°C or Greater Than 1.15 SF (Max. Fuse 125%)	All Other Motors (Max. Fuse 115%)			60°C		75°C		Wire Size (AWG or kcmil)	Conduit (in.)
HP	Amp					TW	THW	TW	THW		
25	34	40	35	60	2	•	•	•	•	6	1
										8	¾
30	40	50	45	60	3	•	•	•	•	6	1
										8	¾
40	52	60*	60*	100	3	•	•	•	•	4	1
										6	1
50	65	80	70	100	3			•	•	3	1¼
										4	1
60	77	90	80	100	4			•	•	1	1¼
										3	1¼
75	96	110	110	200	4	•	•	•	•	1/0	1½
										1	1¼

8-68

3φ 460 V MOTORS AND CIRCUITS—480 V SYSTEM (cont.)

Size of Motor		Motor Overload Protection Low-Peak or Fusetron®		Switch 115% Minimum or HP Rated or Fuse Holder Size	Minimum Size of Starter	Controller Termination Temperature Rating				Minimum Size of Copper Wire and Trade Conduit	
		Motor Less Than 40°C or Greater Than 1.15 SF (Max. Fuse 125%)	All Other Motors (Max. Fuse 115%)			60°C		75°C		Wire Size (AWG or kcmil)	Conduit (in.)
HP	Amp					TW	THW	TW	THW		
100	124	150	125	200	4	•	•	•	•	3/0	2
										2/0	1½
125	156	175	175	200	5	•	•	•	•	4/0	2
										3/0	2
150	180	225	200*	400	5	•	•	•	•	300	2½
										4/0	2
200	240	300	250	400	5	•	•	•	•	500	3
										350	2½
250	302	350	325	400	6	•	•	•	•	4/0-2φ	2-2
										3/0-2φ	2-2
300	361	450	400*	600	6	•	•	•	•	300-2φ	2-1½
										4/0-2φ	2-2

*Fuse reducers required.

8-69

DC MOTORS AND CIRCUITS

Size of Motor		Motor Overload Protection Low-Peak or Fusetron®		Switch 115% Minimum or HP Rated or Fuse Holder Size	Minimum Size of Starter	Controller Termination Temperature Rating				Minimum Size of Copper Wire and Trade Conduit	
		Motor Less Than 40°C or Greater Than 1.15 SF (Max. Fuse 125%)	All Other Motors (Max. Fuse 115%)			60°C		75°C		Wire Size (AWG or kcmil)	Conduit (in.)
HP	Amp					TW	THW	TW	THW		
90 V											
¼	4.0	5	4½	30	0	•	•	•	•	14	½
⅓	5.2	6¼	5⁶/₁₀	30	0	•	•	•	•	14	½
½	6.8	8	7.5	30	0	•	•	•	•	14	½
¾	9.6	12	10	30	0	•	•	•	•	14	½
1	12.2	15	12	30	0	•	•	•	•	14	½
120 V											
¼	3.1	3½	3½	30	0	•	•	•	•	14	½
⅓	4.1	5	4½	30	0	•	•	•	•	14	½
½	5.4	6¼	6	30	0	•	•	•	•	14	½
¾	7.6	9	8	30	0	•	•	•	•	14	½
1	9.5	10	10	30	0	•	•	•	•	14	½

DC MOTORS AND CIRCUITS (cont.)

Size of Motor		Motor Overload Protection Low-Peak or Fusetron®		Switch 115% Minimum or HP Rated or Fuse Holder Size	Minimum Size of Starter	Controller Termination Temperature Rating				Minimum Size of Copper Wire and Trade Conduit	
		Motor Less Than 40°C or Greater Than 1.15 SF (Max. Fuse 125%)	All Other Motors (Max. Fuse 115%)			60°C		75°C		Wire Size (AWG or kcmil)	Conduit (in.)
HP	Amp					TW	THW	TW	THW		
120 V (cont.)											
1½	13.2	15	15	30	1	•	•	•	•	14	½
2	17	20	17½	30	1	•	•	•	•	12	½
5	40	50	45	60	2	•	•	•		6	¾
										8	¾
10	76	90	80	100	3	•		•		2	1
										3	1
180 V											
¼	2	2½	2¼	30	0	•	•	•	•	14	½
⅓	2.6	3²/₁₀	2⁸/₁₀	30	0	•	•	•	•	14	½
½	3.4	4	3½	30	0	•	•	•	•	14	½
¾	4.8	6	5	30	0	•	•	•	•	14	½

*Fuse reducers required.

DC MOTORS AND CIRCUITS (cont.)

Size of Motor		Motor Overload Protection Low-Peak or Fusetron®		Switch 115% Minimum or HP Rated or Fuse Holder Size	Minimum Size of Starter	Controller Termination Temperature Rating				Minimum Size of Copper Wire and Trade Conduit	
		Motor Less Than 40°C or Greater Than 1.15 SF (Max. Fuse 125%)	All Other Motors (Max. Fuse 115%)			60°C		75°C			
HP	Amp					TW	THW	TW	THW	Wire Size (AWG or kcmil)	Conduit (in.)
180 V (cont.)											
1	6.1	7½	7	30	0	•		•		14	½
1½	8.3	10	9	30	1	•		•		14	½
2	10.8	12	12	30	1	•		•		14	½
3	16	20	17½	30	1	•		•	•	12	½
5	27	30*	30*	60	1				•	8	¾

*Fuse reducers required.

8-72

CONTROL RATINGS

| Size | Load (V) | Maximum HP | | | | Cont. Amps | Service Limit Amps | Tungsten & Ballast Type Lamp Amps 480 V Max. | Resistance Heating (kW) | | Transformer Switching 50–60 Hz kVA Rating Inrush Peak Time Continuous Amps | | | | Capacitor kVA Switching Rating 3φ KVAR |
| | | Normal Duty | | Plugging & Jogging Duty | | | | | | | 20 Times | | 20–40 Times | | |
		1φ	3φ	1φ	3φ				1φ	3φ	1φ	3φ	1φ	3φ	
00	115	½	—	—	—	9	11	—	1.15	2.0	—	—	—	—	—
	200	—	1½	—	—	9	11	—	2.0	3.46	—	—	—	—	—
	230	1	1½	—	—	9	11	—	2.3	4.0	—	—	—	—	—
	380	—	1½	—	—	9	11	—	—	6.5	—	—	—	—	—
	460	—	2	—	—	9	11	—	4.6	8.0	—	—	—	—	—
	575	—	2	—	—	9	11	—	5.8	10.0	—	—	—	—	—
0	115	1	—	½	—	18	21	20	2.3	4.0	0.6	—	0.3	—	—
	200	—	3	—	1½	18	21	20	4.0	6.92	—	1.8	—	0.9	—
	230	2	3	1	1½	18	21	20	4.6	8.0	1.2	2.1	0.6	1.0	—
	380	—	5	—	1½	18	21	20	—	13.1	2.4	4.2	1.2	2.1	—
	460	—	5	—	2	18	21	20	9.2	15.9	3.0	5.2	1.5	2.6	—
	575	—	5	—	2	18	21	—	11.5	19.9	—	—	—	—	—
1	115	2	—	1	—	27	32	30	3.5	6.0	1.2	—	0.6	—	—
	200	—	7½	—	3	27	32	30	6	10.4	—	3.6	—	1.8	—
	230	3	7½	2	3	27	32	30	6.9	11.9	2.4	4.3	1.2	2.1	—
	380	—	10	—	5	27	32	30	—	19.7	—	—	—	—	—
	460	—	10	—	5	27	32	30	13.8	23.9	4.9	8.5	2.5	4.3	—
	575	—	10	—	5	27	32	—	17.3	29.8	6.2	11.0	3.1	5.3	—

CONTROL RATINGS (cont.)

Size	Load (V)	Maximum HP Normal Duty 1φ	Normal Duty 3φ	Plugging & Jogging Duty 1φ	Plugging & Jogging Duty 3φ	Cont. Amps	Service Limit Amps	Tungsten & Ballast Type Lamp Amps 480 V Max.	Resistance Heating (kW) 1φ	Resistance Heating (kW) 3φ	Transformer Switching 20 Times 1φ	20 Times 3φ	20-40 Times 1φ	20-40 Times 3φ	Capacitor kVA Switching Ratings 3φ kVAR
1P	115	3	—	1½	—	35	42	45	5.8	—	—	—	—	—	—
	230	5	—	3	—	35	42	45	11.5	—	—	—	—	—	—
1¾	115	—	—	—	—	40	40	45	5.8	9.9	1.6	—	0.8	—	—
	200	—	10	—	5	40	40	45	10	17.3	—	4.9	—	2.4	—
	230	—	10	—	5	40	40	45	11.5	19.9	3.2	5.75	1.6	2.8	—
	380	—	15	—	7½	40	40	45	—	32.9	—	—	—	—	—
	460	—	15	—	7½	40	40	45	23	39.8	6.6	11.2	3.3	5.7	—
	575	—	15	—	7½	40	40	—	28.8	49.7	8.1	14.5	4.1	7.1	—
2	115	3	—	2	—	45	52	60	8.1	13.9	2.1	—	1.0	—	—
	200	—	10	—	7½	45	52	60	14	24.2	—	6.3	—	3.1	—
	230	7½	10	5	10	45	52	60	16.1	27.8	4.1	7.2	2.1	3.6	8
	380	—	15	—	15	45	52	60	—	46.0	—	—	—	—	—
	460	—	25	—	15	45	52	60	32.2	55.7	8.3	14	4.2	7.2	16
	575	—	25	—	15	45	52	—	40.3	69.6	10.0	18	5.2	8.9	20
2½	115	5	—	—	—	60	65	75	10.4	17.9	3.1	—	1.5	—	—
	200	—	15	—	10	60	65	75	18	31.1	—	9.1	—	4.6	—
	230	10	20	—	15	60	65	75	20.7	35.8	6.1	10.6	3.1	5.3	17.5
	380	—	30	—	20	60	65	75	—	59.2	—	—	—	—	—
	460	—	30	—	20	60	65	75	41.4	71.6	12	21	6.1	10.6	34.5
	575	—	30	—	20	60	65	—	51.8	89.5	15	26.5	7.6	13.4	43.5

CONTROL RATINGS (cont.)

Size	Load (V)	Normal Duty 1φ	Normal Duty 3φ	Plugging & Jogging Duty 1φ	Plugging & Jogging Duty 3φ	Cont. Amps	Service Limit Amps	Tungsten & Ballast Type Lamp Amps 480 V Max.	Resistance Heating (kW) 1φ	Resistance Heating (kW) 3φ	20 Times 1φ	20 Times 3φ	20-40 Times 1φ	20-40 Times 3φ	Capacitor kVA Switching Ratings 3φ kVAR
3	115	7½	–	–	–	90	104	100	14.4	24.8	4.1	–	2.0	–	–
	200	–	25	–	15	90	104	100	25	43.3	–	12	–	6.1	–
	230	15	30	–	20	90	104	100	28.8	50.0	8.1	14	4.1	7.0	27
	380	–	50	–	30	90	104	100	–	82.2	–	28	–	14	–
	460	–	50	–	30	90	104	100	57.5	99.4	16	28	8.1	14	53
	575	–	50	–	30	90	104	–	71.9	124	20	35	10	18	67
3½	115	–	–	–	–	115	125	150	18.4	31.8	–	–	–	–	–
	200	–	30	–	20	115	125	150	32	55.4	–	16	–	8	–
	230	–	60	–	25	115	125	150	36.8	63.7	11	18.5	5.4	9.5	33.5
	380	–	60	–	30	115	125	150	–	105	–	37.5	–	18.5	–
	460	–	75	–	40	115	125	150	73.6	127	21.5	37.5	11.0	18.5	66.5
	575	–	75	–	40	115	125	–	92	159	37	47	13.5	23.5	83.5
4	200	–	40	–	25	135	156	200	39	67.5	–	20	–	10	–
	230	–	50	–	30	135	156	200	44.9	77.6	14	23	6.8	12	40
	380	–	75	–	50	135	156	200	–	128	–	47	–	23	–
	460	–	100	–	60	135	156	200	89.7	155	27	47	14	23	80
	575	–	100	–	60	135	156	–	112	194	34	59	17	29	100

CONTROL RATINGS (cont.)

Size	Load (V)	Normal Duty 1φ	Normal Duty 3φ	Plugging & Jogging Duty 1φ	Plugging & Jogging Duty 3φ	Cont. Amps 3φ	Cont. Amps	Service Limit Amps	Tungsten & Ballast Type Lamp Amps 480 V Max.	Resistance Heating (kW) 1φ	Resistance Heating (kW) 3φ	Transformer Switching 20 Times 1φ	Transformer Switching 20 Times 3φ	Transformer Switching 20–40 Times 1φ	Transformer Switching 20–40 Times 3φ	Capacitor kVA Switching Ratings 3φ kVAR
4½	200	—	50	—	—	30	210	225	250	53	91.7	—	30.5	—	15	—
	230	—	75	—	—	40	210	225	250	60.9	105	20.5	35	10.4	18	60
	380	—	100	—	—	75	210	225	250	—	174	—	—	—	—	—
	460	—	150	—	—	100	210	225	250	122	211	40.5	70.5	20.5	35	120
	575	—	150	—	—	100	210	225	—	152	264	51	88	25.5	44	150

STANDARD MOTOR SIZES

Classification	Size (hp)
Milli	1, 1.5, 2, 3, 5, 7.5, 10, 15, 25, 35
Fractional	1/20, 1/12, 1/8, 1/6, 1/4, 1/3, 1/2, 3/4
Full	1, 1½, 2, 3, 5, 7½, 10, 15, 20, 25, 30, 40, 50, 60, 75, 100, 125, 150, 200, 250, 300
Full—special order	350, 400, 450, 500, 600, 700, 800, 900, 1,000, 1,250, 1,500, 1,750, 2,000, 2,250, 2,500, 3,000, 3,500, 4,000, 4,500, 5,000, 5,500, 6,000, 7,000, 8,000, 9,000, 10,000, 11,000, 12,000, 13,000, 14,000, 15,000, 16,000, 17,000, 18,000, 19,000, 20,000, 22,500, 30,000, 32,500, 35,000, 37,500, 40,000, 45,000, 50,000

CHAPTER 9
Conversion Factors and Units of Measurement

MISCELLANEOUS HVAC CONVERSIONS

1 cu. ft. of water = 62.4 lbs.

1 horsepower per ton is obtained at 40°F evaporator saturation temperature and 100°F condensing temperature.

Atmospheric pressure at sea level = 14.696 psia
Gauge pressure = psia minus 14.7
psia = psig plus 14.7

1 psi = 2.31 ft. of water
1 ft. of water = .433 psi
Specific heat of air = .24

HEAT, ENERGY, AND WORK

1 ft. lb.	= .001285 Btu	= 0.13826 kg-m
1 joule	= 1 watt-sec.	= .000948 Btu
1 Btu	= 778.1 ft. lb.	= .252 kcal
1 KCAL	= 3.968 Btu	= 1000 cal
1 hp-hr.	= .746 kw-hr.	= 2544.7 Btu
1 kw-hr.	= 1.341 hp-hr.	= 3413 Btu
1 boiler horsepower	= 33479 Btu/h	= Evaporation of 34.5 water/hr at 212°F

SOLID AND LIQUID EXPENDABLE REFRIGERANTS

Evaporating temperature of dry ice (solid CO_2) at 1 atmosphere	$= -109°F$
Heat of sublimation of dry ice at $-109°F$	$= 246.3$ Btu/lb.
Specific heat of CO_2 gas	$= .2$ Btu/lb./°F
Refrigerating effect of solid CO_2 to gas at 32°F ($246.3 + .2 [109 + 32]$)	$= 274.5$ Btu/lb.
Evaporating temperature of liquid carbon dioxide (CO_2) at 1 atmosphere	$= -70°F$
Heat of vaporization of liquid CO_2 at $-70°F$	$= 149.7$ Btu/lb.
Specific heat of CO_2 gas	$= .2$ Btu/lb./°F
Refrigerating effect of liquid CO_2 to gas at 32°F ($149.7 + .2 [70 + 32]$)	$= 170.1$ Btu/lb.
Evaporating temperature of liquid nitrogen (N_2) at 1 atmosphere	$= -320°F$
Heat of vaporization of liquid N_2 at $-320°F$	$= 85.67$ Btu/lb.
Specific heat of N_2 gas	$= .248$ Btu/lb./°F
Refrigerating effect of liquid nitrogen to gas at 32°F ($85.67 + .248 [320 + 32]$)	$= 172.97$ Btu/lb.

VELOCITY

1 ft./sec. =	.682 miles/hr.	= .3048 m/sec.
1 mile/hr. =	1.467 ft. sec.	= .447 m/sec.
1 mile/hr. =	.868 knots	= 1.609 km/hr.

1 m/sec. =	3.6 km/hr.	= 3.28 ft/sec.
1 km/hr. =	.2778 m/sec.	= .621 miles/hr.
1 knot =	1.152 miles/hr.	= 1 nautical mile/hr.

THERM-HOUR CONVERSION

$$1 \text{ therm-hr.} = 100{,}000 \text{ Btu per hr.}$$
$$1 \text{ brake horsepower} = 2{,}544 \text{ Btu per hr.}$$

$$1 \text{ brake horsepower} = \frac{2{,}544}{100{,}000} = 0.02544 \text{ therm hr.}$$

$$1 \text{ therm-hr.} = \frac{100{,}000}{2{,}544} = 39.3082 \text{ brake horsepower (40 hp)}$$

$$1 \text{ therm-hr.} = \frac{100{,}000}{33{,}475} = 2.9873 \text{ boiler horsepower (3 hp)}$$

CONVERTING INCHES OF MERCURY TO POUNDS PER SQUARE INCH ABSOLUTE

Inches of Hg	mm of Hg	psia	Ft. of Water
30	—	15	—
(29.92)	760	(14.7)	33.40
29	—	14.5	—
28	711	14	32.2
27	—	13.5	—
26	660	13	29.9
25	—	12.5	—
24	610	12	27.6
23	—	11.5	—
22	559	11	25.3
21	—	10.5	—
20	508	10	23.0
19	—	9.5	—
18	457	9	20.7
17	—	8.5	—
16	408	8	18.4
15	—	7.5	—
14	356	7	16.1
13	—	6.5	—
12	305	6	13.8
11	—	5.5	—
10	254	5	11.5
9	—	4.5	—
8	203	4	9.2
7	—	3.5	—
6	152	3	6.9
5	—	2.5	—
4	102	2	4.6
3	—	1.5	—
2	51	1	2.3
1	—	0.5	—
0	0	0	0

PRESSURE CONVERSION CHART

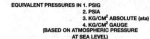

EQUIVALENT PRESSURES IN 1. PSIG
2. PSIA
3. KG/CM² ABSOLUTE (ata)
4. KG/CM² GAUGE
(BASED ON ATMOSPHERIC PRESSURE
AT SEA LEVEL)

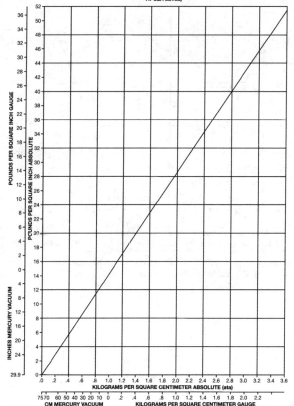

9-5

PRESSURE CONVERSION FACTORS

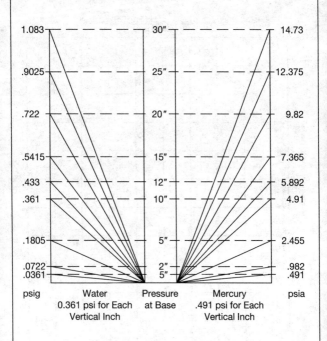

1.083		30"		14.73
.9025		25"		12.375
.722		20"		9.82
.5415		15"		7.365
.433		12"		5.892
.361		10"		4.91
.1805		5"		2.455
.0722		2"		.982
.0361		5"		.491
psig	Water 0.361 psi for Each Vertical Inch	Pressure at Base	Mercury .491 psi for Each Vertical Inch	psia

POUNDS PER SQUARE FOOT TO KILOPASCALS

Pounds per Square Foot	Kilopascals	Pounds per Square Foot	Kilopascals
1	.0479	8	.3832
2	.0958	9	.4311
3	.1437	10	.4788
4	.1916	25	1.1971
5	.2395	50	2.3940
6	.2874	75	3.5911
7	.3353	100	4.7880

POUNDS PER SQUARE INCH TO KILOPASCALS

Pounds per Square Inch	Kilopascals	Pounds per Square Inch	Kilopascals
1	6.895	8	55.160
2	13.790	9	62.055
3	20.685	10	68.950
4	27.580	25	172.375
5	34.475	50	344.750
6	41.370	75	517.125
7	48.265	100	689.500

COMMONLY USED CONVERSION FACTORS

Multiply	By	To Obtain
Acres	43,560	Square feet
Acre-feet	43,560	Cubic feet
Amperes per sq. cm	6.452	Amperes per sq. in.
Amperes per sq. in.	0.1550	Amperes per sq. cm
Ampere-turns per cm	2.540	Ampere-turns per in.
Ampere-turns per in.	0.3937	Ampere-turns per cm
Atmospheres	76.0	Cm of mercury
Atmospheres	29.92	Inches of mercury
Atmospheres	1,033.29	Cm of water
Atmospheres	33.90	Feet of water
Atmospheres	101.325	Kilopascals
Atmospheres	101325	Pascals
Atmospheres	14.70	Pounds per sq. in.
British thermal units	252.0	Calories
British thermal units	777.649	Foot pound-force
British thermal units	3.930×10^{-4}	Horsepower-hours
British thermal units	0.2520	Kilogram-calories
British thermal units	107.514	Kilogram-meters
British thermal units	2.931×10^{-4}	Kilowatt-hours
British thermal units	1,054.35	Watt-seconds
Btu per hour	2.931×10^{-4}	Kilowatts
Btu per minute	0.02358	Horsepower
Btu per minute	0.01758	Kilowatts
Calories, g	0.003971	Btu
Calories, g	3.086	Foot pound-force
Calories, kg	3.968	Btu
Calories, kg	3.085.96	Foot pound-force
Centimeters	0.3937	Inches
Centimeters	0.03281	Feet
Centimeters	0.01	Meters
Circular mils	5.067×10^{-6}	Square centimeters
Circular mils	0.7854×10^{-6}	Square inches
Circular mils	0.7854	Square mils
Cords	128	Cubic feet
Cubic centimeters	0.06102	Cubic inches
Cubic centimeters	3.5315×10^{-5}	Cubic feet
Cubic centimeters	2.6×10^{-4}	Gallons
Cubic feet	0.02832	Cubic meters
Cubic feet	7.481	Gallons
Cubic feet	28.317	Liters

Multiply	By	To Obtain
Cubic inches	16.3871	Cubic centimeters
Cubic inches	5.79×10^{-4}	Cubic feet
Cubic inches	0.00433	Gallons
Cubic meters	35.3147	Cubic feet
Cubic meters	1.308	Cubic yards
Cubic meters	264.172	Gallons
Cubic yards	0.7646	Cubic meters
Cubic yards	201.974	Gallons
Decimeters	10	Centimeters
Decimeters	0.32808	Feet
Decimeters	3.937	Inches
Degrees (angle)	0.00278	Circles
Degrees (angle)	60	Minutes
Degrees (angle)	0.01745	Radians
Dynes	2.248×10^{-6}	Pound-force
Ergs	1	Dyne-centimeters
Ergs	7.376×10^{-8}	Foot pound-force
Ergs	10^{-7}	Joules
Fathoms	6	Feet
Feet	30.48	Centimeters
Feet of air	3.608×10^{-5}	Atmospheres
Feet of air	0.0009	Feet of mercury
Feet of air	0.00122	Feet of water
Feet ot air	0.00108	Inches of mercury
Feet of air	0.00053	Pound per square in.
Feet of mercury	30.48	Centimeters of mercury
Feet of mercury	13.6086	Feet of water
Feet of mercury	163.30	Inches of water
Feet of mercury	5.8938	Pound per square in.
Feet of water	0.0295	Atmospheres
Feet of water	2.2419	Centimeters of mercury
Feet of water	29,888.9	Dynes/square cm
Feet of water	0.8826	Inches of mercury
Feet of water	304.78	Kg. per square meter
Feet of water	2,988.888	Pascals
Feet of water	62.424	Pounds per square ft.
Feet of water	0.4335	Pounds per square in.
Foot pound-force	0.001286	British thermal units
Foot pound-force	5.050×10^{-7}	Horsepower-hours
Foot pound-force	1.356	Joules

COMMONLY USED CONVERSION FACTORS *(cont.)*

Multiply	By	To Obtain
Foot pound-force	0.1383	Kilogram-meters
Foot pound-force	3.766×10^{-7}	Kilowatt-hours
Gallons	3785.41	Cubic centimeters
Gallons	0.1337	Cubic feet
Gallons	231	Cubic inches
Gallons	0.003785	Cubic meters
Gallons	0.00495	Cubic yards
Gallons	3.7854	Liters
Gallons	128	Ounces
Gallons	8	Pints
Gallons	4	Quarts
Gallons	8.33	Weight of water (lbs.)
Gallons per hour	0.13368	Cubic feet per hour
Gallons per minute	8.0208	Cubic feet per hour
Gallons per minute	0.002228	Cubic feet per sec.
Horsepower	42.375	Btu per min.
Horsepower	2,542.5	Btu per hour
Horsepower	550	Foot pounds per sec.
Horsepower	33,000	Foot pounds per min.
Horsepower	1.014	Horsepower (metric)
Horsepower	10.686	Kg. calories per min.
Horsepower	746	Watts
Horsepower	0.7457	Kilowatts
Horsepower (boiler)	33,445.7	Btu per hour
Horsepower-hours	2,546.1	British thermal units
Horsepower-hours	1.98×10^6	Foot pound-force
Horsepower-hours	2.737×10^5	Kilogram-meters
Horsepower-hours	0.7457	Kilowatt-hours
Inches	2.540	Centimeters
Inches	0.08333	Feet
Inches of mercury	0.03342	Atmospheres
Inches of mercury	1.133	Feet of water
Inches of mercury	3,386.39	Pascals
Inches of mercury	70.526	Pounds per square ft.
Inches of mercury	0.4912	Pounds per square in.
Inches of water	0.002458	Atmospheres
Inches of water	2,490.8	Dynes per square cm
Inches of water	0.07355	Inches of mercury
Inches of water	25.398	Kg. per square meter

COMMONLY USED CONVERSION FACTORS *(cont.)*

Multiply	By	To Obtain
Inches of water	0.5781	Ounces per square in.
Inches of water	5.202	Pounds per square ft.
Inches of water	0.03613	Pounds per square in.
Joules	9.478×10^{-4}	British thermal units
Joules	0.2388	Calories
Joules	10^7	Ergs
Joules	0.7376	Foot-pounds
Joules	2.778×10^{-7}	Kilowatt-hours
Joules	0.1020	Kilogram-meters
Joules	1	Watt-seconds
Kilograms	2.205	Pounds
Kilogram-calories	3.968	British thermal units
Kilogram meters	7.233	Foot pound-force
Kg per square meter	0.003281	Feet of water
Kg per square meter	0.2048	Pounds per square ft.
Kg per square meter	0.001422	Pounds per square in.
Kilometers	3,280.84	Feet
Kilometers	0.6214	Miles
Kilopascals	0.009869	Atmospheres
Kilopascals	0.2952	Inches of mercury
Kilopascals	4.021	Inches of water
Kilopascals	0.010197	Kg. per square cm
Kilopascals	1,000	Pascals
Kilopascals	20.88542	Pounds per square ft.
Kilopascals	0.1450377	Pounds per square in.
Kilowatts	56.8725	Btu per min.
Kilowatts	737.56	Foot pounds per sec.
Kilowatts	1.341	Horsepower
Kilowatts-hours	3409.5	British thermal units
Kilowatts-hours	2.655×10^6	Foot pound-force
Knots	1.151	Miles
Liters	1,000	Cubic centimeters
Liters	0.03531	Cubic feet
Liters	61.0237	Cubic inches
Liters	0.001	Cubic meters
Liters	0.00131	Cubic yards
Liters	0.2642	Gallons
Log N_e or in N	0.4343	Log_{10} N
Log N	2.303	Log_e N or in N

COMMONLY USED CONVERSION FACTORS *(cont.)*

Multiply	By	To Obtain
Lumens per square ft.	1	Footcandles
Megohms	10^6	Ohms
Meters	100	Centimeters
Meters	0.54681	Fathoms
Meters	3.281	Feet
Meters	39.37	Inches
Meters	0.001	Kilometers
Meters	1,000	Millimeters
Meters	1.0936	Yards
Meter-Kilograms	7.233	Foot pounds
Microhms	10^{-6}	Ohms
Miles	5,280	Feet
Miles	1.609	Kilometers
Miner's inch	1.5	Cubic feet per min.
Minutes (angle)	0.016667	Degrees
Minutes (angle)	1.85×10^{-4}	Quadrants
Minutes (angle)	2.909×10^{-4}	Radians
Newtons	10^5	Dynes
Newtons	1.0	Joules per meter
Newtons	7.233	Poundals
Newtons	0.22481	Pound-force
Ohms	10^{-6}	Megohms
Ohms	10^6	Microhms
Ohms per mil foot	0.1662	Microhms per cm. cube
Ohms per mil foot	0.06524	Microhms per in. cube
Pascals	9.8692×10^{-6}	Atmospheres
Pascals	3.3455×10^{-4}	Foot of water
Pascals	0.000145	Foot pounds per sq. in.
Pascals	2.953×10^{-4}	Inches of mercury
Pasclas	0.0040146	Inches of water
Pascals	0.101972	Kg per sq. meter
Poundals	0.03108	Pound-force
Pound-force	32.174	Poundals
Pound feet	1.488	Kilograms per meter
Pounds	7,000	Grains
Pounds of water	0.01602	Cubic feet
Pounds of water	0.1198	Gallons
Pounds per cubic foot	16.02	Kg. per cubic meter
Pounds per cubic foot	5.787×10^{-4}	Pounds per cubic in.

COMMONLY USED CONVERSION FACTORS *(cont.)*

Multiply	By	To Obtain
Pounds per cubic inch	27.68	Grams per cubic cm.
Pounds per cubic inch	2.768×10^4	Kg. per cubic meter
Pounds per cubic inch	1,728	Pounds per cubic ft.
Pounds per square foot	0.01602	Feet of water
Pounds per square foot	4.8824	Kg. per square meter
Pounds per square foot	0.006944	Pounds per sq. in.
Pounds per square inch	2.3067	Feet of water
Pounds per square inch	2.036	Inches of mercury
Pounds per square inch	144	Pounds per sq. ft.
Radians	57.296	Degrees
Square centimeters	1.973×10^5	Circular mils
Square feet	2.296×10^{-5}	Acres
Square feet	0.0929	Square meters
Square inches	1.273×10^6	Circular mils
Square inches	6.4516	Square centimeters
Square kilometers	0.3861	Square miles
Square meters	10.764	Square feet
Square miles	640	Acres
Square miles	2.590	Square kilometers
Square millimeters	1,973	Circular mils
Square mils	1.273	Circular mils
Therms	100,000	Btu
Ton (cooling)	288,000	Btu/day
Ton (cooling)	12,000	Btu/hr.
Ton (cooling)	200	Dtu/min
Tons (long)	2,240	Pounds
Tons (metric)	2,204.6	Pounds
Tons (short)	2,000	Pounds
Watts	3.414	Btu
Watts	0.0568	Btu per minute
Watts	10^7	Ergs per sec.
Watts	44.2537	Foot pounds per min.
Watts	0.001341	Horsepower
Watts	14.34	Calories per min.
Watt-hours	3.4144	British thermal units
Watt-hours	2,655.22	Foot pounds
Watt-hours	0.001341	Horsepower-hours
Watt-hours	0.8604	Kilogram-calories
Watt-hours	367.098	Kilogram-meters
Webers	10^8	Maxwells

COMMON ENGINEERING UNITS AND THEIR RELATIONSHIP

Quantity	SI Metric Units/Symbols	Customary Units	Relationship of Units
Acceleration	meters per second squared (m/s²)	feet per second squared (ft./s²)	m/s² = ft./s² × 3.281
Area	square meter (m²) square millimeter (mm²)	square foot (ft.²) square inch (in.²)	m² = ft.² × 10.764 mm² = in.² × 0.00155
Density	kilograms per cubic meter (kg/m³) grams per cubic centimeter (g/cm³)	pounds per cubic foot (lb./ft.³) pounds per cubic inch (lb./in.³)	kg/m³ = lb./ft.² × 16.02 g/cm³ = lb./in.² × 0.036
Work	Joule (J)	foot pound force (ft. lbf. or ft. lb.)	J = ft. lbf. × 1.356
Heat	Joule (J)	British thermal unit (Btu) Calorie (Cal)	J = Btu × 1.055 J = cal × 4.187
Energy	kilowatt (kW)	Horsepower (Hp)	kW = Hp × 0.7457

	Newton (N)	Pound-force (lbf, lb.-f., or lb.)	$N = lbf. \times 4.448$
Force	Newton (N)	kilogram-force (kgf, kg · f., or kp)	$N = \dfrac{kgf}{9.807}$
Length	meter (m)	foot (ft.)	$m = ft. \times 3.281$
	millimeter (mm)	inch (in.)	$mm = \dfrac{in.}{25.4}$
Mass	kilogram (kg)	pound (lb.)	$kg = lb. \times 2.2$
	gram (g)	ounce (oz.)	$g = \dfrac{oz.}{28.35}$
Stress	Pascal = Newton per second (Pa = N/s)	pounds per square inch (lb./in.^{2} or psi)	$Pa = lb./in.^2 \times 6{,}895$
Temperature	degree Celsius (°C)	degree Fahrenheit (°F)	$°C = \dfrac{°F - 32}{1.8}$
Torque	Newton meter (N · m)	foot-pound (ft. lb.)	$N \cdot m = ft. \; lbf. \times 1.356$
		inch-pound (in. lb.)	$N \cdot m = in. \; lbf. \times 0.113$
Volume	cubic meter (m^3)	cubic foot (ft.^{3})	$m^3 = ft.^3 \times 35.314$
	cubic centimeter (cm^3)	cubic inch (in.^{3})	$cm^3 = \dfrac{in.^3}{16.387}$

CONVERSION TABLE FOR TEMPERATURE—°F/°C

°F	°C	°F	°C	°F	°C	°F	°C	°F	°C
-459.4	-273	-22.0	-30	35.6	2	93.2	34	150.8	66
-418.0	-250	-18.4	-28	39.2	4	96.0	36	154.4	68
-328.0	-200	-14.8	-26	42.8	6	100.4	38	158.0	70
-238.0	-150	-11.2	-24	46.4	8	104.0	40	161.6	72
-193.0	-125	-7.6	-22	50.0	10	107.6	42	165.2	74
-148.0	-100	-4.0	-20	53.6	12	111.2	44	168.8	76
-130.0	-90	-0.4	-18	57.2	14	114.8	46	172.4	78
-112.0	-80	3.2	-16	60.8	16	118.4	48	176.0	80
-94.0	-70	6.8	-14	64.4	18	122.0	50	179.6	82
-76.0	-60	10.4	-12	68.0	20	125.6	52	183.2	84
-58.0	-50	14.0	-10	71.6	22	129.2	54	186.8	86
-40.0	-40	17.6	-8	75.2	24	132.8	56	190.4	88
-36.4	-38	21.2	-6	78.8	26	136.4	58	194.0	90
-32.8	-36	24.8	-4	82.4	28	140.0	60	197.6	92
-29.2	-34	28.4	-2	86.0	30	143.6	62	201.2	94
-25.6	-32	32.0	0	89.6	32	147.2	64	204.8	96

208.4	98	347.0	175	590	310	1,004	540	6,332	3,500
212.0	100	356.0	180	608	320	1,040	560	7,232	4,000
221.0	105	365.0	185	626	330	1,076	580	4,500	8,132
230.0	110	374.0	190	644	340	1,112	600	9,032	5,000
239.0	115	383.0	195	662	350	1,202	650	9,932	5,500
248.0	120	392.0	200	680	360	1,292	700	10,832	6,000
257.0	125	410	210	698	370	1,382	750	11,732	6,500
266.0	130	428	220	716	380	1,472	800	12,632	7,000
275.0	135	446	230	734	390	1,562	850	13,532	7,500
284.0	140	464	240	752	400	1,652	900	14,432	8,000
293.0	145	482	250	788	420	1,742	950	15,332	8,500
302.0	150	500	260	824	440	1,832	1,000	16,232	9,000
311.0	155	518	270	860	460	2,732	1,500	17,132	9,500
320.0	160	536	280	896	480	3,632	2,000	18,032	10,000
329.0	165	554	290	932	500	4,532	2,500		
338.0	170	572	300	968	520	5,432	3,000		

1°F is 1/180 of the difference between the temperature of melting ice and boiling water.
1°C is 1/100 of the difference between the temperature of melting ice and boiling water.
Absolute zero = −273.16°C = −459.69°F.

DECIMAL EQUIVALENTS OF FRACTIONS

8ths	32nds	64ths	64ths
1/8 = .125	1/32 = .03125	1/64 = .015625	33/64 = .515625
1/4 = .250	3/32 = .09375	3/64 = .046875	35/64 = .546875
3/8 = .375	5/32 = .15625	5/64 = .078125	37/64 = .57812
1/2 = .500	7/32 = .21875	7/64 = .109375	39/64 = .609375
5/8 = .625	9/32 = .28125	9/64 = .140625	41/64 = .640625
3/4 = .750	11/32 = .34375	11/64 = .171875	43/64 = .671875
7/8 = .875	13/32 = .40625	13/64 = .203128	45/64 = .703125
16ths	15/32 = .46875	15/64 = .234375	47/64 = .734375
1/16 = .0625	17/32 = .53125	17/64 = .265625	49/64 = .765625
3/16 = .1875	19/32 = .59375	19/64 = .296875	51/64 = .796875
5/16 = .3125	21/32 = .65625	21/64 = .328125	53/64 = .828125
7/16 = .4375	23/32 = .71875	23/64 = .359375	55/64 = .859375
9/16 = .5625	25/32 = .78125	25/64 = .390625	57/64 = .890625
11/16 = .6875	27/32 = .84375	27/64 = .421875	59/64 = .921875
13/16 = .8125	29/32 = .90625	29/64 = .453125	61/64 = .953125
15/16 = .9375	31/32 = .96875	31/64 = .484375	63/64 = .984375

MILLIMETER AND DECIMAL INCH EQUIVALENTS

mm in.	mm in.	mm in.	mm in.	mm in.
$\frac{1}{50}$ = .00079	$\frac{30}{50}$ = .02362	11 = .43307	41 = 1.61417	71 = 2.79527
$\frac{2}{50}$ = .00157	$\frac{31}{50}$ = .02441	12 = .47244	42 = 1.65354	
$\frac{3}{50}$ = .00236	$\frac{32}{50}$ = .02520	13 = .51181	43 = 1.69291	72 = 2.83464
$\frac{4}{50}$ = .00315	$\frac{33}{50}$ = .02598	14 = .55118	44 = 1.73228	73 = 2.87401
	$\frac{34}{50}$ = .02677			74 = 2.91338
$\frac{5}{50}$ = .00394		15 = .59055	45 = 1.77165	75 = 2.95275
$\frac{6}{50}$ = .00472	$\frac{35}{50}$ = .02756	16 = .62992	46 = 1.81102	76 = 2.99212
$\frac{7}{50}$ = .00551	$\frac{36}{50}$ = .02835	17 = .66929	47 = 1.85039	
$\frac{8}{50}$ = .00630	$\frac{37}{50}$ = .02913	18 = .70866	48 = 1.88976	77 = 3.03149
$\frac{9}{50}$ = .00709	$\frac{38}{50}$ = .02992	19 = .74803	49 = 1.92913	78 = 3.07086
	$\frac{39}{50}$ = .03071			79 = 3.11023
		20 = .78740	50 = 1.96850	80 = 3.14960
$\frac{10}{50}$ = .00787	$\frac{40}{50}$ = .03150	21 = .82677	51 = 2.00787	81 = 3.18897
$\frac{11}{50}$ = .00866	$\frac{41}{50}$ = .03228	22 = .86614	52 = 2.04724	
$\frac{12}{50}$ = .00945	$\frac{42}{50}$ = .03307	23 = .90551	53 = 2.08661	82 = 3.22834
$\frac{13}{50}$ = .01024	$\frac{43}{50}$ = .03386	24 = .94488	54 = 2.12598	83 = 3.26771
$\frac{14}{50}$ = .01102	$\frac{44}{50}$ = .03465			84 = 3.30708
		25 = .98425	55 = 2.16535	85 = 3.34645
$\frac{15}{50}$ = .01181	$\frac{45}{50}$ = .03543	26 = 1.02362	56 = 2.20472	86 = 3.38582
$\frac{16}{50}$ = .01260	$\frac{46}{50}$ = .03622	27 = 1.06299	57 = 2.24409	
$\frac{17}{50}$ = .01339	$\frac{47}{50}$ = .03701	28 = 1.10236	58 = 2.28346	87 = 3.42519
$\frac{18}{50}$ = .01417	$\frac{48}{50}$ = .03780	29 = 1.14173	59 = 2.32283	88 = 3.46456
$\frac{19}{50}$ = .01496	$\frac{49}{50}$ = .03858			89 = 3.50393
		30 = 1.18110	60 = 2.36220	90 = 3.54330
$\frac{20}{50}$ = .01575	1 = .03937	31 = 1.22047	61 = 2.40157	91 = 3.58267
$\frac{21}{50}$ = .01654	2 = .07874	32 = 1.25984	62 = 2.44094	
$\frac{22}{50}$ = .01732	3 = .11811	33 = 1.29921	63 = 2.48031	92 = 3.62204
$\frac{23}{50}$ = .01811	4 = .15748	34 = 1.33858		93 = 3.66141
$\frac{24}{50}$ = .01890			64 = 2.51968	94 = 3.70078
	5 = .19685	35 = 1.37795	65 = 2.55905	95 = 3.74015
	6 = .23622	36 = 1.41732	66 = 2.59842	96 = 3.77952
$\frac{25}{50}$ = .01969	7 = .27559	37 = 1.45669		
$\frac{26}{50}$ = .02047	8 = .31496	38 = 1.49606	67 = 2.63779	97 = 3.81889
$\frac{27}{50}$ = .02126	9 = .35433	39 = 1.53543	68 = 2.67716	98 = 3.85826
$\frac{28}{50}$ = .02205			69 = 2.71653	99 = 3.89763
$\frac{29}{50}$ = .02283	10 = .39370	40 = 1.57480	70 = 2.75590	100 = 3.93700

AREA OF CIRCLES					
Dia.	Area	Dia.	Area	Dia.	Area
1/8	.0123	4 1/2	15.904	16 1/2	213.82
1/4	.0491	5	19.635	17	226.98
3/8	.1104	5 1/2	23.758	17 1/2	240.52
1/2	.1963	6	28.274	18	254.46
5/8	.3067	6 1/2	33.183	18 1/2	268.80
3/4	.4417	7	38.484	19	283.52
7/8	.6013	7 1/2	44.178	19 1/2	298.64
1	.7854	8	50.265	20	314.16
1 1/8	.9940	8 1/2	56.745	20 1/2	330.06
1 1/4	1.227	9	63.617	21	346.36
1 3/8	1.484	9 1/2	70.882	21 1/2	363.05
1 1/2	1.767	10	78.54	22	380.13
1 5/8	2.073	10 1/2	86.59	22 1/2	397.60
1 3/4	2.405	11	95.03	23	415.47
1 7/8	2.761	11 1/2	103.86	23 1/2	433.73
2	3.141	12	113.09	24	452.39
2 1/4	3.976	12 1/2	122.71	24 1/2	471.43
2 1/2	4.908	13	132.73	25	490.87
2 3/4	5.939	13 1/2	143.13	26	530.93
3	7.068	14	153.93	27	572.55
3 1/4	8.295	14 1/2	165.13	28	615.75
3 1/2	9.621	15	176.71	29	660.52
3 3/4	11.044	15 1/2	188.69	30	706.86
4	12.566	16	201.06	31	754.76

AREA OF CIRCLES *(cont.)*					
Dia.	Area	Dia.	Area	Dia.	Area
32	804.24	56	2463.0	80	5026.5
33	855.30	57	2551.7	81	5153.0
34	907.92	58	2642.0	82	5281.0
35	962.11	59	2733.9	83	5410.6
36	1017.8	60	2827.4	84	5541.7
37	1075.2	61	2922.4	85	5674.5
38	1134.1	62	3019.0	86	5808.8
39	1194.5	63	3117.2	87	5944.6
40	1256.6	64	3216.9	88	6082.1
41	1320.2	65	3318.3	89	6221.1
42	1385.4	66	3421.2	90	6361.7
43	1452.2	67	3525.6	91	6503.8
44	1520.5	68	3631.6	92	6647.6
45	1590.4	69	3739.2	93	6792.9
46	1661.9	70	3848.4	94	6939.7
47	1734.9	71	3959.2	95	7088.2
48	1809.5	72	4071.5	96	7238.2
49	1885.7	73	4185.3	97	7389.8
50	1963.5	74	4300.8	98	7542.9
51	2042.8	75	4417.8	99	7697.7
52	2123.7	76	4536.4	100	7854.0
53	2206.1	77	4656.0	101	8011.8
54	2290.2	78	4778.3	102	8171.3
55	2375.8	79	4901.6	103	8332.3

COMMONLY USED GEOMETRICAL RELATIONSHIPS

Diameter of a circle × 3.1416 = Circumference

Radius of a circle × 6.283185 = Circumference

Square of the radius of a circle × 3.1416 = Area

Square of the diameter of a circle × 0.7854 = Area

Square of the circumference of a circle × 0.07958 = Area

Half the circumference of a circle × half its diameter = Area

Circumference of a circle × 0.159155 = Radius

Square root of the area of a circle × 0.56419 = Radius

Circumference of a circle × 0.31831 = Diameter

Square root of the area of a circle × 1.12838 = Diameter

Diameter of a circle × 0.866 = Side of an inscribed equilateral triangle

Diameter of a circle × 0.7071 = Side of an inscribed square

Circumference of a circle × 0.225 = Side of an inscribed square

Circumference of a circle × 0.282 = Side of an equal square

Diameter of a circle × 0.8862 = Side of an equal square

Base of a triangle × one-half the altitude = Area

Multiplying both diameters and .7854 together = Area of an ellipse

Surface of a sphere × one sixth of its diameter = Volume

Circumference of a sphere × its diameter = Surface

Square of the diameter of a sphere × 3.1416 = Surface

Square of the circumference of a sphere × 0.3183 = Surface

Cube of the diameter of a sphere × 0.5236 = Volume

Cube of the circumference of a sphere × 0.016887 = Volume

Radius of a sphere × 1.1547 = Side of an inscribed cube

Diameter of a sphere divided by $\sqrt{3}$ = Side of an inscribed cube

Area of its base × one third of its altitude = Volume of a cone or pyramid whether round, square, or triangular

Area of one of its sides × 6 = Surface of the cube

Altitude of trapezoid × one half the sum of its parallel sides = Area

CHAPTER 10
Materials, Tools, and Safety

Pipe Size (in.)	Wrench Length (in.)	Weight of Wrench (lbs.)		
		Steel Straight and End Wrench	Aluminum Straight Wrench	Aluminum End Wrench
¾	6	.50	—	—
1	8	.75	—	—
1½	10	1.75	1.00	—
2	12	2.75	—	—
2	14	3.50	2.34	1.75
2½	18	5.75	3.67	3.50
3	24	9.75	6.00	5.75
5	36	19.00	11.00	—
6	48	34.25	18.50	—
8	60	51.25	—	—

WRENCHES FOR USE WITH VARIOUS PIPE SIZES

FLARE NUT WRENCH SIZES

Tube Size OD (in.)	Wrench Size across Flats	
	Old (in.)	New (in.)
¼	¾	⅝
⅜	⅞	13/16
½	1	15/16

SOLDER ALLOYS USED IN REFRIGERATION

Solder	Melting Point (°F)	Flow Point (°F)	Shear Strength (psi)
50–50 Tin–Lead	358	414	83.4
95–5 Tin–Antimony	450	465	327.0
Silver Solder 45 Ag, 15 Cu, 24 Cd, 16 Zn	1,120	1,145	8,340
Phosphorous–Copper	1,310	1,650	8,340

BRAZING TIP SIZES, OXY-ACETYLENE TORCHES

Pipe Size (in.)	Tip Size	Drill Size	Oxygen Press	Acetylene Press
¾ or smaller	2	56	4 to 6 psig	4 to 6 psig
½ to 1⅛	3	53	5 to 8 psig	4 to 7 psig
⅞ to 2	4	49	6 to 11 psig	5 to 8 psig
2 to 6	5	43	7 to 13 psig	6 to 9 psig

PROPERTIES OF WELDING GASES

Gas Type	Gas Characteristics	Tank Sizes (cu. ft.)
Acetylene	C_2H_2, explosive gas, flammable, garlic-like odor, colorless, dangerous if used in pressures over 15 psig (30 psig absolute)	10, 40, 75 100, 300
Argon	Ar, nonexplosive inert gas, tasteless, odorless, colorless	131, 330 4754 (liquid)
Carbon dioxide	CO_2, nonexplosive inert gas, tasteless, odorless, colorless (in large quantities is toxic)	20 lbs., 50 lbs.
Helium	He, nonexplosive inert gas, tasteless, odorless, colorless	221
Hydrogen	H_2, explosive gas, tasteless, odorless, colorless	191
Nitrogen	N_2, nonexplosive inert gas, tasteless, odorless, colorless	20, 40, 80 113, 225
Oxygen	O_2, nonexplosive gas, tasteless, odorless, colorless, supports combustion	20, 40, 80 122, 244 4500 (liquid)

WELDING RODS—36" LONG

Rod Size (in.)	Number of Rods per Pound			
	Aluminum	Brass	Cast Iron	Steel
3/8	—	1.0	.25	1.0
5/16	—	—	.50	1.33
1/4	6.0	2.0	2.25	2.0
3/16	9.0	3.0	5.50	3.5
5/32	—	—	—	5.0
1/8	23.0	7.0	—	8.0
3/32	41.0	13.0	—	14.0
1/16	91.0	29.0	—	31.0

TYPES OF SOLDERING FLUX

To Solder	Use
Cast iron	Cuprous oxide
Galvanized iron, galvanized, steel, tin, zinc	Hydrochloric acid
Pewter and lead	Organic
Brass, copper, gold, iron, silver, steel	Borax
Brass, bronze, cadmium, copper, lead, silver	Resin
Brass, copper, gun metal, iron, nickel, tin, zinc	Ammonia chloride
Bismuth, brass, copper, gold, silver, tin	Zinc chloride
Silver	Sterling
Pewter and lead	Tallow
Stainless steel	Stainless steel (only)

HARD SOLDER ALLOYS

To Hard Solder	Copper %	Gold %	Silver %	Zinc %
Gold	22	67	11	—
Silver	20	—	70	10
Hard brass	45	—	—	55
Soft brass	22	—	—	78
Copper	50	—	—	50
Cast iron	55	—	—	45
Steel and iron	64	—	—	36

SOFT SOLDER ALLOYS

To Soft Solder	Lead %	Tin %	Zinc %	Bism %	Other %
Gold	33	67	—	—	—
Silver	33	67	—	—	—
Brass	34	66	—	—	—
Copper	40	60	—	—	—
Steel and iron	50	50	—	—	—
Galvanized steel	42	58	—	—	—
Tinned steel	36	64	—	—	—
Zinc	45	55	—	—	—
Block Tin	1	99	—	—	—
Lead	67	33	—	—	—
Gun metal	37	63	—	—	—
Pewter	25	25	—	50	—
Bismuth	33	33	—	34	—
Aluminum	—	70	25	—	5

BAND SAW TEETH PER INCH AND SPEED

Type of Material to Be Cut	Size of Material (in.)			
	½–1	1–2	½–1	1–2
	Teeth per inch		Speed (fpm)	
Steels				
Angle iron	14	14	190	175
Armor plate	14	12	100	75
Cast iron	12	10	200	185
Cast steels	14	12	150	75
Graphic steel	14	12	150	125
High-speed steel	14	10	100	75
I-beams and channels	14	14	250	200
Pipe	14	12	250	225
Stainless steel	12	10	60	50
Tubing (thinwall)	14	14	250	200
Nonferrous Metals				
Aluminum (all types)	8	6	250	250
Beryllium	10	8	175	150
Brass	8	8	250	250
Bronze (cast)	10	8	185	125
Bronze (rolled)	12	10	175	125
Copper	10	8	250	225
Magnesium	8	8	250	250

STANDARD TAPS AND DIES (IN.)

Thread Size	Coarse			Fine		
	Drill Size	Threads per in.	Decimal Size	Drill Size	Threads per in.	Decimal Size
4	3	4	3.75	—	—	—
3¾	3	4	3.5	—	—	—
3½	3	4	3.25	—	—	—
3¼	3	4	3.0	—	—	—
3	2	4	2.75	—	—	—
2¾	2	4	2.5	—	—	—
2½	2	4	2.25	—	—	—
2¼	2	4.5	2.0313	—	—	—
2	1	4.5	1.7813	—	—	—
1¾	1	2	1.5469	—	—	—
1½	1	6	1.3281	1²⁷⁄₆₄	12	1.4219
1⅜	1	6	1.2188	1¹⁹⁄₆₄	12	1.2969
1¼	1	7	1.1094	1¹¹⁄₆₄	12	1.1719
1⅛	⁶³⁄₆₄	7	.9844	1³⁄₆₄	12	1.0469
1	⅞	8	.8750	¹⁵⁄₁₆	14	.9375
⅞	⁴⁹⁄₆₄	9	.7656	¹³⁄₁₆	14	.8125
¾	²¹⁄₃₂	10	.6563	¹¹⁄₁₆	16	.6875
⅝	¹⁷⁄₃₂	11	.5313	³⁷⁄₆₄	18	.5781
⁹⁄₁₆	³¹⁄₆₄	12	.4844	³³⁄₆₄	18	.5156
½	²⁷⁄₆₄	13	.4219	²⁹⁄₆₄	20	.4531
⁷⁄₁₆	U	14	.368	²⁵⁄₆₄	20	.3906
⅜	⁵⁄₁₆	16	.3125	Q	24	.332
⁵⁄₁₆	F	18	.2570	I	24	.272
¼	#7	20	.201	#3	28	.213
#12	#16	24	.177	#14	28	.182
#10	#25	24	.1495	#21	32	.159
³⁄₁₆	#26	24	.147	#22	32	.157
#8	#29	32	.136	#29	36	.136
#6	#36	32	.1065	#33	40	.113
#5	#38	40	.1015	#37	44	.104
⅛	³⁄₃₂	32	.0938	#38	40	.1015
#4	#43	40	.089	#42	48	.0935
#3	#47	48	.0785	#45	56	.082
#2	#50	56	.07	#50	64	.07
#1	#53	64	.0595	#53	72	.0595
#0	—	—	—	³⁄₆₄	80	.0469

TAPS AND DIES—METRIC CONVERSIONS

Thread Pitch (mm)	Fine Thread Size		Tap Drill Size	
	(in.)	(mm)	(in.)	(mm)
4.5	1.6535	42	1.4567	37.0
4.0	1.5748	40	1.4173	36.0
4.0	1.5354	39	1.3779	35.0
4.0	1.4961	38	1.3386	34.0
4.0	1.4173	36	1.2598	32.0
3.5	1.3386	34	1.2008	30.5
3.5	1.2992	33	1.1614	29.5
3.5	1.2598	32	1.1220	28.5
3.5	1.1811	30	1.0433	26.5
3.0	1.1024	28	.9842	25.0
3.0	1.0630	27	.9449	24.0
3.0	1.0236	26	.9055	23.0
3.0	.9449	24	.8268	21.0
2.5	.8771	22	.7677	19.5
2.5	.7974	20	.6890	17.5
2.5	.7087	18	.6102	15.5
2.0	.6299	16	.5118	14.0
2.0	.5512	14	.4724	12.0
1.75	.4624	12	.4134	10.5
1.50	.4624	12	.4134	10.5
1.50	.3937	11	.3780	9.6
1.50	.3937	10	.3386	8.6
1.25	.3543	9	.3071	7.8
1.25	.3150	8	.2677	6.8
1.0	.2856	7	.2362	6.0
1.0	.2362	6	.1968	5.0
.90	.2165	5.5	.1811	4.6
.80	.1968	5	.1653	4.2
.75	.1772	4.5	.1476	3.75
.70	.1575	4	.1299	3.3
.75	.1575	4	.1279	3.25
.60	.1378	3.5	.1142	2.9
.60	.1181	3	.0945	2.4
.50	.1181	3	.0984	2.5
.45	.1124	2.6	.0827	2.1
.45	.0984	2.5	.0787	2.0
.40	.0895	2.3	.0748	1.9
.40	.0787	2	.0630	1.6
.45	.0787	2	.0590	1.5
.35	.0590	1.5	.0433	1.1

RECOMMENDED DRILLING SPEEDS IN RPM

Material	Bit Sizes (in.)	RPM Speed Range
Glass	Special metal tube drilling	700
Plastics	$7/16$ and larger $3/8$ $5/16$ $1/4$ $3/16$ $1/8$ $1/16$ and smaller	500–1,000 1,500–2,000 2,000–2,500 3,000–3,500 3,500–4,000 5,000–6,000 6,000–6,500
Woods	1 and larger $3/4$ to 1 $1/2$ to $3/4$ $1/4$ to $1/2$ $1/4$ and smaller carving/routing	700–2,000 2,000–2,300 2,300–3,100 3,100–3,800 3,800–4,000 4,000–6,000
Soft metals	$7/16$ and larger $3/8$ $5/16$ $1/4$ $3/16$ $1/8$ $1/16$ and smaller	1,500–2,500 3,000–3,500 3,500–4,000 4,500–5,000 5,000–6,000 6,000–6,500 6,000–6,500
Steel	$7/16$ and larger $3/8$ $5/16$ $1/4$ $3/16$ $1/8$ $1/16$ and smaller	500–1,000 1,000–1,500 1,000–1,500 1,500–2,000 2,000–2,500 3,000–4,000 5,000–6,500
Cast iron	$7/16$ and larger $3/8$ $5/16$ $1/4$ $3/16$ $1/8$ $1/16$ and smaller	1,000–1,500 1,500–2,000 1,500–2,000 2,000–2,500 2,500–3,000 3,500–4,500 6,000–6,500

METALWORKING LUBRICANTS

Material	Threading	Drilling	Lathing
Machine steel	Dissolvable oil Mineral oil Lard oil	Dissolvable oil Sulpherized oil Mineral lard oil	Dissolvable oil
Tool steel	Lard oil Sulpherized oil	Dissolvable oil Sulpherized oil	Dissolvable oil
Steel alloys	Lard oil Sulpherized oil	Dissolvable oil Sulpherized oil Mineral lard oil	Dissolvable oil
Cast iron	Sulpherized oil Dry Mineral lard oil	Dissolvable oil Dry Air jet	Dissolvable oil Dry
Malleable iron	Soda water Lard oil	Soda water Dry	Soda water Dissolvable oil
Monel metal	Lard oil	Dissolvable oil Lard oil	Dissolvable oil
Aluminum	Kerosene Dissolvable oil Lard oil	Kerosene Dissolvable oil	Dissolvable oil
Brass	Dissolvable oil Lard oil	Kerosene Dissolvable oil Dry	Dissolvable oil
Bronze	Dissolvable oil Lard oil	Dissolvable oil Dry Mineral oil Lard oil	Dissolvable oil
Copper	Dissolvable oil Lard oil	Kerosene Dissolvable oil Dry Mineral lard oil	Dissolvable oil

TORQUE LUBRICATION EFFECTS (FT./LBS.)

Lubricant	5/16" – 18 Thread	1/2" – 13 Thread	Torque Decrease
Graphite	13	62	49–55%
Mily film	14	66	45–52%
White grease	16	79	35–45%
Sae 30	16	79	35–45%
Sae 40	17	83	31–41%
Sae 20	18	87	28–38%
Plated	19	90	26–34%
No lube	29	121	0%

TIGHTENING TORQUE (FT./LBS.)

Wire Size, AWG/kcmil	Driver	Bolt	Other
18-16	1.67	6.25	4.2
14-8	1.67	6.25	6.125
6-4	3.0	12.5	8.0
3-1	3.2	21.00	10.40
0-2/0	4.22	29	12.5
3/0-200	–	37.5	17.0
250-300	–	50.0	21.0

SCREW TORQUES

Screw Size, Inches Across, Hex Flats	Torque (ft./lbs.)
1/8	4.2
5/32	8.3
3/16	15.0
7/32	23.25
1/4	42.0

SHEET METAL SCREW CHARACTERISTICS

Screw Size	Screw Diameter (in.)	Diameter of Pierced Hole (in.)	Hole Size	Metal Gauge
4	.112	.086	44	28
		.086	44	26
		.093	42	24
		.098	42	22
		.100	40	20
6	.138	.111	39	28
		.111	39	26
		.111	39	24
		.111	38	22
		.111	36	20
7	.155	.121	37	28
		.121	37	26
		.121	35	24
		.121	33	22
		.121	32	20
		—	31	18
8	.165	.137	33	26
		.137	33	24
		.137	32	22
		.137	31	20
		—	30	18
10	.191	.158	30	26
		.158	30	24
		.158	30	22
		.158	29	20
		.158	25	18
12	.218	—	26	24
		.185	25	22
		.185	24	20
		.185	22	18
14	.251	—	15	24
		.212	12	22
		.212	11	20
		.212	9	18

STANDARD WOOD SCREW CHARACTERISTICS

Screw Size	Wood Screw Lengths (in.)	Pilot Hole		Shank Hole	
		Softwood Bit #	Hardwood Bit #	Clearance Bit #	Hole Diameter (in.)
0	¼	75	66	52	.060
1	¼ to ⅜	71	57	47	.073
2	¼ to ½	65	54	42	.086
3	¼ to ⅝	58	53	37	.099
4	⅜ to ¾	55	51	32	.112
5	⅜ to ¾	53	47	30	.125
6	⅜ to 1½	52	44	27	.138
7	⅜ to 1½	51	39	22	.151
8	½ to 2	48	35	18	.164
9	⅝ to 2¼	45	33	14	.177
10	⅝ to 2½	43	31	10	.190
11	¾ to 3	40	29	4	.203
12	⅞ to 3½	38	25	2	.216
14	1 to 4½	32	14	D	.242
16	1¼ to 5½	29	10	I	.268
18	1½ to 6	26	6	N	.294
20	1¾ to 6	19	3	P	.320
24	3½ to 6	15	D	V	.372

ALLEN HEAD AND MACHINE SCREW
BOLT AND TORQUE CHARACTERISTICS

Number of Threads per Inch	Allen Head and Mach. Screw Bolt Size (in.)	Allen Head Case H Steel 160,000 psi	Mach. Screw Yellow Brass 60,000 psi	Mach. Screw Silicone Bronze 70,000 psi
		Torque in Foot-Pounds or Inch-Pounds		
4.5	2	8,800	—	—
5	1¾	6,100	—	—
6	1½	3,450	655	595
6	1⅜	2,850	—	—
7	1¼	2,130	450	400
7	1⅛	1,520	365	325
8	1	970	250	215
9	⅞	640	180	160
10	¾	400	117	104
11	⅝	250	88	78
12	⁹⁄₁₆	180	53	49
13	½	125	41	37
14	⁷⁄₁₆	84	30	27
16	⅜	54	20	17
18	⁵⁄₁₆	33	125 in#	110 in#
20	¼	16	70 in#	65 in#
24	#10	60	22 in#	20 in#
32	#8	46	19 in#	16 in#
32	#6	21	10 in#	8 in#
40	#5	—	7.2 in#	6.4 in#
40	#4	—	4.9 in#	4.4 in#
48	#3	—	3.7 in#	3.3 in#
56	#2	—	2.3 in#	2 in#

For fine thread bolts, increase by 9%.

HEX HEAD BOLT AND TORQUE CHARACTERISTICS

BOLT MAKEUP IS STEEL WITH COARSE THREADS

Number of Threads per Inch	Hex Head Bolt Size (in.)	SAE 0-1-2 74,000 psi	SAE Grade 3 100,000 psi	SAE Grade 5 120,000 psi
		Torque (ft./lbs.)		
4.5	2	2,750	5,427	4,550
5	1¾	1,900	3,436	3,150
6	1½	1,100	1,943	1,775
6	1⅜	900	1,624	1,500
7	1¼	675	1,211	1,105
7	1⅛	480	872	794
8	1	310	551	587
9	⅞	206	372	382
10	¾	155	234	257
11	⅝	96	145	154
12	9/16	69	103	114
13	½	47	69	78
14	7/16	32	47	54
16	⅜	20	30	33
18	5/16	12	17	19
20	¼	6	9	10

For fine thread bolts, increase by 9%.

10-14

HEX HEAD BOLT AND TORQUE CHARACTERISTICS *(cont.)*

BOLT MAKEUP IS STEEL WITH COARSE THREADS

Number of Threads per Inch	Hex Head Bolt Size (in.)	 SAE Grade 6 133,000 psi	 SAE Grade 7 133,000 psi	 SAE Grade 8 150,000 psi
		Torque (ft./lbs.)		
4.5	2	7,491	7,500	8,200
5	1¾	5,189	5,300	5,650
6	1½	2,913	3,000	3,200
6	1⅜	2,434	2,500	2,650
7	1¼	1,815	1,825	1,975
7	1⅛	1,304	1,325	1,430
8	1	825	840	700
9	⅞	550	570	600
10	¾	350	360	380
11	⅝	209	215	230
12	⁹⁄₁₆	150	154	169
13	½	106	110	119
14	⁷⁄₁₆	69	71	78
16	⅜	43	44	47
18	⁵⁄₁₆	24	25	29
20	¼	12.5	13	14

For fine thread bolts, increase by 9%.
For special alloy bolts, obtain torque rating from the manufacturer.

WHITWORTH HEX HEAD BOLT AND TORQUE CHARACTERISTICS

BOLT MAKEUP IS STEEL WITH COARSE THREADS

Number of Threads per Inch	Whitworth Type Hex Head Bolt Size (in.)	Grades A & B 62,720 psi	Grade S 112,000 psi	Grade T 123,200 psi	Grade V 145,600 psi
			Torque (ft./lbs.)		
8	1	276	497	611	693
9	⅞	186	322	407	459
11	¾	118	213	259	287
11	⅝	73	128	155	175
12	9/16	52	94	111	128
12	½	36	64	79	89
14	7/16	24	43	51	58
16	⅜	15	27	31	36
18	5/16	9	15	18	21
20	¼	5	7	9	10

For fine thread bolts, increase by 9%.

10-16

METRIC HEX HEAD BOLT AND TORQUE CHARACTERISTICS

BOLT MAKEUP IS STEEL WITH COARSE THREADS

Thread Pitch (mm)	Bolt Size (mm)	(5D) Standard 5D 71,160 psi	(8G) Standard 8G 113,800 psi	(10K) Standard 10K 142,000 psi	(12K) Standard 12K 170,674 psi
			Torque (ft./lbs.)		
3.0	24	261	419	570	689
2.5	22	182	284	394	464
2.0	18	111	182	236	183
2.0	16	83	132	175	208
1.25	14	55	89	117	137
1.25	12	34	54	70	86
1.25	10	19	31	40	49
1.0	8	10	16	22	27
1.0	6	5	6	8	10

For fine thread bolts, increase by 9%.

10-17

PULLEY AND GEAR FORMULAS

For single reduction or increase of speed by means of belting where the speed at which each shaft should run is known, and one pulley is in place:

Multiply the diameter of the pulley which you have by the number of revolutions per minute that its shaft makes; divide this product by the speed in revolutions per minute at which the second shaft should run. The result is the diameter of pulley to use.

Where both shafts with pulleys are in operation and the speed of one is known:

Multiply the speed of the shaft by diameter of its pulley and divide this product by diameter of pulley on the other shaft. The result is the speed of the second shaft.

Where a countershaft is used, to obtain size of main driving or driven pulley, or speed of main driving or driven shaft, it is necessary to calculate, as above, between the known end of the transmission and the countershaft, and then repeat this calculation between the countershaft and the unknown end.

A set of gears of the same pitch transmits speeds in proportion to the number of teeth they contain. Count the number of teeth in the gear wheel and use this quantity instead of the diameter of pulley, mentioned above, to obtain number of teeth cut in unknown gear, or speed of second shaft.

Formulas for Finding Pulley Sizes:

$$d = \frac{D \times S}{s'} \qquad D = \frac{d \times s'}{S}$$

d = Diameter of driven pulley

D = Diameter of driving pulley

s' = Number of revolutions per minute of driven pulley

S = Number of revolutions per minute of driving pulley

PULLEY AND GEAR FORMULAS *(cont.)*

Formulas for Finding Gear Sizes:

$$n = \frac{N \times S}{s'} \quad N = \frac{n \times s'}{S}$$

n = Number of teeth in pinion (driving gear)

N = Number of teeth in gear (driven gear)

s' = Number of revolutions per minute of gear

S = Number of revolutions per minute of pinion

Formula to Determine Shaft Diameter:

$$\text{Diameter of shaft in inches} = \sqrt[3]{\frac{K \times HP}{RPM}}$$

HP = Horsepower to be transmitted

RPM = Speed of shaft

K = Factor which varies from 50 to 125 depending on type of shaft and distance between supporting bearings

For line shaft having bearings 8′ apart:

K = 90 for turned shafting

K = 70 for cold-rolled shafting

Formula to Determine Belt Length:

$$\text{Length of belt} = \frac{3.14\,(D + d)}{2} + 2\left(\sqrt{X^2 + \left(\frac{D - d}{2} \right)^2} \right)$$

D = Diameter of large pulley

d = Diameter of small pulley

X = Distance between centers of shafting

STANDARD "V" BELT LENGTHS (IN.)

A BELTS			B BELTS			C BELTS		
Standard Belt No.	Pitch Length	Outside Length	Standard Belt No.	Pitch Length	Outside Length	Standard Belt No.	Pitch Length	Outside Length
A26	27.3	28.0	B35	36.8	38.0	C51	53.9	55.0
A31	32.3	33.0	B38	39.8	41.0	C60	62.9	64.0
A35	36.3	37.0	B42	43.8	45.0	C68	70.9	81.0
A38	39.3	40.0	B46	47.8	49.0	C75	77.9	79.0
A42	43.3	44.0	B51	52.8	54.0	C81	83.9	85.0
A46	47.3	48.0	B55	56.8	58.0	C85	87.9	89.0
A51	52.3	53.0	B60	61.8	63.0	C90	92.9	94.0
A55	56.3	57.0	B68	69.8	71.0	C96	98.9	100.0
A60	61.3	62.0	B75	76.8	78.0	C105	107.9	109.0
A68	69.3	70.0	B81	82.8	84.0	C112	114.9	116.0
A75	76.3	77.0	B85	86.8	88.0	C120	122.9	124.0
A80	81.3	82.0	B90	91.8	93.0	C128	130.9	132.0
A85	86.3	87.0	B97	98.8	100.0	C136	138.9	140.0
A90	91.3	92.0	B105	106.8	108.0	C144	146.9	148.0
A96	97.3	98.0	B112	113.8	115.0	C158	160.9	162.0
A105	106.3	107.0	B120	121.8	123.0	C162	164.9	166.0
A112	113.3	114.0	B128	129.8	131.0	C173	175.9	177.0
A120	121.3	122.0	B136	137.8	139.0	C180	182.9	184.0
A128	129.3	130.0	B144	145.8	147.0	C195	197.9	199.0

D BELTS			B158	159.8	161.0	C210	212.9	214.0
Standard Belt No.	Pitch Length	Outside Length	B173	174.8	176.0	C240	240.9	242.0
			B180	181.8	183.0	C270	270.9	272.0
D120	123.3	125.0	B195	196.8	198.0	C300	300.9	302.0
D128	131.3	133.0	B210	211.8	213.0	C360	360.9	362.0
D144	147.3	149.0	B240	240.3	241.5	C390	390.9	392.0
D158	161.3	163.0	B270	270.3	271.5	C420	420.9	422.0
D162	165.3	167.0	B300	300.3	301.5			
D173	176.3	178.0						
D180	183.3	185.0	E BELTS			E BELTS		
D195	198.3	200.0	Standard Belt No.	Pitch Length	Outside Length	Standard Belt No.	Pitch Length	Outside Length
D210	213.3	215.0						
D240	240.8	242.0	E180	184.5	187.5	E360	361.0	364.0
D270	270.8	272.5	E195	199.5	202.5	E390	391.0	394.0
D300	300.8	302.5	E210	214.5	217.5	E420	421.0	424.0
D330	330.8	332.5	E240	241.0	244.0	E480	481.0	484.0
D360	360.8	362.5	E270	271.0	274.0	E540	541.0	544.0
D390	390.8	392.5	E300	301.0	304.0	E600	601.0	604.0
D420	420.8	422.5	E330	331.0	334.0			
D480	480.8	482.5						
D540	540.8	542.5						
D600	600.8	602.5						

STANDARD "V" BELT LENGTHS (IN.) *(cont.)*

3V Belts		5V Belts		8V Belts	
3V250	25.0	5V500	50.0	8V1000	100.0
3V265	26.5	5V530	53.0	8V1060	106.0
3V280	28.0	5V560	56.0	8V1120	112.0
3V300	30.0	5V600	60.0	8V1180	118.0
3V315	31.5	5V630	63.0	8V1250	125.0
3V335	33.5	5V670	67.0	8V1320	132.0
3V355	35.5	5V710	71.0	8V1400	140.0
3V375	37.5	5V750	75.0	8V1500	150.0
3V400	40.0	5V800	80.0	8V1600	160.0
3V425	42.5	5V850	85.0	8V1700	170.0
3V450	45.0	5V900	90.0	8V1800	180.0
3V475	47.5	5V950	95.0	8V1900	190.0
3V500	50.0	5V1000	100.0	8V2000	200.0
3V530	53.0	5V1060	106.0	8V2120	212.0
3V560	56.0	5V1120	112.0	8V2240	224.0
3V600	60.0	5V1180	118.0	8V2360	236.0
3V630	63.0	5V1250	125.0	8V2500	250.0
3V670	67.0	5V1320	132.0	8V2650	265.0
3V710	71.0	5V1400	140.0	8V2800	280.0
3V750	75.0	5V1500	150.0	8V3000	300.0
3V800	80.0	5V1600	160.0	8V3150	315.0
3V850	85.0	5V1700	170.0	8V3350	335.0
3V900	90.0	5V1800	180.0	8V3550	355.0
3V950	95.0	5V1900	190.0	8V3750	375.0
3V1000	100.0	5V2000	200.0	8V4000	400.0
3V1060	106.0	5V2120	212.0	8V4250	425.0
3V1120	112.0	5V2240	224.0	8V4500	450.0
3V1180	118.0	5V2360	236.0	8V5000	500.0
3V1250	128.0	5V2500	250.0		
3V1320	132.0	5V2650	265.0		
3V1400	140.0	5V2800	280.0		
		5V3000	300.0		
		5V3150	315.0		
		5V3350	335.0		
		5V3550	355.0		

If the 60" "B" section belt shown is made 3/10" longer, it will be code marked 53 rather than 50. If made 3/10" shorter, it will be marked 47. While both have the belt number B60, they cannot be used in a set because of the difference in length.

Typical Code Marking

B60   MANUFACTURER'S NAME   50

Nominal
size and length

Length
code number

WIRE ROPE CHARACTERISTICS
FOR 6 STRAND BY 19 WIRE TYPE

WR Diameter (in.)	Weight (lbs./foot)	Breaking Point (lbs.)	Safe Load (lbs.)
¼	0.10	4,800	675
5/16	0.16	7,400	1,000
⅜	0.23	10,600	1,500
7/16	0.31	14,400	2,000
½	0.40	18,700	2,400
9/16	0.51	23,600	3,300
⅝	0.63	29,000	4,000
¾	0.90	41,400	6,000
⅞	1.23	56,000	8,000
1	1.60	72,800	10,000
1⅛	2.03	91,400	13,000
1¼	2.50	112,400	16,000
1⅜	3.03	135,000	19,000
1½	3.60	160,000	22,000
1¾	4.90	216,000	30,500
2	6.40	278,000	40,000
2½	10.00	424,000	60,000

The above values are for vertical pulls
at average ambient temperatures.

CABLE CLAMPS PER WIRE ROPE SIZE

Wire Rope Diameter (in.)	Number of Clamps Required	Clip Spacing (in.)	Rope Turn-back (in.)
1/8	2	3	3 1/4
3/16	2	3	3 3/4
1/4	2	3 1/4	4 3/4
5/16	2	3 1/4	5 1/4
3/8	2	4	6 1/2
7/16	2	4 1/2	4
1/2	3	5	11 1/2
9/16	3	5 1/2	12
5/8	3	5 3/4	12
3/4	4	6 3/4	18
7/8	4	8	19
1	5	8 3/4	26
1 1/8	6	9 3/4	34
1 1/4	6	10 3/4	37
1 7/16	7	11 1/2	44
1 1/2	7	12 1/2	48
1 5/8	7	13 1/4	51
1 3/4	7	14 1/2	53
2	8	16 1/2	71
2 1/4	8	16 1/2	73
2 1/2	9	17 3/4	84
2 3/4	10	18	100
3	10	18	106

WIRE ROPE SLINGS—STRAIGHT LEG

Wire Rope Diameter (in.)	Straight 1 Leg	Choker 1 Leg	60° Choker 2 Leg	45° Choker 2 Leg	30° Choker 2 Leg
			Capacity (tons)		
1/4	1/2	1/3	2/3	1/2	1/3
3/8	1	3/4	1 1/4	1	3/4
1/2	2	1 1/2	2 1/2	2	1 1/2
5/8	3	2	4	3	2
3/4	4	3	5	4	3
1	7	5	8	7	5
1 1/4	10	7	12	9	7
1 1/2	13	9	16	13	9
2	21	15	27	22	15
2 1/2	28	22	38	31	22
3	36	28	49	40	28
3 1/2	40	34	59	48	34

10-24

WIRE ROPE SLINGS – BASKET

Wire Rope Diameter (in.)	60° Basket 2 Leg	45° Basket 2 Leg	30° Basket 2 Leg	60° Basket 4 Leg	45° Basket 4 Leg	30° Basket 4 Leg
			Capacity (tons)			
1/4	2/3	1/2	1/3	1	1	3/4
3/8	1 1/2	-	3/4	3	2	1 1/2
1/2	2 1/2	2	1 1/2	5	4	3
5/8	4	3	2	7	6	4
3/4	5	4	3	11	9	6
1	9	7	5	18	15	10
1 1/4	13	11	7	26	21	15
1 1/2	17	14	10	35	28	20
2	27	22	15	53	44	31
2 1/2	38	31	22	75	61	43
3	49	40	29	97	80	56
3 1/2	59	49	34	118	97	68

10-25

SAFE LOADS FOR SHACKLES

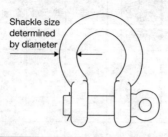

Shackle size determined by diameter

Size (in.)	Safe Load at 90° (tons)
$1/4$	$1/3$
$5/16$	$1/2$
$3/8$	$3/4$
$7/16$	1
$1/2$	$1\frac{1}{2}$
$5/8$	2
$3/4$	3
$7/8$	4
1	$5\frac{1}{2}$
$1\frac{1}{8}$	$6\frac{1}{2}$
$1\frac{1}{4}$	8
$1\frac{3}{8}$	10
$1\frac{1}{2}$	12
$1\frac{3}{4}$	16
2	21
$2\frac{1}{4}$	27
$2\frac{1}{2}$	34
$2\frac{3}{4}$	40
3	50

ROPE CHARACTERISTICS

Rope Diameter (in.)	Safe Load Ratio	Nylon		Polypropylene		Manila	
		Break Lbs.	Lbs./ 100 Feet	Break Lbs.	Lbs./ 100 Feet	Break Lbs.	Lbs./ 100 Feet
3/16	10:1	1,000	1.0	800	0.7	406	1.5
1/4	10:1	1,650	1.5	1,250	1.2	540	2.0
5/16	10:1	2,550	2.5	1,900	1.8	900	2.9
3/8	10:1	3,700	3.5	2,700	2.8	1,220	4.1
7/16	10:1	5,000	5.0	3,500	3.8	1,580	5.3
1/2	9:1	6,400	6.5	4,200	4.7	2,380	7.5
9/16	8:1	8,000	8.3	5,100	6.1	3,100	10.4
5/8	8:1	10,400	10.5	6,200	7.5	3,960	13.3
3/4	7:1	14,200	14.5	8,500	10.7	4,860	16.7
13/16	7:1	17,000	17.0	9,900	12.7	5,850	19.5
7/8	7:1	20,000	20.0	11,500	15.0	6,950	22.4
1	7:1	25,000	26.4	14,000	18.0	8,100	27.0
1 1/16	7:1	28,800	29.0	16,000	20.4	9,450	31.2
1 1/8	7:1	33,000	34.0	18,300	23.8	10,800	36.0
1 1/4	7:1	37,500	40.0	21,000	27.0	12,200	41.6
1 5/16	7:1	43,000	45.0	23,500	30.4	13,500	47.8
1 1/2	7:1	53,000	55.0	29,700	38.4	16,700	60.0
1 5/8	7:1	65,000	66.5	36,000	47.6	20,200	74.5
1 3/4	7:1	78,000	83.0	43,000	59.0	23,800	89.5
2	7:1	92,000	95.0	52,000	69.0	28,000	108
2 1/8	7:1	106,000	109	61,000	80.0		
2 1/4	6:1	125,000	129	69,000	92.0		
2 1/2	6:1	140,000	149	80,000	107		
2 5/8	6:1	162,000	168	90,000	120		
2 7/8	6:1	180,000	189	101,000	137		
3	6:1	200,000	210	114,000	153		
3 1/4	6:1	250,000	264	137,000	190		
3 1/2	6:1	300,000	312	162,000	232		
4	6:1	360,000	380	190,000	276		

Lbs./foot = Rope weight per linear foot
Break lbs. = Tensile strength
Safe load ratio = Break strength to safe load
Example: 7/16" nylon rope break strength = 5,000 lbs.
5000/10 = 500 lbs. safe working load
Note: increased temperatures decrease rope strength.

MANILA ROPE SLINGS—STRAIGHT LEG

Rope Diameter (in.)	Straight 2 Leg	Straight 4 Leg	Choker 2 Leg	60° Choker 4 Leg	45° Choker 4 Leg	30° Choker 4 Leg
			Capacity (tons)			
1/2	1/2	1	1/3	2/3	1/2	1/3
3/4	3/4	1 1/2	3/4	1 1/4	1	3/4
1	1 1/2	3	1 1/4	2	1 1/2	1 1/4
1 1/2	3	6	2	4	3	2
2	5	10	4	7	6	4
2 1/2	7	15	6	10	8	6
3	10	20	8	14	12	8
3 1/2	14	29	11	20	16	11
4	17	34	13	23	19	13

10-28

MANILA ROPE SLINGS—BASKET

Rope Diameter (in.)	60° Basket 4 Leg	45° Basket 4 Leg	30° Basket 4 Leg	60° Basket 6 Leg	45° Basket 6 Leg	30° Basket 6 Leg
			Capacity (tons)			
1/2	3/4	2/3	1/2	1	3/4	2/3
3/4	1 1/2	1	3/4	2 1/4	2	1
1	2 1/2	2	1 1/2	4	3	2
1 1/2	5	4	3	8	6	4
2	9	7	5	13	11	7
2 1/2	13	11	7	19	16	11
3	18	15	10	27	22	15
3 1/2	25	21	15	38	31	22
4	30	24	17	44	36	25

10-29

STANDARD HAND SIGNALS FOR CONTROLLING CRANE OPERATIONS

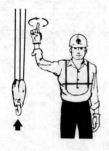

HOIST. Forearm vertical, forefinger pointing up, move hand in small horizontal circles.

LOWER. Arm extended downward, forefinger pointing down, move hand in small horizontal circles.

USE MAIN HOIST. Tap fist on head; then use regular signals.

STANDARD HAND SIGNALS FOR
CONTROLLING CRANE OPERATIONS *(cont.)*

USE WHIPLINE. Tap elbow with one hand; then use regular signals.

RAISE BOOM. Arm extended, fingers closed, thumb pointing upward.

LOWER BOOM. Arm extended, fingers closed, thumb pointing down.

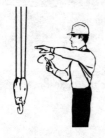

MOVE SLOWLY. One hand gives motion signal, other hand motionless in front of hand giving the motion signal.

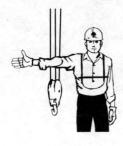

RAISE THE BOOM AND LOWER THE LOAD. Arm extended, thumb pointing up, flex fingers in and out.

LOWER THE BOOM AND RAISE THE LOAD. Arm extended, thumb pointing down, flex fingers in and out.

STANDARD HAND SIGNALS FOR
CONTROLLING CRANE OPERATIONS *(cont.)*

SWING. Arm extended, point with finger in direction of swing.

STOP. Arm extended, palm down, hold.

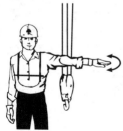

EMERGENCY STOP. Arm extended, palm down, move hand rapidly right and left.

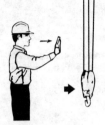

TRAVEL. Arm extended forward, hand open and slightly raised, pushing motion in direction of travel.

EXTEND BOOM. Both fists in front of body with thumbs pointing outward.

RETRACT BOOM. Both fists in front of body with thumbs pointing toward each other.

WIRE AND SHEET METAL GAUGES
(In Approximate Decimals of an Inch)

Gauge Numbers	United States	American or Brown & Sharpe	Washburn & Moen. Am. Steel & Wire Co. Roebling	Trenton Iron Co.	Birmingham or Stubs' Iron Wire	Stubs' Steel Wire	British Imperial
7.0	.500	—	.4900	—	—	—	.500
6.0	.489	—	.4615	—	—	—	.464
5.0	.438	—	.4305	.450	—	—	.432
4.0	.406	.460	.3938	.400	.454	—	.400
000	.375	.410	.3625	.360	.425	—	.372
00	.344	.365	.3310	.330	.380	—	.348
0	.313	.325	.3065	.305	.340	—	.324
1	.281	.289	.2830	.285	.300	.227	.300
2	.266	.258	.2625	.265	.284	.219	.276
3	.250	.229	.2437	.245	.259	.212	.252
4	.234	.204	.2253	.225	.238	.207	.232
5	.219	.182	.2070	.205	.220	.204	.212
6	.203	.162	.1920	.190	.203	.201	.192
7	.188	.144	.1770	.175	.180	.199	.176
8	.172	.128	.1620	.160	.165	.197	.160
9	.156	.144	.1483	.145	.148	.194	.144

WIRE AND SHEET METAL GAUGES (cont.)
(In Approximate Decimals of an Inch)

Gauge Numbers	United States	American or Brown & Sharpe	Washburn & Moen. Am. Steel & Wire Co. Roebling	Trenton Iron Co.	Birmingham or Stubs' Iron Wire	Stubs' Steel Wire	British Imperial
10	.141	.102	.1360	.130	.134	.191	.128
11	.125	.0907	.1206	.1175	.120	.188	.116
12	.109	.0808	.1055	.105	.109	.185	.104
13	.0938	.0720	.0916	.0925	.095	.182	.092
14	.0781	.0641	.0800	.0806	.083	.180	.080
15	.0703	.0571	.0720	.070	.072	.178	.072
16	.0625	.0508	.0625	.061	.065	.175	.064
17	.0563	.0453	.0540	.0525	.058	.172	.056
18	.0500	.0403	.0475	.045	.049	.168	.048
19	.0438	.0359	.0410	.040	.042	.164	.040
20	.0375	.0320	.0348	.035	.035	.161	.036
21	.0344	.0285	.0318	.031	.032	.157	.032
22	.0313	.0253	.0286	.028	.028	.155	.028
23	.0281	.0226	.0258	.025	.025	.153	.024

24	.0250	.0201	.0230	.0225	.022	.151	.022
25	.0219	.0179	.0204	.020	.020	.148	.020
26	.0188	.0159	.0181	.018	.018	.146	.018
27	.0172	.0142	.0173	.017	.016	.143	.0164
28	.0156	.0126	.0162	.016	.014	.139	.0149
29	.0141	.0113	.0150	.015	.013	.134	.0136
30	.0125	.0100	.0140	.014	.012	.127	.0124
31	.0109	.0089	.0132	.013	.010	.120	.0116
32	.0102	.0080	.0128	.012	.009	.115	.0108
33	.0094	.0071	.0118	.011	.008	.112	.010
34	.0086	.0063	.0104	.010	.007	.110	.0092
35	.0078	.0056	.0095	.0095	.005	.108	.0084
36	.0070	.0050	.0090	.009	.004	.106	.00076
37	.0066	.0045	.0085	.0085	—	.103	.0068
38	.0063	.0040	.0080	.008	—	.101	.006
39	.0059	.0035	.0075	.0075	—	.099	.0052
40	.0055	.0031	.0070	.007	—	.097	.0048

CARBON MONOXIDE EXPOSURE LEVELS

Concentration Level	Exposure Time	Result of Exposure
9 ppm	Short-term exposure	Maximum allowable concentration for short-term exposure in living area
35 ppm	8 hr.	Maximum allowable concentration for a continuous exposure, in any 8-hr. period, according to federal law
200 ppm	2–3 hr.	Slight headache, tiredness, dizziness, nausea; maximum CO concentration exposure at any time
400 ppm	1–2 hr. After 3 hr. —	Frontal headaches Life threatening Maximum ppm in flue gas (on a free air basis) allowed
800 ppm	45 min. 2 hr. 2–3 hr.	Dizziness, nausea, and convulsions Unconscious Death
1,600 ppm	20 min. 1 hr.	Headache, dizziness, nausea Death
3,200 ppm	5–10 min. 30 min.	Headache, dizziness, nausea Death
6,400 ppm	1–2 min. 10–15 min.	Headache, dizziness, nausea Death
12,800 ppm	1–3 min.	Death

EFFECTS OF CARBON MONOXIDE EXPOSURE

	Concentration of Carbon Monoxide in ppm		
Hours of Exposure	**Barely Perceptable**	**Sickness**	**Death**
0.5	600	1000	2000
1.0	200	600	1600
2	100	300	1000
3	75	200	700
4	50	150	400
5	35	125	300
6	25	120	200
7	25	100	200
8	25	100	150

SELECTING RESPIRATORS BY HAZARD

Hazard	Respirator
Oxygen deficiency	Self-contained breathing apparatus. Hose mask with blower. Combination air-line respirator with auxiliary self-contained air supply or an air-storage receiver with alarm.
Gas and vapor contaminants immediately dangerous to life and health	Self-contained breathing apparatus. Hose mask with blower. Air-purifying full facepiece respirator (for escape only). Combination air-line respirator with auxiliary self-contained air supply or an air-storage receiver with alarm.
Not immediately dangerous to life and health	Air-line respirator. Hose mask without blower. Air-purifying, half-mask, or mouthpiece respirator with chemical cartridge.
Particulate contaminants immediately dangerous to life and health	Self-contained breathing apparatus. Hose mask with blower. Air-purifying, full facepiece respirator with appropriate filter. Self-rescue mouthpiece respirator (for escape only). Combination air-line respirator with auxiliary self-contained air supply or an air-storage receiver with alarm.
Not immediately dangerous to life and health	Air-purifying, half-mask, or mouthpiece respirator with filter pad or cartridge. Air-line respirator. Air-line abrasive-blasting respirator. Hose-mask without blower.

Immediately dangerous to life and health is a condition that either poses an immediate threat of severe exposure to contaminants (radioactive materials) or those which are likely to have adverse delayed effects on health.

SELECTING RESPIRATORS BY HAZARD *(cont.)*

Hazard	Respirator
Combination gas, vapor, and particulate contaminants immediately dangerous to life and health	Self-contained breathing apparatus. Hose mask with blower. Air-purifying, full facepiece respirator with chemical canister and appropriate filter (gas mask with filter). Self-rescue mouthpiece respirator (for escape only). Combination air-line respirator with auxiliary self-contained air-supply or an air-storage receiver with alarm.
Not immediately dangerous to life and health	Air-line respirator. Hose mask without blower. Air-purifying, half-mask, or mouthpiece respirator with chemical cartridge and appropriate filter.

Immediately dangerous to life and health is a condition that either poses an immediate threat of severe exposure to contaminants (radioactive materials) or those which are likely to have adverse delayed effects on health.

PERMISSIBLE NOISE EXPOSURES

Duration per Day (hr.)	Sound Level in dBA (Slow Response)
8	90
6	92
4	95
3	97
2	100
1½	102
1	105
½	110
¼ or less	115

EYE PROTECTION RECOMMENDATIONS

Operation	Hazards	Recommended Protectors
Acetylene—burning, Acetylene—cutting, Acetylene—welding	Sparks, harmful rays, molten metal, flying particles	7, 8, 9
Chemical handling	Splash, acid burns, fumes	2, 10 (for severe exposure add 10 over 2)
Chipping	Flying particles	1, 3, 4, 5, 6, 7A, 8A
Electric (arc) welding	Sparks, intense rays, molten metal	9, 11 (11 in combination with 4, 5, 6 in tinted lenses advisable)
Furnace operations	Glare, heat, molten metal	7, 8, 9 (for severe exposure add 10)
Grinding—light	Flying particles	1, 3, 4, 5, 6, 10
Grinding—heavy	Flying particles	1, 3, 7A, 8A (for severe exposure add 10)
Laboratory	Chemical splash, glass breakage	2 (10 when in combination with 4, 5, 6)
Machining	Flying particles	1, 3, 4, 5, 6, 10
Molten metals	Heat, glare, sparks, splash	7, 8 (10 in combination with 4, 5, 6 in tinted lenses)
Spot welding	Flying particles, sparks	1, 3, 4, 5, 6, 10

FILTER LENS SHADE NUMBERS FOR PROTECTION AGAINST RADIANT ENERGY

Welding Operation	Shade Number
Gas-shielded arc welding (nonferrous) $1/16$, $3/32$, $1/8$, $5/32$" diameter electrodes	11
Gas-shielded arc welding (ferrous) $1/16$, $3/32$, $1/8$, $5/32$" diameter electrodes	12
Shielded metal-arc welding $1/16$, $3/32$, $1/8$, $5/32$" diameter electrodes	10
Shielded metal-arc welding $3/16$, $7/32$, $1/4$" diameter electrodes	12
Shielded metal-arc welding $5/16$, $3/8$" diameter electrodes	14
Atomic hydrogen welding	10 to 14
Carbon-arc welding	14
Soldering	2
Torch brazing	3 or 4
Light cutting, up to 1"	3 or 4
Medium cutting, 1" to 6"	4 or 5
Heavy cutting, over 6"	5 or 6
Gas welding (light), up to $1/8$"	4 or 5
Gas welding (medium), $1/8$" to $1/2$"	5 or 6
Gas welding (heavy), over $1/2$"	6 or 8

10-43

EYE AND FACE PROTECTORS

1. GOGGLES, flexible fitting – regular ventilation
2. GOGGLES, flexible fitting – hooded ventilation
3. GOGGLES, cushioned fitting – rigid body
4. SPECTACLES, metal frame – with sideshields[1]
5. SPECTACLES, plastic frame – with sideshields
6. SPECTACLES, metal-plastic frame – with sideshields[1]
7. WELDING GOGGLES, eyecup type – tinted lenses
7A. CHIPPING GOGGLES, eyecup type – clear safety lenses
8. WELDING GOGGLES, coverspec type – tinted lenses
8A. CHIPPING GOGGLES, coverspec type – clear safety lenses
9. WELDING GOGGLES, coverspec type – tinted plate lens
10. FACE SHIELD (available with plastic or mesh window)
11. WELDING HELMETS

[1]**Non-side shield spectacles are available for limited hazard use requiring only frontal protection.**

TYPES OF FIRE EXTINGUISHERS

TYPE A: To extinguish fires involving trash, cloth, paper, and other wood- or pulp-based materials. The flames are put out by water-based ingredients or dry chemicals.

TYPE B: To extinguish fires involving greases, paints, solvents, gas, and other petroleum-based liquids. The flames are put out by cutting off oxygen and stopping the release of flammable vapors. Dry chemicals, foams, and halon are used.

TYPE C: To extinguish fires involving electricity. The combustion is put out the same way as with a type B extinguisher but, most importantly, the chemical in a type C <u>MUST</u> be nonconductive to electricity in order to be safe and effective.

TYPE D: To extinguish fires involving combustible metals. Please be advised to obtain important information from your local fire department on the requirements for type D fire extinguishers for your area.

Any combination of letters indicate that an extinguisher will put out more than one type of fire. Type BC will put out two types of fires. The size of the fire to be extinguished is shown by a number in front of the letter such as 100A. For example:

Class 1A will extinguish 25 burning sticks 40 inches long.

Class 1B will extinguish a paint thinner fire 2.5 square feet in size.

Class 100B will put out a fire 100 times larger than type 1B.

Here are some basic guidelines to follow:

- By using a type ABC you will cover most basic fires.
- Use fire extinguishers with a gauge and that are constructed with metal. Also note if the unit is U.L. approved.
- Utilize more than one extinguisher and be sure that each unit is mounted in a clearly visible and accessible manner.
- After purchasing any fire extinguisher always review the basic instructions for its intended use. Never deviate from the manufacturer's guidelines. Following this simple procedure could end up saving lives.

EXTENSION CORD SIZES FOR PORTABLE TOOLS

Cord Length (ft.)	Full-Load Rating of Tool in Amperes at 115 Volts					
	0 to 2.0	2.1 to 3.4	3.5 to 5.0	5.1 to 7.0	7.1 to 12.0	12.1 to 16.0
	Wire Size (AWG)					
25	18	18	18	16	14	14
50	18	18	18	16	14	12
75	18	18	16	14	12	10
100	18	16	14	12	10	8
200	16	14	12	10	8	6
300	14	12	10	8	6	4
400	12	10	8	6	4	4
500	12	10	8	6	4	2
600	10	8	6	4	2	2
800	10	8	6	4	2	1
1,000	8	6	4	2	1	0

CHAPTER 11
Symbols and Abbreviations

PIPE LEGENDS		
Classification	**Color of Band**	**Color of Letters**
Fire protection	Red	White
Dangerous	Yellow	Black
Safe	Green	Black
Protective	Blue	White

Outside Diameter of Pipe or Covering (in.)	**Width of Color Band (in.)**	**Size of Legend Letters (in.)**
¾ to 1¼	8	½
1½ to 2	8	2/4
2¼ to 6	12	1¼
8 to 10	24	2½
More than 10	32	3½

PIPING	
Heating	
——HPS——	High-pressure steam
——MPS——	Medium–pressure steam
——LPS——	Low-pressure steam
——HPC——	High-pressure condensate
——MPC——	Medium-pressure condensate
——LPC——	Low-pressure condensate
——HPR——	High-pressure return
——MPR——	Medium-pressure return
——LPR——	Low-pressure return
——BBD——	Boiler blowdown
—— PC ——	Pumped condensate
——CP——	Condensate pump discharge
——VPD——	Vacuum pump discharge
——MU——	Makeup water

Heating *(cont.)*

——ATV——   Atmospheric vent

——V——   Air reliel line (vent)

——FOD——   Fuel oil discharge

——FOG——   Fuel oil gauge

——FOS——   Fuel oil suction

——FOF——   Fuel oil flow

——FOR——   Fuel oil return

——FOV——   Fuel oil tank vent

——HWS —   Low-temperature hot water supply

——MTWS——   Medium-temperature hot water supply

——HTWS——   High-temperature hot water supply

——HWR——   Low-temperature hot water return

——MTWR——   Medium-temperature hot water return

——HTWR——   High-temperature hot water return

PIPING *(cont.)*

Heating *(cont.)*

—— A ——   Compressed air

—— VAC ——   Vacuum (air)

—(NAME)E—   Existing piping

⚹⚹(NAME)⚹⚹   Pipe to be removed

Air Conditioning and Refrigeration

—— RD ——   Refrigerant discharge

—— RS ——   Refrigerant suction

—— B ——   Brine supply

—— BR ——   Brine return

—— C ——   Condenser water supply

—— CR ——   Condenser water return

—CHWS—   Chilled water supply

—CHWR—   Chilled water return

—— FILL ——   Fill line

—— H ——   Humidification line

—— D ——   Drain

PIPING (cont.)

Air Conditioning and Refrigeration

—HCS—   Hot/chilled water supply

—HCR—   Hot/chilled water return

—RL—   Refrigerant liquid

—HPWS—   Heat pump water supply

—HPWR—   Heat pump water return

Plumbing

—————   Soil, waste, or leader (above grade)

— — —   Soil, waste, or leader (below grade)

— S —   Sanitary drain above floor or grade

····· S ·····   Sanitary drain below floor or grade

— ST —   Storm drain above floor or grade

····· ST ·····   Storm drain below floor or grade

—CD—   Condensate drain above floor or grade

·····CD·····   Condensate drain below floor or grade

— — — —   Vent

—·—·—·—   Cold water

Plumbing *(cont.)*

—··—··—··	Hot water
———··——··	Hot water return
—G—G—	Gas
—ACID—	Acid water
—DWS—	Drinking water supply
—DWR—	Drinking water return
—FOS—	Fuel oil suction
—FOR—	Fuel oil return
—FOV—	Fuel oil tank vent
—VAC—	Vacuum (air)
—A—	Compressed air
—(NAME I)—	Chemical supply pipes
—☐D—	Floor drain
Y	Funnel drain, open

FITTINGS

Fitting connections are often specified with the fitting symbol. For example, an elbow could have the following types of connections:

Flanged   Soldered

Welded   Bell & spigot

Screwed   Solvent cement

Fittings are shown with screwed connections unless specified.

Bushing

Reducer, concentric, straight crown

Reducer, concentric

Reducer, concentric, straight invert

Tee, single sweep

FITTINGS *(cont.)*

Symbol	Description
	Cap
	Connection, bottom
	Connection, top
	Coupling (joint)
	Cross
	Elbow, 90°
	Elbow, 45°
	Elbow, turned up
	Elbow, turned down
	Elbow, reducing, show sizes

FITTINGS (cont.)

	Elbow, base
	Elbow, long radius
	Elbow, double branch
	Elbow, side outlet, outlet up
	Elbow, side outlet, outlet down
	Lateral
	Reducer, concentric
	Reducer, concentric, straight invert
	Reducer, concentric, straight crown
	Tee

FITTINGS (cont.)

$+\!\!-\!\!\bigcirc\!\!-\!\!+$	Tee, outlet up
$+\!\!-\!\!\bigcirc\!\!-\!\!+$	Tee, outlet down
${}_{6}\!\!-\!\!\overset{2}{+}\!\!-\!\!{}_{4}$	Tee, reducing, show sizes
$+\!\!-\!\!\bigcirc\!\!-\!\!+$	Tee, side outlet, outlet up
$+\!\!-\!\!\bigcirc\!\!-\!\!+$	Tee, side outlet, outlet down
▬	Union
─╫─	Union, screwed
─╫╫─	Union, flanged

PIPING SPECIALTIES

△	Air eliminator
Ⓢ	Air separator
□ AV	Air vent, automatic
MV	Air vent, manual
═══	Alignment guide

PIPING SPECIALTIES *(cont.)*

✕ PA	Anchor, intermediate
⊠ PA	Anchor, main
●	Ball joint
▭ EJ - 1	Expansion joint
⎍	Expansion loop
⬙⬙⬙	Flexible connector
▢ FD	Floor drain
‖ OFM-1	Flowmeter, orifice
⬦ VFM-1	Flowmeter, venturi
▱ FS	Flow switch

PIPING SPECIALTIES *(cont.)*

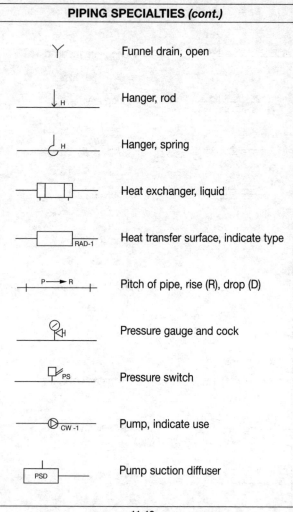

	Funnel drain, open
	Hanger, rod
	Hanger, spring
	Heat exchanger, liquid
	Heat transfer surface, indicate type
	Pitch of pipe, rise (R), drop (D)
	Pressure gauge and cock
	Pressure switch
	Pump, indicate use
	Pump suction diffuser

PIPING SPECIALTIES *(cont.)*

‖————‖	Spool piece, flanged
	Strainer
	Strainer, blow off
—8—	Strainer, duplex
—▭FO	Tank, indicate use
	Thermometer
	Thermometer well, only
🅣	Thermostatic, electric
Ⓣ	Thermostatic, pneumatic
Ⓣ	Thermostatic, self-contained

PIPING SPECIALTIES (cont.)

$\otimes_{\text{F \& T}}$	Traps, steam, indicate type
(water trap symbol)	Trap, water
$\square_{\text{UH 1,2}}$	Unit heater, indicate type

VALVES

**Valves are shown with screwed connections
unless specified**

(gate valve symbol)	Gate valve	(angle gate valve symbol)	Angle gate valve
(globe valve symbol)	Globe valve	(angle globe valve symbol)	Angle globe valve
(check valve symbol)	Check valve	(angle check valve symbol)	Angle check valve
(quick opening valve symbol)	Quick opening valve		
(float opening valve symbol)	Float opening valve	(post indicator gate valve symbol)	Post indicator gate valve

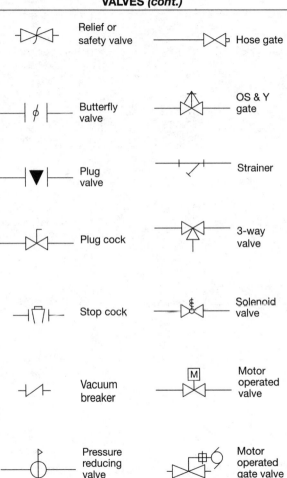

VALVES (cont.)

Relief or safety valve

Hose gate

Butterfly valve

OS & Y gate

Plug valve

Strainer

Plug cock

3-way valve

Stop cock

Solenoid valve

Vacuum breaker

Motor operated valve

Pressure reducing valve

Motor operated gate valve

VALVES (cont.)

Diaphragm operated valve		Gas cock	
Reducing valve (self-actuated)		Shock absorber	
Reducing valve (external pilot connection)		Temperature and pressure relief valve	
Lock shield		Governor	
2-way automatic control		Thermometer	
3-way automatic control		Pressure gauge	

STEAM SYMBOLS

 Steam trap

 Steam trap with integral strainer

 Balanced pressure thermostatic TRP

 Float & thermostatic steam traps

 Inverted bucket trap

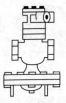

Control valve
with solenoid

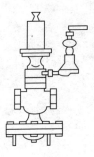

Pressure–temperature
regulator

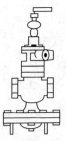

Temperature regulator
with solenoid

STEAM SYMBOLS *(cont.)*

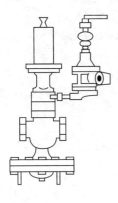

Pressure–temperature regulator with solenoid

H.C. temperature regulator

H.C. temperature regulator

Strainer

Balance master valve

 Balance master valve

 Radiator valve

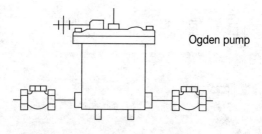 Ogden pump

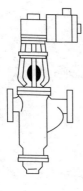

Scraper strainer

Type vs air vent

Air eliminator

CONTROL DIAGRAM SYMBOLS

AC INV	AC Inverter
AFS	Airflow station
AMS	Air measuring station
_ _ _ _ _	Control tubing
__ _ __	Control wiring
DO	Damper operator
DPS	Differential pressure switch
DPT	Differential pressure transmitter
EP	Electric–pneumatic switch
ES	End switch
FZ	Freezestat
FS	Flow switch
FT	Flow transmitter
H	Humidifer

CONTROL DIAGRAM SYMBOLS (cont.)

Symbol	Description
⊏─┤H┤	Humidity sensor
⊏─┤HL┤	High limit switch
HC	Humidity controller
HOA	Hand-off automatic switch
MS | HOA	Combined motor starter and Hand-off automatic switch
M	Motor
MS	Motor starter
NC	Normally closed
NO	Normally opened
PS	Pressure sensor
PC	Pressure controller
PE	Pheumatic–electric switch
PT	Pressure transmitter
R	Relay

Symbol	Description
SDT	Smoke detector
SP	Static pressure w/pitot tube
SPT	Static pressure transmitter
SR	Switching relay
SS	Signal selector
T	Temperature sensor
T	Temperature sensor w/bulb
TC	Temperature controller
TDR	Time delay relay
VSD	Variable speed drive
VP	Velocity pressure sensor
VC	Volume controller
VT	Volume transmitter
⊙	Reference static pressure

◑	Gauge
⊠	Light
⊟	On–off switch
⏷	3-position switch
ATC	Automatic temperature control panel
▤	CRT Display
⊗	PC with printer
Ⓜ	Main air
Ⓗ	Humidistat
Ⓣ	Thermostat

AIR MOVING DEVICES AND COMPONENTS

Fans

 Axial flow

 Centrifugal

 Propeller

 Roof ventilator, intake

 Roof ventilator, exhaust

 Roof ventilator, louvered

Ductwork

 Direction of flow

 Duct size, first figure is side shown

Ductwork

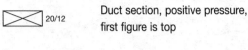

Duct section, positive pressure, first figure is top
20/12

Duct section, negative pressure
500/300
500/300

Change of elevation, rise (R), drop (D)
→R

Access doors, vertical or horizontal
AD 10/10

Acoustical lining (insulation)

Gooseneck hood (cowl)

Cowl, (gooseneck) and flashing

Flexible connection

AIR MOVING DEVICES AND COMPONENTS *(cont.)*

Ductwork

 Flexible duct

 Sound attenuator

 Terminal unit, mixing

 Terminal unit, reheat

 Terminal unit, variable volume

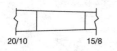 Transition

  Turning vanes

 Detectors, fire and/or smoke

Dampers

	Adjustable blank off
	Back draft damper
	Pneumatic-operated damper
	Electric-operated damper
	Control, electric
	Control, pneumatic
	Fire damper and sleeve, provide access door
	Vertical position
	Horizontal position
	Manual volume

Dampers

	Manual splitter
	Smoke damper, provide access door
	Standard branch, supply or return, no splitter
	Heater, duct, electric
	Point of change in duct construction (by static pressure class)
20 × 12	Duct (1st figure, side shown 2nd figure, side not shown)
	Acoustical lining duct dimensions for net free area

AIR MOVING DEVICES AND COMPONENTS *(cont.)*

Dampers

	Direction of flow
S 30 × 12	Duct section (supply)
E OR R 20 × 12	Duct section (exhaust or return)
R →	Inclined rise (R) or drop (D), arrow in direction of air flow
	Transitions give sizes. Note F.O.T.: flat on top of F.O.B: flat on bottom if applicable
S ← R →	Standard branch for supply & return (no splitter)
	Splitter damper

Dampers

 Volume damper manual operation

 Automatic dampers motor operated

 Access door (AD)
Access panel (AP)

 Fire damper:
Show ◀ vertical pos.
Show ◆ horiz. pos.

 Ceiling damper or alternate
protection for fire rated CLG

 Turning waves

 Back draft damper

AIR MOVING DEVICES AND COMPONENTS *(cont.)*

Dampers

$\dfrac{20 \times 12 \text{ SG}}{700 \text{ CFM}}$	Supply grille (SG)
$\dfrac{20 \times 12 \text{ RG}}{700 \text{ CFM}}$	Return (RG) or exhaust (EG) grille (note at FLR or GLG)
$\dfrac{20 \times 12 \text{ SR}}{700 \text{ CFM}}$	Supply register (SR) (a grille + integral vol. control)

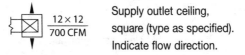

$\dfrac{20 \times 12 \text{ GR}}{700 \text{ CFM}}$	Exhaust or return air inlet ceiling (indicate type)
$\dfrac{20}{700 \text{ CFM}}$	Supply outlet ceiling, round (type as specified). Indicate flow direction.
$\dfrac{12 \times 12}{700 \text{ CFM}}$	Supply outlet ceiling, square (type as specified). Indicate flow direction.

Dampers

 Terminal unit (give type and/or schedule)

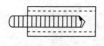 Combination diffuser and light fixture

 Door grille

 Sound trap

 Fan and motor with belt guard and flexible connections

 Ventilating unit (type as specified)

 Unit heater (downblast)

Dampers

Unit heater (horizontal)

Unit heater (centrifugal fan) plan

Thermostat

Power or gravity roof ventilator—exhaust (ERV)

Power or gravity roof ventilator—intake (SRV)

Power or gravity roof ventilator—louvered

Louvers and screen

Grilles, Registers and Diffusers

 Exhaust grille or register

 Supply grille or register

Supply outllet

Exhaust inlet

 Grille or register, sidewall

Grille or register, ceiling

Heat stop for fire rated ceiling

 Louver and screen

 Louver, door or wall

Door grille

Undercut door

 Ceiling diffuser, rectangular

Grilles, Registers and Diffusers

 Ceiling diffuser, round

 Diffuser, linear

 Diffuser and light fixture combination

 Transfer grille assembly

REFRIGERATION

Compressors

 Centrifugal

 Reciprocating

 Rotary

 Rotary screw

REFRIGERATION *(cont.)*

Condensers

 Air cooled

 Evaporate

 Water cooled

Condensing Units

 Air cooled

 Water cooled

Condenser—Evaporator

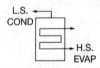 Cascade system

Cooling Towers

 Cooling towers

REFRIGERATION *(cont.)*

Cooling Towers

Spray pond

Evaporators

Finned coil

Forced convection

Immersion cooling unit

Plate coil

Pipe coil

Liquid Chillers (Chillers Only)

Direct expansion

Flooded

REFRIGERATION *(cont.)*	
Cooling Towers	
	Tank, closed
	Tan, open
Chilling Units	
	Absorption
	Centrifugal
	Reciprocating
	Rotary screw
CONTROLS	
Refrigeration Controls	
	Capillary tube
	Expansion valve, hand
	Expansion valve, automatic

Refrigeration Controls

Expansion valve, thermostatic

Float valve, high side

Float valve, low side

Thermal bulb

Solenoid valve

Constant pressure valve, suction

Evaporator pressure regulating valve, thermostatic, throttling type

Evaporator pressure regulating valve, thermostatic, snap action

Evaporator, pressure regulating valve, throttling type, evaporator side

Compressor suction valve, pressure limiting, throttling type, compressor side

Thermo-suction valve

Snap action valve

CONTROLS *(cont.)*

Refrigeration Controls

   Refrigerant reversing valve

**Temperature or Temperature-Actuated
Electrical or Flow Controls**

   Thermostat, self-contained

   Thermostat, remote bulb

**Pressure of Pressure-Actuated
Electrical or Flow Controls**

   Pressure switch

   Pressure switch, dual (high–low)

   Pressure switch, differential oil pressure

   Automatic reducing valve

   Automatic bypass valve

   Valve, pressure reducing

   Valve, condenser water regulating

AUXILIARY EQUIPMENT

Refrigerant

	Filter
	Strainer
	Filter and drier
	Scale trap
	Drier
	Vibration absorber
	Heat exchanger
	Oil separator
	Sight glass
	Fusible plug
	Rupture disc
	Receiver, high pressure horizontal

AUXILIARY EQUIPMENT *(cont.)*

Refrigerant

Receiver, high pressure, vertical

Receiver, low pressure

Intercooler

Intercooler/desuperheater

ENERGY RECOVERY EQUIPMENT

Condenser, Double Bundle

Condenser, double bundle

Air-to-Air Energy Recovery

Rotary heat wheel

Coil loop

ENERGY RECOVERY EQUIPMENT *(cont.)*

Air-to-Air Energy Recovery

  Heat pipe

  Fixed plate

  Plate fin, gross flow

POWER RESOURCES

  Motor, electric, number indicates horsepower

  Motor, electric, (number for identification of description in specifications)

  Engine (indicate fuel)

  Gas turbine

  Steam turbine

  Steam turbine, condensing

MISCELLANEOUS PLAN SYMBOLS

  New connection to existing

  Thermostat

  Temperature sensor

MISCELLANEOUS PLAN SYMBOLS *(cont.)*

Symbol	Description
(H)	Humidity
[H]	Humidity sensor
(S)	Switch
[P]	Pressure sensor
(XX)	Sheet note (number), applies only on the sheet it appears on
[XX]	Coordination point between floor plans and diagrams (number)
⬡X	Demolition note (number)
(L) (L)	Plenum light

Direction of view
Section number
Drawing on which section
or detail is shown

Detail or section number
Drawing from which section
or detail is taken

Equipment symbol (see schedule)
Equipment designation
Equipment reference number
System number if applicable

Equipment symbol (see schedule)
Equipment designation
Equipment reference number

MISCELLANEOUS PLAN SYMBOLS *(cont.)*

Symbol	Description
	Existing to remain
	Existing to be removed
	New work
	Supply air up
	Supply air down
	Exhaust or return air up
	Exhaust or return air down
	Round duct up
	Round duct down
24 × 36	Rectangular duct size (first figure—side shown)
36 × 14∅	Flat oval duct size (first figure—side shown)
→	Direction of flow
UP DN	Duct inclined rise or drop in direction of flow

MISCELLANEOUS PLAN SYMBOLS (cont.)

	90° elbow with turning vanes
	45° elbow (no vanes)
	Supply or return branch connection
	Supply or return with spin collar connection
	Lateral connection round ductwork
	Conical tee round ductwork
	Duct with internal lining

MATHEMATICAL SYMBOLS

a^n	a raised to the power n
$\sqrt{a}$, $a^{0.5}$	Square root of a
∞	Infinity
%	Percent
Σ	Summation of
ln	Natural log
log	Logarithm to base 10

SUBSCRIPTS

a,b,...	Referring to different phases, states or physical conditions of a substance, or to different substances
a	Air
a	Ambient
b	Barometric (pressure)
c	Referring to critical state or critcal value
c	Convection
db	Dry bulb
dp	Dew point
e	Base of natural logarithms
f	Referring to saturated liquid
f	Film
fg	Referring to evaporation or condensation
F	Friction
g	Referring to saturated vapor
h	Referring to change of phase in evaporation
H	Water vapor
i	Referring to saturated solid
i	Internal
if	Referring to change of phase in melting
ig	Referring to change of phase in sublimation
k	Kinetic
L	Latent
m	Mean value

SUBSCRIPTS *(cont.)*	
M	Molar basis
o	Referring to initial or standard states or conditions
p	Referring to constant pressure conditions or processes
p	Potential
r	Refrigerant
r	Radiant or radiation
s	Referring to moist air at saturation
s	Sensible
s	Referring to isentropic conditions or processes
s	Static (pressure)
s	Surface
t	Total (pressure)
T	Referring to isothermal conditions or processes
v	Referring to constant volume conditions or processes
v	Vapor
v	Velocity (pressure)
w	Wall
w	Water
wb	Wet bulb
1,2,...	Different points in a process, or different instants of time

STANDARD PLAN ABBREVIATIONS

A	Compressed air line or area
ABC	Above ceiling
AC	Air chamber, alternating current
A/C	Air conditioning
AFF	Above finished floor
AFG	Above finished grade
AL	Aluminum
AMB	Ambient
AMP	Amphere
AP	Access panel
APPROX	Approximate
ARR	Arrangement
ATC	Automatic temperature control or at ceiling
ATM	Atmosphere
AUTO	Automatic
AUX	Auxiliary
AVG	Average
BBD	Boiler blowdown
BF	Boiler feed

STANDARD PLAN ABBREVIATIONS *(cont.)*

BHP	Boiler horsepower or brake horsepower
BOD	Bottom of duct
BOP	Bottom of pipe
BOT	Bottom
BP	Back pressure
B & S	Bell-and-spigot
BSMT	Basement
BTU	British thermal unit
BV	Butterfly valve
°C	Degrees Celsius
C	Condensate line
C TO C	Center to center
CA	Compressed air
CAL	Calorie
CAP	Capacity
CD	Condensate drain
CF	Chemical feed or cubic foot
CFH	Cubic feet per hour
CFM	Cubic foot (feet) per minute
CI	Cast iron

CIRC	Circular
CL	Center line
CM	Centimeter
CM²	Square centimeter
CM³	Cubic centimeter
CO	Cleanout
COL	Column
CONC	Concrete or concentric
CONN	Connect or connection
CONT	Continuation
CPVC	Chlorinated polyvinyl chloride
CR	Condenser return
CRW	Chemical resistant waste
CS	Condenser supply
CTR	Center
CU	Cubic
CU FT.	Cubic feet
CU IN.	Cubic inches
CV	Check valve
CW	Cold water
CWR	Cold water riser

STANDARD PLAN ABBREVIATIONS (cont.)

D	Drain or deep
DB	Dry bulb
DDC	Direct digital control
DEG	Degree
DELTAT	Temperature difference
DET	Detail
DIA	Diameter
DISC	Disconnect
DM	Decimeter
DM^2	Square decimeter
DM^3	Cubic decimeter
DN	Down
DP	Dew point temperature
DR	Drain
DWG	Drawing
EA	Each or exhaust air
EAT	Entering air temperature

STANDARD PLAN ABBREVIATIONS *(cont.)*

E to C	End to center
EDR	Equivalent direct radiation
EER	Energy efficiency ratio
EFF	Efficiency
EJ	Expansion joint
EL	Elevation
ELB	Elbow
ELEC	Electrical
ENT	Entering
ESP	External static pressure
ET	Expansion tank
EVAP	Evaporator
EWT	Entering water temperature
EXH	Exhaust
EXP	Expansion
EXST	Existing
EXT	External

STANDARD PLAN ABBREVIATIONS *(cont.)*

°F	Degrees Fahrenheit
F	Fahrenheit
F&T	Float & thermostatic
FC	Flexible connector/flexible connection
FCO	Floor cleanout
FD	Floor drain
FDW	Feed water
FEC	Fire extinguisher cabinet
FF	Finish floor
FG	Finish grade
FHC	Fire hose cabinet
FLA	Full load amps
FLR	Floor
FM	Flow meter
FO	Fuel oil
FOV	Flush out valve
FPM	Foot (feet) per minute
FPS	Foot (feet) per second
FS	Flow switch or federal specs

STANDARD PLAN ABBREVIATIONS *(cont.)*

FT	Foot (feet)
FTG	Fitting
FU	Fixture unit
FV	Flush valve
G	Gram or gas line
GA	Gauge
GAL	Gallons
GALV	Galvanized
GC	General contractor
GL.V	Globe valve
GND	Ground
GPD	Gallons per day
GPH	Gallons per hour
GPM	Gallons per minute
GPS	Gallons per second
GR	Grain
GV	Gate valve
GWH	Gas water heater

STANDARD PLAN ABBREVIATIONS *(cont.)*

H_2O	Water
HB	Hose bibb
Hd	Head
HD	Head
Hg	Mercury
HG	Hot gas or hose gare
HGT	Height
HMD	Humidity
HORIZ	Horizontal
HP	Horsepower
HR	Hour
HTD	Heated
HTR	Heater
HW	Hot water
HWH	Hot water heater
HWR	Hot water return or Hot water riser
HWS	Hot water supply
HWT	Hot water tank

STANDARD PLAN ABBREVIATIONS *(cont.)*

HZ	Hertz
ID	Inside diameter
IN.	Inch
INHg	Inches of mercury
INSUL	Insulation
INT	International
INTL	Internal
IPS	Iron pipe size
IV	Indirect vent
IW	Indirect waste
J	Joule
K	Kelvin
KG	Kilogram
KM	Kilometer
KM²	Square kilometer
KPA	Kilopascal
KS	Kitchen sink

STANDARD PLAN ABBREVIATIONS *(cont.)*

KW	Kilowatt
L	Length or liter
LAT	Leaving air temperature
LB.	Pound
LBF	Pound-force
LIN	Lineal
LIQ	Liquid
LP	Low pressure
LRA	Locked rotor amps
LVL	Level
LVR	Louver
LWT	Leaving water temperature
M	Meter
M²	Square meter
M TYPE	Lightest type of rigid copper pipe
MAN	Manual
MAT	Mixed air temperature
MAX	Maximum

STANDARD PLAN ABBREVIATIONS *(cont.)*

MBH	Thousand British thermal units per hour
MFR	Manufacturer
MG	Milligram
MGD	Million gallons per day
MIN	Minimum or minute
ML	Milliliter
MM	Millimeter
MM³	Cubic millimeter
MOD	Model
MOUNT	Mounting
MPT	Male pipe thread
MTD	Mounted
MU	Make up
NA	Not applicable
NC	Normally closed
NEG	Negative
NIC	Not in contract
NO	Normally open

STANDARD PLAN ABBREVIATIONS *(cont.)*

NPHP	Name plate horsepower
NPS	Nominal pipe size
NPSH	Net positive suction head
NTS	Not to scale
O	Oxygen
OA	Outside air
OAT	Outside air temperature
OC	On center
OD	Outside diameter
OED	Open end duct
OF	Overflow
OFCI	Owner furnished contractor installed
OV	Outlet velocity
OZ.	Ounce
PA	Pascal
PC	Plumbing contractor
PCR	Pumped condensate return
PD	Pressure drop

STANDARD PLAN ABBREVIATIONS *(cont.)*

PF	Power factor
PG	Pressure gauge
PL	Plate
PNEU	Pneumatic
PRESS	Pressure
PROP	Propeller
PRV	Pressure-reducing valve
PSI	Pounds per square inch
PSIA	Pounds per square inch atmospheric or absolute
PSIG	Pounds per square in gauge
PV	Plug valve
QTY	Quantity
RA	Return air
RAD	Radius
RAT	Return air temperature
RD	Roof drain
R/E	Return and Exhaust

STANDARD PLAN ABBREVIATIONS *(cont.)*

RECOV	Recovery
RED	Reducer
REF	Reference
RH	Relative Humidity
REQD	Required
REV	Revision
RL	Refrigerant liquid
RM	Room
RS	Refrigerant suction
RTN	Return
RV	Relief valve
S	Switch
SA	Shock absorber or supply air
SAT	Supply air temperature
SCH	Schedule
SD	Smoke damper with access door
SDT	Saturated discharge temperature
SEC	Secondary, seconds

STANDARD PLAN ABBREVIATIONS *(cont.)*

SENS	Sensible
SEP	Separate
SEQ	Sequence
SER	Series
SERV	Service
SF	Service factor
SHT	Sheet
SI	International System of Units (metric)
SOL	Solenoid
SP	Static pressure
SPEC	Specification
SQ.	Square
SQ. FT.	Square feet
SS	Stainless steel
SSH	Static suction head
SST	Saturated suction temperature
STD	Standard
STH	Static total head

STANDARD PLAN ABBREVIATIONS *(cont.)*

STL	Steel
SUCT	Suction
SPLY	Supply
SV	Service
SVH	Static velocity head
SW	Service weight
SWS	Service water
TD	Temperature differential
TDH	Total dynamic head
TEMP	Temperature
TH	Thermometer
THK	Thick
TP	Total pressure
TSP	Total static pressure
UF	Under floor
UH	Unit heater
V	Vent or volt or volume

STANDARD PLAN ABBREVIATIONS *(cont.)*

VAC	Vacuum
VAV	Variable air volume
VB	Vacuum breaker
VCI	Vacuum cleaning inlet
VCL	Vacuum cleaning line
VEL	Velocity
VERT	Vertical
VIB	Vibration
VOL	Volume
VSD	Variable speed drive
VP	Velocity pressure
VTR	Vent thru roof
W	Watt, width, or wide
WB	Wet bulb
WCO	Wall cleanout
WG	Water gauge
WH	Water heater
WH-1	Water heater and number
XH	Extra heavy

EQUIPMENT DESIGNATIONS	
ACC	Air-cooled condenser
ACCU	Air-cooled condensing unit
ACU	Packaged air conditioning unit
AFM	Terminal air flow module
AFS	Air flow station
AHU	Air handling unit
BLR	Boiler
BFP	Boiler feed pump
BFS	Boiler feed set
C	Convector
CC	Cooling coil
CF	Charcoal filter
CFP	Chemical feed pump
CFU	Chemical feed unit
CH	Chiller
CP	Condensate pump
CRAC	Computer room air conditioning unit
CSG	Clean steam generator
CT	Cooling tower
DA	Deaerator
DAC	Door air curtain
DC	Duct coil
DEH	Dehumidifier
DUC	Dust collector
EUH	Electric unit heater
ET	Expansion tank
F	Fan
FCU	Fan coil unit
FM	Flow meter
FMB	Filter mixing box

EQUIPMENT DESIGNATIONS (cont.)

FOP	Fuel oil pump
FT	Flash tank
GC	Glycol cooler
H	Humidifier
HC	Heating coil
HE	Heat-exchanger
HEF	High-efficiency filter
HEPA	High-efficiency particulate air filter
HP	Heat pump
HRC	Heat recovery coil
HRU	Heat recovery unit
HV	Heating & ventilating unit
LFM	Laminar flow module
MB	Mixing box
MAH	Makeup air heater
OAI	Outside air intake
P	Pump
PF	Prefilter
PHC	Preheat coil
PRV	Pressure-reducing valve
PSG	Pure steam generator
PTAC	Packaged terminal air conditioner
RH	Relief hood
RV	Relief valve
SC	Steam coil
UH	Unit heater
V	Variable air volume box
VF	Variable air volume box—fan powered
WF	Wall fin radiation

HVAC AFFILIATION ABBREVIATIONS

AGA	American Gas Association
ANMC	American National Metric Council
ANSI	American National Standards Institute
ASHRAE	American Society of Heating, Refrigerating and Air Conditioning Engineers
ASME	American Society of Mechanical Engineers
ASPE	American Society of Plumbing Engineers
ASSE	American Society of Sanitary Engineers or American Society of Safety Engineers
ASTM	American Society for Testing Material
AWWA	American Water Works Association
BOCA	Building Officials Conference of America
MCA	Mechanical Contractors Association
NAPHCC	National Association of Plumbing, Heating, and Cooling Contractors
NBFU	National Board of Fire Underwriters
NBS	National Bureau of Standards
NFPA	National Fire Protection Association
NSF	National Sanitation Foundation Testing Laboratory

CHAPTER 12
Glossary

GLOSSARY

A

Absolute Humidity: The weight of water vapor per unit; grams per cubic foot or grams per cubic meter.

Absolute Pressure: The sum of gage pressure and atmospheric pressure.

Absolute Temperature: Temperature measured from absolute zero.

Absolute Zero: A temperature equal to $-459.6°F$ or $-273°C$.

Absorbent: The ability to soak up another substance.

Absorber: A solution or surface that is capable of soaking up (taking in) another substance or energy form.

Absorption: The action of a material in extracting one or more substances present in the atmosphere or a mixture of gases or liquids accompanied by physical change, chemical change, or both.

Absorption Chiller: A chiller that uses a brine solution and water to provide refrigeration without the aid of a compressor.

Absorption Refrigerator: Refrigerator that creates low temperatures by using the cooling effect formed when a refrigerant is absorbed by chemical substance.

Accessible Hermetic: Assembly of motor and compressor inside a single bolted housing unit.

Accumulator: A shell placed in a suction line for separating the liquid entrained in the suction gas.

Acrolein: An agent used with methyl chloride to call attention to the escape of refrigerant.

Activated Alumina: A form of aluminum oxide (Al_2O_3) that absorbs moisture readily and is used as a drying agent.

Activated Carbon: Specially processed carbon used as a filter drier; commonly used to clean air.

Add On Heat Pump: Installing a heat pump in conjunction with an existing fossil fuel furnace. The result is a dual fuel system.

Adiabatic Compression: Compressing refrigerant gas without removing or adding heat.

Adsorbent: Substance with the property to hold molecules of fluids without causing a chemical or physical change.

Adsorption: The adhesion of a thin layer of molecules of a gas or liquid to a solid object.

Aeration: Exposing a substance or area to air circulation.

AFUE (Annual Fuel Utilization Efficiency): The ratio of annual output of useful energy or heat to the annual energy input to the furnace. The higher the AFUE, the more efficient the furnace.

Air Ambient: Generally speaking, the air surrounding an object.

Air Break: An inverted opening placed in the chimney of a gas furnace to prevent back pressure from outside wind from reaching the furnace flame or pilot.

Air Change: The amount of air required to completely replace the air in a room or building; not to be confused with recirculated air.

Air Circulation: Natural or imparted motion of air.

Air Cleaner: A device designed for the purpose of removing airborne impurities such as dust, gases, vapors, fumes, and smoke. An air cleaner includes air washers, air filters, electrostatic precipitors, and charcoal filters.

Air Coil: Coil on some types of heat pumps used either as an evaporator or a condenser.

Air-Cooled Condenser: Heat of compression is transferred from condensing coils to surrounding air. This may be done either by convection or by a fan or blower.

Air Conditioning: The simultaneous control of all, or at least the first three, of the following factors affecting the physical and chemical conditions of the atmosphere within a structure— temperature, humidity, motion, distribution, dust, bacteria, odors, toxic gases, and ionization.

Air Conditioning System, Central Fan: A mechanical indirect system of heating, ventilating, or air conditioning in which the air is treated or handled by equipment located outside the rooms

GLOSSARY (cont.)

served, usually at a central location and conveyed to and from the rooms by means of a fan and a system of distributing ducts.

Air Conditioning System, Year-Round: An air conditioning system that ventilates, heats, and humidifies in winter and cools and dehumidifies in summer to provide the desired degree of air motion and cleanliness.

Air Conditioning Unit: A piece of equipment designed as a specific air-treating combination, consisting of a means for ventilation, air circulation, air cleaning, and heat transfer, with a control means for maintaining temperature and humidity within prescribed limits.

Air Core Solenoid: A solenoid that has a hollow core instead of a solid core.

Air Curtain: A system in which a blower is activated when a door is opened to blow across the open area, preventing the transfer of air between outdoors and indoors.

Air Defrosting: Evaporator defrosting that occurs as evaporator warms when the compressor is not running.

Air Diffuser: A circular, square, or rectangular air distribution outlet, generally located in the ceiling, and comprised of deflecting members discharging supply air in various directions and planes, arranged to promote mixing of primary air with secondary room air.

Air, Dry: Air unmixed with or containing no water vapor.

Airflow: The distribution or movement of air.

Air Gap: The space between magnetic poles or between rotating and stationary assemblies in a motor or generator.

Air Handler: Fan-blower, filter, and housing parts of a system

Air Infiltration: The in-leakage of air through cracks, crevices, doors, windows, or other openings caused by wind pressure or temperature difference.

Air, Recirculated: Return air passed through the conditioner before being again supplied to the conditioned space.

Air, Return: Air returned from conditioned or refrigerated space.

Air, Saturated: Moist air in which the partial pressure of the water vapor is equal to the vapor pressure of water at the existing temperature. This occurs when dry air and saturated water vapor coexist at the same dry-bulb temperature.

Air Source Equipment: Heat pumps or air conditioners that use the outdoor air to transfer heat to and from the refrigerant in the unit.

Air, Standard: Air with a density of 0.075 lb./cu. ft. and an absolute viscosity of 0.122 × 10 lb. mass/ft.-sec. This is substantially equivalent to dry air at 70°F and 29.92 in. Hg barometer.

Air-to-Air Heat Pump: A heat pump that uses outdoor air, as opposed to a *geothermal heat pump*.

Air Washer: Device used to clean air while changing its humidity.

Alcohol Brine: Water and alcohol solution that remains a liquid below 32°F (0°C).

Ambient Compensator: An electronic device that provides a small amount of heat to the refrigeration compartment to ensure that the machinery continues to cycle when ambient temperatures are low.

Ambient Temperature: The temperature of the medium surrounding an object. In a domestic system having an air-cooled condenser, it is the temperature of the air entering the condenser.

Ammeter: Electric meter used to measure current.

Ammonia: Chemical combination of nitrogen and hydrogen (NH_3). Ammonia refrigerant is identified as R-717.

Amperage: Current flow past a given point in a circuit.

Ampere: Unit of electric current.

Analyzer: A device used in the high side of an absorption system for increasing the concentration of vapor entering the rectifier or condenser.

Anemometer: An instrument for measuring the velocity of air in motion.

Anticipator: A device used with a start–stop control to reduce the control differential.

Antifreeze, Liquid: A substance added to the refrigerant to prevent formation of ice crystals at the expansion valve. Antifreeze agents in general do not prevent corrosion due to moisture. The use of a liquid should be a temporary measure where large quantities of water are involved, unless a drier is used to reduce the moisture content. Ice crystals may form when moisture is present below the corrosion limits, and in such instances, a suitable noncorrosive antifreeze liquid is often of value.

ARI (Air-Conditioning and Refrigeration Institute): Is a nonprofit, voluntary organization composed of manufacturers that publishes standards for testing and rating heat pumps and air conditioners to provide you with a standardized measure of comparison.

Armature: Part of an electric motor, generator, or other device moved by magnetism.

Articulated Connection Rods: Short connecting rods in a compressor.

ASHRAE: American Society of Heating, Refrigerating and Air Conditioning Engineers.

Aspirating Psychrometer: Device that draws a sample of air through it to measure the humidity.

Aspiration: Movement produced by suction.

Atmospheric Dust Spot Efficiency: Measurement of a device's ability to remove atmospheric air from test air.

Atmospheric Pressure: The pressure exerted by the atmosphere in all directions as indicated by a barometer. Standard atmospheric pressure is considered to be 14.695 psi (pounds per square inch), which is equivalent to 29.92 in. Hg (inches of mercury).

Atomize: To reduce to a fine spray.

Automatic Control: Action reached through self-operated or self-actuated means, not requiring manual adjustment.

Automatic Defrost: System of removing ice and frost from evaporators automatically.

Automatic Expansion Valve: A pressure-actuated device that regulates the flow of refrigerant from the liquid line into the evaporator to maintain a constant evaporator pressure.

Auxiliary Evaporator: Small evaporator consisting of coils of tinned tubing below the shelves in a display case.

Azeotropic Mixture: A liquid mixture having constant maximum and minimum boiling points. Refrigerants comprising the azeotropic mixture do not combine chemically, yet the mixture provides constant characteristics.

B

Back Pressure: Pressure in low side of refrigerating systems; also called suction pressure or low-side pressure.

Back Seating: Fluid opening/closing, such as a gauge opening, to seat the joint where the valve stem goes through the valve body.

Baffle: A partition used to divert the flow of air or a fluid.

Balance Point: The lowest outdoor temperature at which the refrigeration cycle of a heat pump will supply the heating requirements without the aid of a supplementary heat source.

Balanced Pressure: The same pressure in a system or container that exists outside the system or container.

Balancing: The process of adjusting the flow of air in duct systems or water flow in hot water heating systems. Proper balancing is performed using accurate instrumentation to

deliver the right amount of heating or cooling to each area or room of the home.

Bar: Unit of pressure. One bar equals .9869 atmosphere (approximately one atmosphere, 14.51 psi).

Barometer: An instrument for measuring atmospheric pressure.

Baudelot Cooler: Heat exchanger in which water flows by gravity over the outside of tubes or plates.

Bellows: Corrugated cylindrical container that moves as pressures change, or provides a seal during movement of parts.

Bellows Seal: Method of sealing the valve stem. The ends of the sealing material are fastened to the bonnet and to the stem. Seal expands and contracts with the stem level.

Bending Spring: Coil spring that is placed on inside or outside of tubing to keep it from collapsing while bending.

Bernoulli's Theorem: In a stream of liquid, the sum of elevation head, pressure head, and velocity remains constant along any line of flow, provided no work is done by or upon liquid on course of its flow; decreases in proportion to energy lost in flow.

Bimetal Strip: Temperature regulating or indicating device that works on the principle that two dissimilar metals with unequal expansion rates, welded together, will bend as temperatures change.

Bioaerosals: Airborne microorganisms derived from viruses, bacteria, fungi, protozoa, mites, and pollen.

Blast Freezer: Low-temperature evaporator that uses a fan to force air over the evaporator surface.

Bleeder: A pipe sometimes attached to a condenser to lead off liquid refrigerant parallel to the main flow.

Bleeding: Slowly reducing the pressure of liquid or gas from a system by opening a valve slightly.

Blend: A mixture of various refrigerants.

Blower (Fan): An air handling device for moving air in a distribution system.

Boiler: A closed vessel in which liquid is heated or vaporized.

Boiler, High-Pressure: A boiler operating with water

temperature and water pressure above low-pressure boiler ratings.

Boiler Horsepower: The equivalent evaporation of 34.5 lbs. of water per hr. from and at 212°F, equal to 33,475 Btu.

Boiler, Low-Pressure: A boiler operating with up to 250°F (121°C) water temperature and 160 psi water pressure or less.

Boiling Point: The temperature at which a liquid is vaporized upon the addition of heat, dependent on the refrigerant and the absolute pressure at the surface of the liquid and vapor.

Boiling Temperature: Temperature at which a fluid changes from a liquid to a gas.

Bonnet: In a furnace, the sheet metal chamber where heat collects before being distributed.

Booster: Common term applied to the use of a compressor as the first stage in a cascade refrigerating system.

Bourdon Tube: Thin-walled tube of elastic metal flattened and bent into a circular shape that tends to straighten as inside pressure is increased. Used in pressure gauges.

Boyle's Law: The volume and pressure of a gas vary inversely if the temperature remains the same. Example: If the pressure is doubled on a quantity of gas, its volume is reduced one half. It the volume is doubled, gas has its pressure reduced by one half.

Brazing: Method of joining metals with nonferrous (without iron) filler using heat between 800°F (427°C) and the melting point of base metals.

Breaker Strip: Strip of wood or plastic used to cover the joint between the outside case and inside liner of a refrigerator.

Breeching: Space in hot water or steam boilers between the end of the tubing and the jacket.

Brine: A liquid cooled by a refrigerant and used for transmission of heat without a change in its state.

Brine System: A system whereby brine cooled by a refrigerating system is circulated through pipes to the point where the refrigeration is needed.

British Thermal Unit (Btu): The amount of heat required to raise the temperature of 1 lb. of water 1°F.

Btu/h (British Thermal Units Per Hour): Btus.

Building-Related Illness (BRI): An illness caused by an airborne virus in a building.

Built-Up Terminal: Electrical terminal attached to a compressor dome.

Bulb, Sensitive: Part of a sealed fluid device that reacts to temperature. Used to measure temperature or to control a mechanism.

Bunker: Space where ice or a cooling element is placed in commercial installations.

Burner: Device in which burning of fuel takes place.

Butane: A hydrocarbon, flammable refrigerant used in small units.

Bypass Cycle: A cycle using a bypass line with either hot gas or liquid used to defrost an evaporator or for low-pressure control.

C

Cabinet: The housing of a refrigerator.

Calcium Chloride: A chemical having the formula $CaCl_2$, which, in granular form, is used as a drier.

Calcium Sulfate: A solid chemical of the formula $CaSO_4$, which may be used as a drying agent.

Calibrate: To position indicators to determine accurate measurements.

Calibration: The process of dividing and numbering the scale of an instrument.

Calorie: Heat required to raise the temperature of 1g of water 1°C.

Calorimeter: Device used to measure quantities of heat or determine specific heats.

Capacitance (C): Property of a nonconductor (condenser or capacitor) that permits storage of electrical energy in an electrostatic field.

Capacitive Reactance: The opposition, or resistance, to an alternating current as a result of capacitance.

Capacity: The output or producing ability of a piece of cooling or heating equipment. Cooling and heating capacity are normally referred to in Btus.

Capacity Reducer: In a compressor, a device, such as a clearance pocket, movable cylinder head, or suction bypass, by which compressor capacity can be adjusted without otherwise changing the operating conditions.

Capacity, Refrigerating: The ability of a refrigerating system to remove heat.

Capillarity: The action by which the surface of a liquid in contact with a solid (as in a slender tube) is raised or lowered.

Capillary Tube: A tube of small internal diameter used as a liquid refrigerant-flow control or expansion device between high and low sides; also used to transmit pressure from the bulb of some temperature controls to the operating element.

Capillary Tube System: A refrigerant control system in which pressure difference is maintained through the use of a thin capillary tube.

Carbon Dioxide (CO_2): Compound of carbon and oxygen that is sometimes used as a refrigerant. Refrigerant number is R-744.

Carbon Filter: Air filter using activated carbon as an air cleansing agent.

Carbon Monoxide (CO): Colorless, odorless, and poisonous gas produced when carbon fuels are burned with too little air.

Carbon Tetrachloride (CCl₄): Colorless, nonflammable, and toxic liquid used as a solvent.

Carrene: Refrigerant in Group A1 (R-11). Chemical combination of carbon, chlorine, and fluorine.

Cascade Systems: Arrangement in which two or more refrigerating systems are used in a series; uses the evaporator of one machine to cool the condenser of other machine. Produces ultralow temperatures.

Cavitation: Localized gaseous condition within a liquid stream.

Celsius: A system in which the freezing point of water is called 0°C and its boiling point 100°C at normal pressure.

Central Air Conditioning: A system capable of providing heating, cooling, humidifying, and dehumidifying.

Central Forced-Air Heating System: A piece of equipment that produces heat in a centralized area, then distributes it throughout the home through a duct system.

Central Station: Central location of condensing unit

with either wet or air-cooled condenser. Evaporator located as needed and connected to the central condensing unit.

Centrifugal Compressor: A compressor employing centrifugal force for compression.

Centrifugal Force: Force that pushes a rotating object away from the center of its rotation.

Centrifugal Switch: An electrical switch that is opened and closed by centrifugal force.

Ceramic Ignitor: Electric ignition system used in a water glycol solution, forced-air furnace. Electrically heated to create ignition of the gas–air mixture in the combustion chamber.

CFC (Chlorofluorocarbon): A class of refrigerants. Generally refers to the chlorofluorocarbon family of refrigerants. Sometimes called Freon.

CFM (Cubic Feet per Minute): A standard of airflow measurement. A typical system produces 400 CFM per ton of air conditioning.

Change of Air: Introduction of new, cleansed, or

recirculated air to a conditioned space, measured by the number of complete changes per unit time.

Charge: The amount of refrigerant in a system.

Charging Board: Specially designed panel or cabinet fitted with gauges, valves, and refrigerant cylinders used for charging refrigerant and oil into refrigerating mechanisms.

Charles' Law: At a constant pressure, mass and temperature of a gas are inversely proportional.

Chemical Refrigeration: System of cooling using a disposable refrigerant. Also called *expendable refrigerant system*.

Chiller: Air conditioning system that circulates chilled water to various cooling coils in an installation.

Chimney: Vertical shaft enclosing one or more flues for carrying flue gases to the outside atmosphere.

Chimney Connector: Conduit (pipe) connecting the furnace to the vertical flue.

Chimney Effect: The tendency of air or gas to rise when heated, due to lower density.

Chimney Flue: Flue gas passageway in a chimney.

Chlorodifluoromethane: Refrigerant better known as R-22. Chemical formula is $CHClF_2$. Cylinder color code is green.

Choke Tube: Throttling device used to maintain correct pressure difference between the high side and the low side in a refrigerating mechanism. Capillary tubes are sometimes called choke tubes.

Circuit, Parallel: The current divides and travels through two or more paths and then returns through a common path.

Circuit, Pilot: Secondary circuit used to control a main circuit or a device in the main circuit.

Circuit, Series: Current flow travels, in turn, through all devices connected together.

Clean Room: A room in which special efforts are made to eliminate dust and other contaminants.

Clearance Space: Small space in a cylinder from which compressed gas is not completely expelled. Compressors are designed to have as small a clearance space as possible.

Climate: The average weather conditions for a region.

Coefficient of Conductivity: Measure of the relative rate at which different materials conduct heat. A good conductor of heat has a high coefficient of conductivity.

Coefficient of Expansion: The fractional increase in length or volume of a material per degree rise in temperature.

Coefficient of Performance (Heat Pump): Ratio of heating effect produced to the energy supplied, each expressed in the same thermal units.

Cogeneration: Using waste energy as a primary heat source.

Coil: Any heating or cooling element made of pipe or tubing connected in series.

Cold Ban: A plastic trim piece used to reduce heat flow between the outer and inner shell of a refrigerator door.

Cold Junction: The part of a thermoelectric system that absorbs heat as the system operates.

Cold Wall: Refrigerator construction that has the inner lining of the refrigerator serving as the cooling surface.

Colloids: Miniature cells peculiar to meats, fish, and poultry, which, if disrupted, cause food to become rancid. Low temperatures minimize this action.

Combined Annual Efficiency (CAE) Ratio: Rating system used for combined heating systems, which heat both air and water.

Combustible Liquids: Liquid having a flash point above 140°F (60°C); known as Class 3 liquids.

Combustion: The process of igniting and burning.

Comfort Chart: Chart used in air conditioning to show the dry-bulb temperature, humidity, and air movement for human comfort.

Comfort Cooler: System used to reduce the temperature in the living space in homes. These systems are not complete air conditioners.

Comfort Zone: The range of temperatures, humidities, and air velocities at which the greatest percentage of people feel comfortable.

Compound Gauge: Instrument for measuring pressure.

Compound Pump: A rotary pump that has two rotors in series.

Compound Refrigerating Systems: A system that has several compressors or compressor cylinders in series. The system is used to pump low-pressure vapors to condensing pressures.

Compression: Term used to denote increase of pressure on a fluid by using mechanical energy.

Compression Chiller: A chiller that achieves the required pressure difference through the use of a compressor.

Compression Gauge: Instrument used to measure positive pressures (pressures above atmospheric pressures) only. Gauge dial usually runs from 0 to 300 psig (101.3–2200 kPa)

Compression Ratio: Ratio of the absolute high-side (compressor discharge) pressure to the absolute low-side (compressor suction) pressure.

Compression Ring: Upper piston ring.

Compressor: That part of a mechanical refrigerating system that receives the refrigerant vapor at low pressure and compresses it into a smaller volume at higher pressure.

Compressor Booster: A compressor for very low pressures, usually discharging into the suction line of another compressor.

Compressor, Centrifugal: A nonpositive displacement compressor that depends on centrifugal effect, at least in part, for pressure rise.

Compressor, Hermetic: A compressor in which the driving motor is sealed in the same dome or housing as the compression.

Compressor, Multiple Stage: A compressor having two or more compressive steps. Discharge from each step is the intake pressure of the next in series.

Compressor, Open-Type: A compressor with a shaft or other moving part, extending through a casing, to be driven by an outside source of power, thus requiring a stuffing box, shaft seals, or equivalent rubbing contact between a fixed and moving part.

Compressor, Reciprocating: A positive displacement compressor with a piston or pistons moving in a straight line but alternately in opposite directions.

Compressor, Rotary: One in which compression is attained in a cylinder by rotation of a positive displacement member.

Compressor Seal: Leakproof seal between crankshaft and compressor body in open compressors.

Compressor, Single Stage: Compressor having only one compressive step between low-side pressure and high-side pressure.

Condensate: A fluid formed when a gas is cooled to its liquid state.

Condensate Pump: Device to remove water condensate that collects beneath an evaporator.

Condensation: Liquid or droplets that form when a gas or vapor is cooled below its dew point.

Condense: Action of changing a gas or vapor to a liquid.

Condenser: A heat-transfer device that receives high-pressure vapor at temperatures above that of the cooling medium, such as air or water, to which the condenser passes latent heat

from the refrigerant, causing the refrigerant vapor to liquefy.

Condenser, Air Cooled: Heat exchanger that transfers heat to surrounding air.

Condenser, Water Cooled: Heat exchanger designed to transfer heat from hot gaseous refrigerant to water.

Condenser Comb: Comb-like device, metal or plastic, used to straighten the metal fins on condensers and evaporators.

Condenser Fan: Forced-air device used to move air through air-cooled condenser.

Condensing: The process of giving up latent heat of vaporization to liquefy a vapor.

Condensing Furnace: High-efficiency, gas, forced-air furnace that extracts the latent heat lost in conventional gas forced-air furnaces.

Condensing Pressure: Pressure inside a condenser at which refrigerant vapor gives up its latent heat of vaporization and becomes a liquid. This varies with the temperature.

Condensing Temperature: Temperature inside a condenser at which refrigerant vapor gives up its latent heat of vaporization and becomes a liquid. This varies with the pressure.

Condensing Unit: Part of a refrigerating mechanism which pumps vaporized refrigerant from the evaporator, compresses it, liquefies it in the condenser, and returns it to the refrigerant control.

Condensing Unit, Sealed: A mechanical condensing unit in which the compressor and compressor motor are enclosed in the same housing, with no external shaft or shaft seal, the compressor motor operating in the refrigerant atmosphere.

Condensing Unit Service Valves: Shutoff valves mounted on the condensing unit to enable service technicians to install and service the unit.

Conduction, Thermal: Passage of heat from one point to another by transmission of molecular energy from particle to particle through a conductor.

Conductivity, Thermal: The ability of a material to pass heat from one point to another, generally expressed in terms of Btu per hour per square foot of material per inch of thickness per degree temperature difference.

Conductor: Substance or body capable of transmitting electricity or heat.

Configuration: This describes the direction in which a furnace outputs heat. A furnace may have an upflow, downflow, or crossflow (horizontal) configuration.

Constant-Pressure Valve: A valve of the throttling type, responsive to pressure, located in the suction line of an evaporator to maintain a desired constant pressure in the evaporator higher than the main suction line pressure.

Constant-Temperature Valve: A valve of the throttling type, responsive to the temperature of a thermostatic bulb. This valve is located in the suction line of an evaporator to reduce the refrigerating effect on the coil to just maintain a desired minimum temperature.

Constrictor: Tube or orifice used to restrict the flow of a gas or a liquid.

Contaminant: Substance such as dirt, moisture, or other

matter foreign to refrigerant or refrigerant oil in system.

Continuous Absorption System: System that has a continuous flow of energy input.

Control: Automatic or manual device used to stop, start, or regulate the flow of gas, liquid, or electricity.

Control, Defrosting: Device used to automatically defrost the evaporator. It may operate by means of a clock or door-cycling mechanism, or during the off cycle.

Control, Low-Pressure: Cycling device connected to the low-pressure side of system.

Control Point: The condition being maintained by a proportional control.

Control, Pressure Motor: High- or low-pressure control connected into the electrical circuit and used to start and stop motor. It is activated by demand for refrigeration or for safety.

Control, Refrigerant: Device used to regulate flow of liquid refrigerant into evaporator. Can be a capillary tube, expansion valve, or high-side and low-side float valves.

Control, Temperature: Temperature-operated thermostatic device that automatically opens or closes a circuit.

Control Valve: Valve that regulates the flow or pressure of a medium that affects a controlled process. Control valves are operated by remote signals from independent devices using any of a number of control media such as pneumatic, electric, or electrohydraulic.

Convection: The circulatory motion that occurs in a fluid at a nonuniform temperature owing to the variation of its density and the action of gravity.

Convection, Forced: Convection resulting from forced circulation of a fluid as by a fan.

Convection, Natural: Circulation of a gas or liquid due to difference in density resulting from temperature differences.

Cooler: Heat exchanger that removes heat from a substance.

Cooling Coil: Coils cooled by a fluid that does not evaporate (such as brine). The evaporator is sometimes incorrectly referred to as a *cooling coil*.

Cooling Tower, Water: An enclosed device for evaporatively cooling water.

COP (Coefficient of Performance): COP compares the heating capacity of a heat pump to the amount of electricity required to operate the heat pump in the heating mode. COPs vary with the outside temperature: as the temperature falls, the COP falls also, because the heat pump is less efficient at lower temperatures. ARI standards compare equipment at two temperatures, 47°F and 17°F, to give you an idea of the COP in both mild and colder temperatures. Geothermal equipment is compared at 32°F enter water temperature. COP and HSPF cannot be compared equally. Air source equipment is rated by HSPF or COP and geothermal equipment is rated by COP.

Copper Plating: Abnormal condition developing in some units in which copper is electrolytically deposited on compressor surfaces.

Core Valves: Schrader valve used to gain access to a hermetic unit.

Corrosion: Deterioration of materials from chemical action.

Counterflow: Flow in opposite direction.

Crankshaft Seal: Leakproof joint between crankshaft and compressor body.

Crankthrow: Distance between centerline of main bearing journal and centerline of the crankpin or eccentric.

Critical Pressure: The vapor pressure corresponding to the critical temperature.

Critical Temperature: The temperature above which a vapor cannot be liquefied, regardless of pressure.

Critical Velocity: The velocity above which fluid flow is turbulent.

Cross-Charged: Sealed container of two fluids that, together, create a desired pressure–temperature curve.

Cryogenic Fluid: A substance that exists as a liquid or gas at temperatures of −250°F (−157°C) or lower.

Cryogenics: Refrigeration that deals with producing temperatures of −250°F (−157°C) and lower.

Current Relay: Device that opens or closes a circuit.

Cut-In: The temperature or pressure at which the control circuit closes.

Cut-Out: The temperature or pressure at which the control circuit opens.

Cycle: A series of events or operations that repeat.

Cycle, Defrosting: The portion of a refrigeration operation that permits the cooling unit to defrost.

Cycle, Refrigeration: A complete course of operation of a refrigerant back to the starting point measured in themodynamic terms. Also used in general for any repeated process for any system.

Cylinder: 1—Device that converts fluid power into linear mechanical force and motion. This usually consists of movable elements such as a piston and piston rod, plunger or ram, operating within a cylindrical bore. 2—Closed container for fluids.

Cylinder Head: Plate or cap that encloses compression end of compressor cylinder.

Cylinder, Refrigerant: Cylinder in which refrigerant is stored and dispensed. Color code painted on cylinder indicates kind of refrigerant.

D

Dalton's Law of Partial Pressure: The sum of the individual pressures of the constituents equals the total pressure of the mixture.

Damper: A device that is located in ductwork to adjust airflow. This movable plate opens and closes to control airflow. Dampers are used effectively in zoning to regulate airflow to certain rooms. There are basically two types of dampers: manual and motorized. A manual damper generally consists of a sheet metal (or similar material) flap, shaped to fit the inside of a round or rectangular duct. By rotating a handle located outside of the duct a technician can adjust (see Balancing) airflow to match the needs of a particular area or room. A motorized damper is generally used in a zoned system (see Zoning) to automatically deliver conditioned air to specific rooms or zones.

Dasher: Stirring mechanism in a dispensing freezer.

Deaeration: Act of separating air from a substance.

Deck (Coil Deck): Insulated horizontal partition between refrigerated space and evaporator space.

Defrost Cycle: The process of removing ice or frost buildup from the outdoor coil during the heating season.

Defrost Timer: Shuts unit off long enough to permit ice and frost accumulation on evaporator to melt.

Defrosting Evaporator: Evaporator operating at such temperatures that ice and frost on surface melts during the off cycle.

Degree Day: A unit based on temperature difference and time used to specify the nominal heating load in winter.

Dehumidification: The reduction of vater vapor in air by cooling the air below the dew point; removal of water vapor from air by chemical means, refrigeration, etc.

Dehumidifier: An air cooler used for lowering the moisture content of the air passing through it.

Dehydrated Oil: Lubricant that has had most of its water content removed (dry oil).

Dehydrator: A device used to remove moisture from the refrigerant.

Dehydrator-Receiver: Small tank that serves as liquid

refrigerant reservoir and also contains a desiccant to remove moisture.

Density: The mass or weight per unit of volume.

Deodorizer: Adsorbs various odors.

Desiccant: Substance used to collect and hold moisture. A drying agent. Common desiccants are activated alumina and silica gel.

Design Pressure: Highest pressure expected during operation. Sometimes calculated as operating pressure plus a safety allowance.

Dew Point, Air: The temperature at which air, on being cooled, gives up moisture, or dew.

Diac: A two load alternating current semiconductor that allows current to flow in both directions at a preset voltage.

Diagnostics: The process of identifying or determining the nature and circumstances of an existing condition.

Diaphragm: Flexible material usually made of thin metal, rubber, or plastic.

Dichlorodifluoromethane: Refrigerant commonly known as R-12.

Dielectric Fluid: Fluid with high electrical resistance.

Differential (of a Control): The difference between the cut-in and cut-out temperature or pressure.

Diffuser: Attachments for duct openings that distribute the air in wide flow patterns.

Diode: Two-element electron tube that will allow more electron flow in one direction in a circuit than in the other.

Direct Connected: Driver and driven, as motor and compressor, positively connected in line to operate at the same speed.

Direct Expansion: A system in which the evaporator is located in the material or space refrigerated or in the air-circulating passages communicating with such space.

Direct Expansion Evaporator: Evaporator using either an automatic expansion valve (AEV) or a thermostatic expansion valve (TEV) refrigerant control.

Direct-Spark Ignition: A furnace control in which a spark is used to ignite the gas–air mixture. There is no constantly burning pilot light.

Displacement, Actual: The volume of gas at the compressor inlet actually moved in a given time.

Displacement, Theoretical: The total volume displaced by all the pistons of a compressor for every stroke during a definite interval.

Distilling Apparatus: Fluid-reclaiming device used to reclaim used refrigerants.

Distribution Controls: Systems that help evenly and efficiently transfer the heating or cooling medium to the area where it is needed.

District Heating and Cooling: Use of a central utility system designed to provide heating and cooling.

Dome-Hat: Sealed metal container for the motor compressor of a refrigerating unit.

Door Heater: A heater located around the door opening of a freezer.

Double-Duty Case: Commercial refrigerator in which a part of space is for refrigerated storage and part is equipped with glass windows for display purposes.

Double-Thickness Flare: Copper, aluminum, or steel tubing end that has been formed into two-wall thickness, 37° to 45° bell mouth or flare.

Downflow Furnace: A furnace that pulls in return air from the top and expels warm air at the bottom.

Draft Gauge: Instrument used to measure air movement by measuring air pressure differences.

Draft Indicator: Instrument used to indicate or measure chimney draft or combustion gas movement. Draft is measured in units of 1" of water column.

Draft Regulator: Device that maintains a desired draft in a combustion-heated appliance.

Drier: Substance or device used to remove moisture from a refrigeration system.

Drip Pan: Pan-shaped panel or trough used to collect condensate from evaporator.

Dry Bulb: An instrument with a sensitive element to measure ambient air temperature.

Dry-Bulb Temperature: Air temperature as indicated by an ordinary thermometer.

Dry Ice: Refrigerating substance made of solid carbon dioxide, which changes directly from a solid to a gas (sublimates). Its subliming temperature is −109°F (−78°C).

Dry System: Refrigeration system that has the evaporator liquid refrigerant mainly in the atomized or droplet condition.

Dry-Type Evaporator: An evaporator of the continuous-tube type where the refrigerant from a pressure-reducing device is fed into one end and the suction line connected to the outlet end.

Dual Fuel System: A dual heating system, for example, a heat pump and a fossil fuel furnace.

Dual-Pressure Regulator: A combination of a high-pressure and a low-pressure regulator.

Duct: A passageway made of sheet metal or other suitable material used for conveying air at low pressure.

Ductwork: Pipes or channels that carry air.

Dust: An air suspension (aerosol) of solid particles of earthy material.

GLOSSARY *(cont.)*

Dynamometer: Device for measuring power output or power input of a mechanism.

E

Eccentric: Circle or disk mounted off center on a shaft.

Economizer: A mechanism that removes flash gas from the evaporator.

EER (Energy Efficiency Ratio): A ratio calculated by dividing the cooling capacity in Btus per hour (Btu/h) by the power input in watts at any given set of rating conditions, expressed in Btu/h per watt (Btu/h/watt). EER and SEER cannot be compared equally. Air source equipment is rated by SEER and geothermal equipment is rated by EER.

Effective Area: Actual flow area of an air inlet or outlet. Gross area minus area of vanes or grille bars.

Effective Latent Heat: The amount of heat absorbed from the cabinet and evaporator.

Effectiveness (Absorption Systems): Method of evaluating absorption cooling systems, in which the cooling effect is divided by the work equivalent to the heat supplied to the absorber.

Effective Temperature: Overall effect of air temperature, humidity, and air movement on human comfort.

Efficiency: For furnaces, it is the rate at which a furnace maximizes fuel use. This rate is numerically described as a ratio called AFUE (see AFUE). As of January 1991, no furnaces can be manufactured with efficiencies lower than 78% AFUE. High efficiency furnaces will be rated 85 to 95% AFUE.

Efficiency, Mechanical: The ratio of the output of a machine to the input in equivalent units.

Efficiency, Volume: The ratio of the volume of gas actually pumped by a compressor or pump to the theoretical displacement of the compressor.

Ejector: A device that utilizes static pressure to build up a high fluid velocity in a restricted area to obtain a lower static pressure at that point so that fluid from another source may be drawn in.

Electric Defrosting: Use of electric resistance heating coils to melt ice and frost off evaporators during defrosting.

GLOSSARY (cont.)

Electric Heating: System in which heat from electrical resistance units is used to heat a building.

Electrodeposition: Process in which metallic particles are applied to another metal surface through the use of an electric current.

Electrolysis: A chemical change in a substance caused by movement of electricity.

Electrolytic Condenser-Capacitor: Plate or surface capable of storing small electrical charges.

Electronic Control Diagnostics: Trouble codes that may be referenced on an automatic climate control system to diagnose problems.

Electronic Leak Detector: Electronic instrument that measures electronic flow across a gas gap. Electronic flow changes indicate presence of refrigerant gas molecules.

Electronic Sight Glass: Device that sends an audible signal when system is low in refrigerant.

Electrostatic Air Filter: A filter that gives dust particles an electric charge. This

causes particles to be attracted to a plate so they can be removed from air.

Element, Bimetallic: An element formed of two metals having different coefficients of thermal expansion, such as used in temperature-indicating and -controlling devices.

Emergency Heat (Supplementary Electric Heat): The backup electric heat built into a heat pump system.

Emulsion: A relatively stable suspension of small, but not colloidal, particles of a substance in a liquid.

End Bell: End structure of the plate of an electric motor, which usually holds the motor bearings.

Endothermal: Chemical reaction in which heat is absorbed.

End Play: Slight movement of shaft along its centerline.

Energy: Actual or potential ability to do work.

Energy Audit: Process of accurately determining the current energy consumption for a given area.

Energy Utilization Index (EUI): A number used to compare energy usage for different areas. It is calculated by dividing the energy consumption by the square footage of the conditioned area.

Enthalpy: Total amount of heat in 1 lb. of a substance calculated from accepted temperature base. Temperature of 32°F (0°C) is accepted base for water vapor calculation. For refrigerator calculations, accepted base is −40°F (−40°C).

Entropy: Engineering calculation used to determine heat available. Measured in BTU per pound degree change for a substance.

Environment: The surrounding conditions.

Enzyme: Complex organic substance, originating from living cells, that speeds up chemical changes in foods. Enzyme action is slowed by cooling.

Equalizer: A piping arrangement to maintain a common liquid level or pressure between two or more chambers.

Equivalent Length: Length of piping plus pressure losses due to bends, fixtures, etc.

ERV (Energy Recovery Ventilator): This device

preheats incoming outside air during the winter and precools incoming air during the summer to reduce the impact of heating and/or cooling the indoor air.

Ethane (R-170): Refrigerant sometimes added to other refrigerants to improve oil circulation.

Eutectic: That unique mixture of two substances providing the lowest possible melting temperature.

Evacuation: Removal of air (gas) and moisture from a refrigeration or air conditioning system.

Evaporation: The changing of a liquid to a gas. Heat is absorbed in this process.

Evaporative Condenser: A refrigerant condenser utilizing the evaporation of water by air at the condenser surface as a means of dissipating heat.

Evaporative Cooling: The process of cooling by means of the evaporation of water in air.

Evaporator: A device in which the refrigerant evaporates while absorbing heat.

Evaporator Coil: The coil that is inside your house in a split system. In the evaporator, refrigerant evaporates and absorbs heat from air passed over the coil.

Evaporator, Dry: Evaporator in which the refrigerant is in droplet form.

Evaporator Fan: Fan that increases airflow over the heat exchange surface of evaporators.

Evaporator, Flooded: Evaporator containing liquid refrigerant at all times.

Exfiltration: Flow of air from a building to the outdoors.

Exhaust Valve: A movable port that provides an outlet for the cylinder gases in a compressor or engine.

Exothermal: Chemical reaction in which heat is released.

Expansion Joint: Device in piping designed to allow movement of the pipe caused by thermal expansion and contraction.

Expansion Tank: A tank used to allow water to expand and contract with temperature changes.

Expansion Valve, Automatic: A device that regulates the flow of refrigerant from the liquid line into the evaporator to maintain a constant pressure.

Expansion Valve, Thermostatic: A device that regulates the flow of refrigerant into an evaporator so as to maintain an evaporation temperature in a definite relationship to the temperature of a thermostatic bulb.

Expendable Refrigerant System: A system that discards the refrigerant after it has evaporated.

External Drive: Term used to indicate a compressor driven directly from the shaft or by a belt using an external motor.

External Equalizer: In a thermostatic expansion valve, a tube connection from the chamber containing the pressure-actuated element of the valve to the outlet of the evaporator coil.

F

Fahrenheit: A system in which 32°F denotes the freezing point of water and 212°F the boiling point.

Fail-Safe Control: A device that opens a circuit when a sensing element loses its pressure.

Fan: Radial or axial flow device used for moving or producing flow of gases.

Farad: Unit of electrical capacity. Capacity of a condenser which, when charged with 1 coulomb of electricity, gives a difference of potential of 1 V.

Faraday Experiment: Silver chloride absorbs ammonia when cool and releases it when heated. This is the basis on which some absorption refrigerators operate.

Fast Food Freezing: Method that uses liquid nitrogen or carbon dioxide to turn fresh food into long-lasting frozen food. It is often referred to as *cryogenic food freezing*.

Feedback Control System: Control system that is constantly correcting the condition. Also called a "closed loop system."

Fill (Cooling Tower): Material in a cooling tower over which water flows.

Filter: A device for removing dust particles from air or unwanted elements from liquids.

Firepot: Refractory-lined combustion chamber.

Flammable Liquids: Liquids having a flash point below 140°F (60°C) and a vapor pressure not exceeding 40 psia (276 kPa) at 100°F (38°C).

Flapper Valve: Thin metal valve used in refrigeration compressors that allows gaseous refrigerants to flow in only one direction.

Flare: An enlargement at the end of a piece of flexible tubing by which the tubing is connected to a fitting or another piece of tubing.

Flare Fitting: A type of connector for soft tubing that involves the flaring of the tube to provide a mechanical seal.

Flash Gas: Instantaneous evaporation of some liquid refrigerant in an evaporator, which cools the remaining liquid refrigerant to the desired evaporation temperature.

Flash Point: Temperature at which flammable liquid will give off sufficient vapor to support a flash flame but will not support continuous combustion.

Flash Weld: Resistance weld in which mating parts are brought together under considerable pressure while a heavy electrical current is passed through the joint to be welded.

Flooded System: Type of refrigerating system in which the liquid refrigerant fills most of the evaporator at all times.

Flooded System, High-Side Float: Refrigeration system that has a float operated by the level of the high-side liquid refrigerant.

Flooded System, Low-Side Float: Refrigerating system that has a low-side float refrigerant control.

Flow Check Piston: Piston assembly, with an orifice in the center, that can operate as an expansion valve.

Flow Meter: Instrument used to measure velocity or volume of fluid movement.

Flue: Gas or air passage that usually depends on natural convection to cause the combustion gases to flow.

Flush: Operation to remove any material or fluids from refrigeration system parts by purging them to the atmosphere using refrigerant or other fluids.

Foaming: Formation of a foam in an oil refrigerant, due to rapid boiling out of the refrigerant dissolved in the oil when the pressure is suddenly reduced.

Foam Leak Detector: System of soap bubbles or special foaming liquids brushed over joints and connections to locate leaks.

Foot-Pound: Unit of work. A foot-pound is the amount of work done in lifting 1 lb. 1 ft.

Forced Air: Uses a blower motor to move air through the furnace and into the ductwork.

Forced Convection: Movement of fluid by mechanical force such as fans or pumps.

Forced-Air Heating: A heating system that uses a fan to circulate the heated air.

Forced-Circulation Evaporator: An evaporator that uses a fan to circulate air.

Force-Feed Oiling: Lubrication system which uses a pump to force oil to surfaces of moving parts.

Freeze Drying: Process of food preservation wherein food is frozen and ice content changed rapidly into a vapor, which is then absorbed on an evaporator.

Freezeup: Failure of a refrigeration unit to operate normally due to formation of ice at the expansion valve.

Freezing-Point Depression: Temperature at which ice will form in solution of water and salt.

Freon-12: The common name for dichlorodifluoromethane (CC2F2).

Frostback: The flooding of liquid from an evaporator into the suction line, accompanied by frost formation on the suction line in most cases.

Frost Control, Automatic: Control that automatically cycles refrigerating system to remove frost on evaporator.

Frost Control, Semiautomatic: Control that starts defrost part of a cycle manually and then returns system to normal operation automatically.

Frost-Free Refrigerator: Refrigerated cabinet that operates with an automatic defrost during each cycle.

Frosting Evaporator: Refrigerating system that maintains the evaporator at frosting temperatures during all phases of cycle.

Frozen: 1—Water in its solid state. 2—Preserved by freezing. 3—Seized (as in machine parts) due to lack of lubrication.

Furnace: That part of an environmental system which converts gas, oil, electricity,

or other fuel into heat for distribution within a structure.

Fusible Plug: A safety plug used in vessels containing refrigerant. The plug is designed to melt at high temperatures (usually about 165°F) to prevent excessive pressure from bursting the vessel.

G

Galvanic Action: Corrosion of two unlike metals due to electrical current passing between them.

Gas: Vapor phase of a substance.

Gasket: Resilient (spongy) or flexible material used between mating surfaces of parts to give a leakproof seal.

Gas, Noncondensable: Gas that will not form into a liquid under the operating pressure–temperature conditions.

Gauge: An instrument used for measuring various pressures or liquid levels.

Gauge, High Pressure: Instrument for measuring pressures in range of 0 psig to 500 psig (101.3 kPa to 3600 kPa).

Gauge, Low Pressure: Instrument for measuring

pressures in range of 0 psia to 50 psia (0 kPa to 350 kPa).

Gauge Manifold: Chamber device constructed to hold both compound and high-pressure gauges.

Gauge Port: Opening or connection provided for installing a gauge.

Gauge, Vacuum: Instrument used to measure pressures below atmospheric pressure.

Generator: A basic component of any absorption-refrigeration system.

Geothermal: An underground or underwater temperature source used for the operation of a heating and cooling system (heat pump).

Geothermal Equipment: Heat pumps that use the ground to transfer heat to and from the refrigerant in the unit. The unit circulates water through a heat exchanger to a closed loop buried in the ground or by pumping water from a well through the unit.

Global Warming Potential (GWP): Numeric value assigned to refrigerants to express the risk each refrigerant poses to the environment.

Glyco–Water Solution Forced-Air Furnace: Furnace with 50% glycol and 50% distilled water solution, which passes through a tube-and-fin heat exchanger to distribute heat through the furnace duct system.

Grain: Unit of weight equal to 1/7000 lb., used to indicate the amount of moisture in the air.

Gravity Air: Air that naturally rises when heated and flows through warm air ducts. When it cools, it becomes denser (heavier) and flows down.

Gravity Flow: The tendency of liquids to flow downward and rest at the lowest possible point.

Gravity Heating: Heating system in which heated air is distributed by natural rising (no fans are used for circulation).

Gravity, Specific: The density of a standard material usually compared to that of water or air.

Grille: Ornamental or louvered opening placed in a room at the end of an air passageway.

Grommet: Plastic, metal, or rubber doughnut-shaped protectors, which line holes where wires or tubing pass through panels.

Ground Coil: Heat exchanger buried in the ground. May be used either as an evaporator or as a condenser.

Gun Burner: Furnace burner that atomizes oil by pushing it through an orifice into the combustion chamber.

H

Halide Refrigerants: Family of refrigerants containing halogen chemicals.

Halide Torch: A leak tester generally using alcohol and burning with a blue flame; when the sampling tube draws in halocarbon refrigerant vapor, the color of flame changes to bright green.

Halogens: Substances containing fluorine, chlorine, bromine, or iodine.

HCFC (Hydrochlorofluorocarbon): A class of refrigerants. Generally refers to halogenated chlorofluorocarbon family of refrigerants.

Head: Pressure, usually expressed in feet of water, inches of mercury, or millimeters of mercury.

Header: Length of pipe or vessel, to which two or more pipelines are joined, that carries fluid from a common source to various points of use.

Head Friction: Head required to overcome friction of the interior surface of a conductor and between fluid particles in motion.

Head Pressure: Pressure that exists in condensing side of refrigerating system.

Head Pressure Control: Pressure-operated control that opens electrical circuit if high-side pressure becomes too high.

Head Pressure Safety Cutout: Motor protection device wired in series with motor; will shut off the motor when excessive head pressures occur.

Head, Static: Pressure of fluid expressed in terms of height of column of the fluid, such as water or mercury.

Head Velocity: Height of fluid equivalent to its velocity pressure in flowing fluid.

Heat: Basic form of energy which may be partially converted into other forms and into which all other forms may be entirely converted.

Heat Absorber: The low-pressure side of a

refrigeration system. The evaporator absorbs heat.

Heat Anticipators: A thermostatic anticipator.

Heat Dissipator: The high-pressure side of a refrigeration system. The condenser dissipates heat.

Heat Exchanger: This is a device that enables furnaces to transfer heat from combustion safely into breathable air. The primary heat exchanger transfers heat from combustion gases to the air blowing through the ductwork. In high efficiency furnaces, secondary heat exchangers recover heat that used to be vented up the chimney with the exhaust gases. Part of the heat recovered causes the water and acid to condense out of the exhaust gas. Because this liquid is corrosive, secondary heat exchangers must be designed to prevent deterioration. Usually they are made of stainless steel.

Heat Gain: The amount of heat gained, measured in Btus, from a space to be conditioned, at the local summer outdoor design temperature and a specified indoor design condition.

Heating Coil: Heat transfer device consisting of a coil of piping that releases heat.

Heating System, Electric: Heating produced by the rise of temperature caused by the passage of an electric current through a conductor having a high resistance to the current flow.

Heating System, Steam: A heating system in which heat is transferred from a boiler or other source to the heating units by steam at, above, or below atmospheric pressure.

Heating System, Vacuum: A two-pipe steam heating system equipped with the necessary accessory apparatus to permit operating the system below atmospheric pressure.

Heating System, Warm Air: A warm-air heating plant consisting of a heating unit (fuel burning furnace) enclosed in a casing from which the heated air is distributed to various rooms of the building through ducts.

Heating Value: Amount of heat that may be obtained by burning a fuel. The heating value is usually expressed in Btu per lb., Btu per gal., or kJ/kg.

Heat Input Method: Method of sizing motor in which the required energy from the motor is the amount of heat added to the vapor in the compressor.

Heat Lag: The time it takes for heat to travel through a substance heated on one side.

Heat Leakage Load: Total amount of heat that leaks from a structure.

Heat Load: Amount of heat removed during a period of 24 hr.

Heat Loss: The amount of heat lost, measured in Btus, from a space to be conditioned, at the local winter outdoor design temperature and a specified indoor design condition.

Heat of Compression: Additional temperature produced by increased pressure.

Heat of Fusion: Latent heat involved in changing between the solid and the liquid states.

Heat of Vaporization: Latent heat involved in the change between liquid and vapor states.

Heat Pipe: High-efficiency gas furnace that uses vertical liquid-filled pipes. The pipes are heated by a burner at their base, and the liquid boils and vaporizes within the pipe. The furnace blower circulates air over the pipes for heating.

Heat Pump: Compression cycle system used to supply heat to a temperature controlled space. Same system can also remove heat from the same space.

Heat Recovery System: Produces and stores hot water by transferring heat from condenser to cooler water.

Heat Sink: Relatively cold surface capable of absorbing heat.

Heat Transfer: Movement of heat from one body or substance to another.

Heat-Transfer Coefficient (U-value): A measure of the amount of heat that a material or combination of materials will allow through.

Heat-Transfer Module: Primary system of heat transfer in a glycol–water solution forced-air furnace. The heat transfer module contains the ignitor, burner, and primary solution circulating coil.

Heat-Transfer Rate (Q): The amount of heat transfer through a given material per unit time.

Heat, Sensible: Heat that is associated with a change in temperature; specific heat exchange of temperature, in contrast to a heat interchange in which a change of state (latent heat) occurs.

Heat, Specific: The ratio of the quantity of heat required to raise the temperature of a given mass of any substance one degree to the quantity required to raise the temperature of an equal mass of a standard substance (usually water at 59°F) one degree.

Hermetically Sealed Unit: A refrigerating unit containing the motor and compressor in a sealed container.

Hermetic Compressor: Compressor that has the driving motor sealed inside the compressor housing. The motor operates in an atmosphere of the refrigerant.

HFC (Hydrofluorocarbon): A class of refrigerants. Generally refers to hydrofluorocarbon family of refrigerants.

Hg (Mercury): Heavy silver-white metallic element; only metal that is liquid at ordinary room temperature.

High-Efficiency Gas Furnace: Furnace that uses recycling of combustion gases or pulse combustion to obtain operating efficiencies from 85% to 95%.

High-Limit Control: Control that stops the flow of gas when the bonnet on a furnace is too hot. Also called a *safety stat.*

High-Pressure Cutout: A control device connected into the high-pressure part of a refrigerating system to stop the machine when the pressure becomes excessive.

High Side: The parts of a refrigerating system subject to the condenser pressure.

High-Side Float: Refrigerant control mechanism that controls the level of the liquid refrigerant in the high-pressure side of the mechanism.

Horizontal Furnace: A furnace that lies on its side, pulling in return air from one side and expelling warm air from the other.

Horsepower: A unit of power. Work done at the rate of 33,000 ft.-lbs. per min., or 550 ft.-lbs. sec.

Hot Gas: High-temperature gas taken from the compressor used to defrost the evaporator.

Hot Gas Bypass: Piping system in refrigerating unit that moves hot refrigerant gas from the condenser into the low-pressure side.

Hot Gas Defrost: Defrosting system in which hot refrigerant gas from the high side is directed through the evaporator for a short period of time at predetermined intervals to remove frost.

Hot Junction: The part of thermoelectric circuit that releases heat.

Hot Surface Ignition System: Furnace ignition system in which a silicon carbide element is heated to light the main burner. No pilot light is needed.

Hot Water Heating System: System in which water is circulated through heating coils.

HRV (Heat Recovery Ventilator): This device brings fresh, outside air into a home while simultaneously exhausting stale indoor air outside. In the process of doing this, an HRY removes heat from the exhaust air and transfer it to the incoming air, preheating it.

HSPF (Heating Seasonal Performance Factor): Heating seasonal performance factor is similar to SEER, but it measures the efficiency of the heating portion of your heat pump. Like SEER, industry minimums have been raised recently, and the minimum is now 6.80 HSPF. The total heating output of a heat pump during its normal annual usage period for heating divided by the total electric power input in watt-hours during the same period. COP and HSPF cannot be compared equally. Air source equipment is rated by HSPF or COP and geothermal equipment is rated by COP. ARI standards compares air source equipment at two temperatures. 47°F and 17°F. Geothermal equipment is compared at 32°F enter water temperature.

Humidifier: A device to add moisture to air.

Humidistat: A device designed to regulate humidity input by reacting to changes in the moisture content of the air.

Humidity: The amount of moisture in the air. Air conditioners remove moisture for added comfort.

Humidity, Absolute: The definite amount of water

contained in a definite quantity of air (usually measured in grains of water per pound or per cubic foot of air).

Humidity, Relative: The ratio of the water-vapor pressure of air compared to the vapor pressure it would have if saturated at its dry-bulb temperature.

Humidity, Specific: The weight of vapor associated with 1 lb. of dry air; also termed humidity ratio.

Hunting: The cycling above and below the set point.

Hydrometer: Floating instrument used to measure the specific gravity of a liquid.

Hydronic: Heating system that circulates a heated fluid, usually water, through baseboard coils by means of a circulating pump controlled by a thermostat.

Hydrostatic Pressure: The pressure due to liquid in a container that contains no gas space.

Hygrometer: Instrument used to measure amount of moisture in the air.

Hygroscopic: Ability of a substance to absorb and release moisture and change physical dimensions as its moisture content changes.

I

IAQ: Indoor air quality.

Ice Bank: Refrigerating systems that form a bank of ice around the evaporator to provide reserve cooling capacity.

Ice-Melting Effect: Amount of heat absorbed by melting ice at 32°F (0°C) is 144 Btu per pound of ice or 288,000 Btu per ton.

Idler: Pulley used on some belt drives to provide proper belt tension and to eliminate belt vibration.

Ignition System: Method of lighting a furnace burner.

Ignition Transformer: Transformer designed to provide a high-voltage current. Used in many heating systems to ignite fuel.

Immersion Freezing: Freezing of articles by dipping them into liquid refrigerant.

Incomplete Combustion: Combustion with insufficient oxygen.

Indoor Coil: Refrigerant containing portion of a fan coil unit similar to a car radiator, typically made of several rows of copper tubing with aluminum fins.

Indoor Coils: The outdoor unit (air conditioner or heat pump) and the indoor unit (coil or blower coil). The term "most popular coil" indicates the actual tested combinations; other ratings may be simulated and unrealistic. Be sure that the efficiency ratings you are comparing are for "most popular coil." You'll know the ratings are attainable and close to reality.

Indoor Unit: Contains the indoor coil, fan, motor, and filtering device, sometimes called the air handler.

Inductance: Inducing voltage in a coil due to the change in the rate of flow of current in the coil.

Induction Motor: An AC motor that operates on the principle of a rotating magnetic field.

Infiltration: Air flow inward into a space through walls, leaks around doors and windows, or through the building materials used in the structure.

Inhibitor: Substance that prevents a chemical reaction.

Insulation, Thermal: Material that is a poor conductor of heat; used to retard flow of heat through wall.

Integrated Circuit Board: Electronic circuit made from transistors, resistors, etc., all placed into a package referred to as a "chip," since all circuits are on one base of semiconductor material.

Interlocked: Controlled by a switch that does not allow a component to operate when a hazardous condition exists.

Intermittent Cycle: Cycle which repeats itself at varying time intervals.

Intrinsic Semiconductor: Material that is neither conductor nor insulator, such as silicon and germanium, often used in temperature sensing devices.

Isothermal: Changes of volume or pressure under conditions of constant temperature.

IWC (Inches of Water Column): Commonly used in the USA.

J

Jet Cooling System: Jet pump is used to produce a vacuum so water or refrigerant may evaporate at relatively low temperatures. These systems usually require a large condenser and have a low efficiency.

GLOSSARY *(cont.)*

Jet Pump: A centrifugal pump combined with an ejector, which can replace the compressor in some refrigeration systems.

Joint: Connecting point as between two pipes.

Joule: Metric unit of heat.

Joule–Thomson Effect: The change in the temperature of a gas on its expansion through a porous plug from a higher pressure to a lower pressure.

Journal, Crankshaft: Part of shaft that contacts the bearing on the large end of the piston rod.

K

Kata Thermometer: Large-bulb alcohol thermometer used to measure air speed or atmospheric conditions by means of cooling effect.

Kelvin Scale (K): Thermometer scale on which unit of measurement equals the Celsius degree and according to which absolute zero is 0°, the equivalent of −273.16°C. Water freezes at 273.16°K and boils at 373.16°K.

Kilocalorie: Great calorie (1000 calories) used in engineering science.

Kilopascal (kPa): Metric unit of pressure equal to 1000 Pascals.

Kinetic Energy: Energy of motion.

King Valve: Liquid receiver service valve.

KW (Kilowatt): A kilowatt equals 1,000 watts.

KWh (Kilowatt Hour): A kilowatt hour (kWh) is the amount of kilowatts of electricity used in 1 hr. of operation of any equipment.

L

Lamp, Steri: Lamp that has a high-intensity ultraviolet ray used to kill bacteria. Also used in food storage cabinets and in air ducts.

Latent Heat: Heat, that when added or removed, causes a change in state but no change in temperature.

Latent Heat of Condensation: Amount of heat released (lost) to change from a vapor (gas) to a liquid.

Latent Heat of Vaporization: Amount of heat required to change from a liquid to a vapor (gas).

Leak Detector: A device used to detect refrigerant leaks in a refrigerating system.

Legionnaire's Disease Bacterium (LDB): Thought to be transmitted by airborne routes, possibly by open air cooling towers or evaporative condensers in commercial systems.

Limit Control: Control used to open or close electrical circuits as temperature or pressure limits are reached.

Liquefied Gases: A gas below a certain temperature and above a certain pressure that becomes liquid.

Liquid Absorbent: Chemical in liquid form that has the property to absorb other fluids.

Liquid Desuperheater: Valve that permits small flow of refrigerant to enter low side of systems to cool suction gas.

Liquid Floodback: A surge of liquid returning to the compressor.

Liquid Indicator: Device located in liquid line that provides a glass window through which liquid flow may be watched.

Liquid Line: Tube that carries liquid refrigerant from the condenser or liquid receiver to the refrigerant control mechanism.

Liquid Nitrogen: Nitrogen in liquid form used as a low-temperature refrigerant in expendable or chemical refrigerating systems.

Liquid Receiver: Cylinder (container) connected to condenser outlet for storage of liquid refrigerant in a system.

Liquid Receiver Service Valve: Two- or three-way manual valve located at the outlet of the receiver and used for installation and service purposes. It is sometimes called the king valve.

Liquid Transfer Method: Method of liquid refrigerant recovery in which the air conditioning unit is pressurized and refrigerant is removed by the created pressure difference.

Liquid-Vapor Valve Refrigerant Cylinder: Dual hand valve on refrigerant cylinders that is used to release either gas or liquid refrigerant from the cylinder.

Lithium Bromide: Chemical commonly used as the absorbent in absorption cooling system. Water would then be the refrigerant.

Load: The required rate of heat removal.

Lockout Relay: A device that shuts down a circuit whenever a safety control device is open.

GLOSSARY (cont.)

Low Side: The parts of a refrigeration system subject to the evaporator pressure.

Low-Pressure Safety Cutout: Motor protection device that senses low-side pressure. Control is wired in series with the motor and will shut off during periods of excessively low suction pressure. Also called *low-side pressure indicator.*

Low-Side Float Valve: Refrigerant control valve operated by level of liquid refrigerant in low-pressure side of system.

Low-Side Pressure: Pressure in cooling side of refrigerating cycle.

Low-Side Pressure Control: Device used to keep low-side evaporating pressure from dropping below a certain pressure.

LP Fuel: Liquefied petroleum used as a fuel gas.

M

Machine Room: Area where commercial and industrial refrigeration machinery—except evaporators—is located.

Magnetic Gasket: Door-sealing material that keeps a door tightly closed with small magnets inserted in gasket.

Main: A pipe or duct for distributing to or collecting conditioned air from various branches.

Make-Up Air Units: An air unit used to create a slight positive pressure in homes, reducing infiltration.

Manifold, Service: Chamber equipped with gauges and manual valves, used by service technicians to service refrigerating systems.

Manometer: A U-shaped liquid-filled tube for measuring pressure differences.

Mass: Quantity of matter held together so as to form one body.

Matched: In refrigeration systems, the correct balancing of the following items: heat load, condensing unit capacity, evaporator capacity, and total system capacity.

MBH: Thousands of British thermal units (1 MBH = 1000 Btu).

McLeod Gauge: Instrument used to measure high vacuums.

Mean Effective Pressure (MEP): Average pressure on a

surface when a changing pressure condition exists.

Mechanical Efficiency: The ratio of work done by a machine to the work done on it or energy used by it.

Mechanical Equivalent of Heat: An energy-conversion ratio of 778.18 ft. lbs.-1 Btu.

Melting Point: Temperature at which a substance will melt at atmospheric pressure.

Micrometer: Precision-measuring instrument used for making measurements accurate to .001 to .0001".

Micron: Unit of length in metric system equal to one millionth of a meter.

Micron Gauge: Instrument for measuring vacuums very close to a perfect vacuum.

Milli: Prefix denoting one thousandth (1/1000); for example, millivolt means one thousandth of a volt.

Minerals: Refers to substances found in water (carbonate, sulfate, lime, iron, etc.) that produce scale formation inside tubing.

Minimum Stable Signal (MSS): Correct setting for an expansion valve where it is utilizing the evaporator

efficiently but remains free from hunting.

Modulating Controls: A control capable of gradual adjustments, rather than simple on–off control.

Modulating Refrigeration Cycle: Refrigerating system of variable capacity.

Modules: Thermoelectric cooling units.

Moisture Indicator: Instrument used to measure moisture content of a refrigerant.

Mollier's Diagram: Graph of refrigerant pressure, heat, and temperature properties.

Muffler: Sound absorber chamber in refrigeration system. Used to reduce sound of gas pulsations.

Mullion Heater: Electrical heating element mounted in the mullion. Used to keep mullion from sweating or frosting.

Multiple System: Refrigerating mechanism in which several evaporators are connected to one condensing unit.

Multiple-Pass Recycling Machine: A refrigerant recycling machine that cycles the refrigerant through a filter drier several times to separate and remove oil.

Multistage System: A system used to produce very low temperatures.

N

Natural Gas: A mixture of methane and other hydrocarbons used as fuel.

Needle Point Valve: Valve having a needle point plug and a small seat orifice for low-flow metering.

Negative Temperature Coefficient Thermistor (NTC): Electronic thermistor that decreases in resistance as temperature increases.

Net Capacity: The interior volume of a refrigeration cabinet.

Neutralizer: Substance used to counteract acids in a refrigeration system.

Newton: Force required to accelerate an object that has a mass of 1 kg to 1 m/sec^2.

Noise Dosimeter: Instrument used to measure sound.

Noncondensables: Foreign gases mixed with a refrigerant, which cannot be condensed into liquid form at the temperatures and pressures at which the refrigerant condenses.

Nonfrosting Evaporator: Evaporator that never collects frost or ice on its surface.

Noninductive Load: An electrical load consisting of resistance that does not affect the power factor.

Normal Charge: Thermal element charge that is part liquid and part gas under all operating conditions.

O

Octyl Alcohol–Ethyl-Hexanol: Additive in absorption machines that reduces surface tension in absorber.

Off Cycle: Segment of refrigeration cycle when the system is not operating.

Offset: In a proportional control system, the deviation between the set point and the control point.

Oil Binding: Condition in which an oil layer on top of the refrigerant liquid may prevent it from evaporating at normal pressure and temperature.

Oil Burner: A device for burning vaporized oil (gas) to produce heat.

Oilless Bushing: Assembly in which a shaft passes through a sintering bushing (which is impregnated with oil). Bushing is permanently lubricated.

Oil Level Regulator: A device that controls the oil level in the compressor.

Oil Pressure Safety Cutout: Motor protection device that senses oil pressure in the compressor. It is wired in series with the compressor and will shut it off during periods of low oil pressure.

Oil, Refrigeration: Specially prepared oil used in refrigerator mechanism, which circulates, to some extent, with refrigerant.

Oil Reservoir: Container that stores the compressor's oil supply.

Oil Ring: Lower piston ring.

Oil Separator: Device used to remove oil from gaseous refrigerant.

Oil Slugging: Oil being pumped out of the compressor.

Open Compressor: Term used to indicate an external drive compressor.

Open-Cycle Refrigeration: Refrigeration system in which the refrigerant is released to the atmosphere after evaporation.

Open System: Refrigerating system which uses a belt-driven or a coupling-driven compressor.

Operating Differential: The actual temperature or pressure difference in the conditioned area.

Operating Pressure: Actual pressure at which the system works under normal conditions. This pressure may be positive or negative (vacuum).

Orifice: Accurate size opening for controlling fluid flow.

Orifice Tube: Metering device consisting of a restricting tube with inlet and outlet screens.

O-Rings: Sealing devices used between parts where there may be some motion.

Outdoor Coil: Refrigerant containing portion of a fan coil unit similar to a car radiator, typically made of several rows of copper tubing with aluminum fins.

Overload: Load greater than that for which the system was intended.

Overload Protector: Device, either temperature, pressure, or current operated, that will stop operation of unit if dangerous conditions arise.

Oxidation: The chemical combining of oxygen with a

GLOSSARY (cont.)

specific material, resulting in deterioration of that material.

Ozone: The O_3 form of oxygen, sometimes used in air conditioning or cold-storage rooms to eliminate odors.

Ozone Depletion Potential (ODP): Numeric value assigned to each refrigerant to express the risk to the ozone layer that the given refrigerant may cause.

P

Packaged Terminal Air Conditioning: A combination heating and cooling unit designed for a single room or zone.

Package Unit or Package System: A self-contained unit that is outside the home and connected to a duct system by a penetration through the home's foundation, except for geothermal, which is a self-contained indoor unit that is placed in a closet, attached garage, basement, or mechanical room.

Packing: The stuffing around a shaft to prevent fluid leakage between the shaft and parts around the shaft.

Partial Pressures: Condition where two or more gases occupy a space and each one creates part of the total pressure.

Parts per Million (ppm): Unit of concentration of one element in another.

Pascal (Pa): Unit of pressure in the metric system.

Pascal's Law: Pressure imposed upon a fluid is transmitted equally in all directions.

Passive Solar Heating System: A solar energy system that is dependent upon the radiation striking directly on the surface to be heated.

Peltier Effect: When direct current is passed through two adjacent metals, one junction will become cooler and the other will become warmer.

Perimeter Drier: An electrical resistance heat wire located in a freezer door to prevent condensation on the exterior of the cabinet and around the freezer door.

Perimeter Hot Gas Tube System: System that has a tube located on the surface of the outer portion of the cabinet to prevent condensation from forming.

Permeable: Having openings that allow the passage of liquid or gas.

pH: Measurement of the free hydrogen ion concentration in an aqueous solution. A pH of 7 is neutral.

Phase Loss Monitor: Motor protection device for polyphase motors that measures current flow to detect phase loss.

Phial: Term sometimes used to denote the sensing element on a thermostatic expansion valve.

Piercing Valve: A type of service valve used on hermetic units.

Piezoelectric: Property of quartz crystal that causes it to vibrate when a high frequency (500 kHz or higher) voltage is applied. Used to atomize water in a humidifier.

Piston: Close-fitting part or plug that moves up and down in a cylinder.

Pitot Tube: Tube used to measure air velocities.

Planck's Constant: Constant value (6.626×10^{-34} J·sec) which, when multiplied by the frequency of radiation, determines the amount of energy in a photon.

Plenum Chamber: Chamber or container for moving air or other gas under a slight positive pressure.

Pneumatic System: An air conditioning system in which pneumatic motors are operated by pressurized air lines.

Polychlorinated Biphenyl (PCB): Dielectric fluid used in capacitors and transformers that is toxic. Use of PCB in transformers and capacitors is strictly regulated by the Environmental Protection Agency.

Polystyrene: Plastic used as an insulation in some refrigerated structures.

Polyurethane: Plastic used in insulation and molded products.

Ponded Roof: Flat roof designed to hold a quantity of water, which acts as a cooling device.

Portable Service Cylinder: Container used to store refrigerant. Two most common types are disposable and refillable.

Positive Pressure: A pressure greater than atmospheric.

Positive Temperature Coefficient Thermistor (PTC): Electronic thermistor that increases in resistance as temperature increases.

Potassium Permanganate: Chemical used in carbon filters to help reduce odors.

Potentiometer: Instrument for measuring or controlling by sensing small changes in electrical resistance.

Pound-Force: Force applied to a 1-lb. mass to give it an acceleration of 32.173 ft./sec^2 (gravitational acceleration).

Pour Point: Lowest temperature at which a liquid will pour or flow.

Power: 1—Time rate at which work is done or energy emitted. 2—Source or means of supplying energy.

Power Burner: A burner that has air blown into it by a blower.

Power Saver Switch: A switch that disconnects heaters in a refrigeration cabinet.

Precooler Condenser: Used to cool the refrigerant prior to entering the main condenser.

Pressure: The force exerted per unit of area.

Pressure Cycling Switch: Pressure-controlled switch located on the inlet line of the evaporator to prevent rapid cycling of the compressor.

Pressure Drop: Loss in pressure, as from one end of a refrigerant line to the other, due to friction, static head, etc.

Pressure Gauge: Instrument for measuring the pressure exerted by the contents on its container.

Pressure, Gauge: Pressure above atmospheric pressure.

Pressure, Head: Force caused by the weight of a column or body of fluids.

Pressure-Heat Diagram: Graph of refrigerant pressure, heat, and temperature properties (Mollier's diagram).

Pressure Limiter: Device that remains closed until a certain pressure is reached, then opens and releases fluid to another part of system or breaks an electric circuit.

Pressure Motor Control: Device that opens and closes an electrical circuit as pressures change.

Pressure-Operated Altitude (POA) Valve: Device that maintains a constant low-side pressure, independent of altitude of operation.

Pressure, Operating: Pressure at which a system is operating.

GLOSSARY (cont.)

Pressure Regulator, Evaporator: Automatic pressure-regulating valve mounted in the suction line between the evaporator outlet and the compressor inlet. Its purpose is to maintain a predetermined pressure and temperature in the evaporator.

Pressure-Relief Valve: A valve or rupture member designed to relieve excessive pressure automatically.

Pressure, Suction: Pressure in low-pressure side of a refrigerating system.

Pressure Switch: Switch operated by a change in pressure.

Pressure Water Valve: Device used to control water flow. It is responsive to head pressure of refrigerating system.

Primary Air: In a combustion system, the air mixed with fuel prior to ignition.

Primary Coil: A tube-and-fin circular coil that contains a water–glycol solution, which surrounds the ignitor and burner. This coil is used in a water–glycol gas forced-air furnace.

Primary Control: Device that directly controls operation of heating system.

Process Tube: Length of tubing fastened to hermetic unit dome, used for servicing unit.

Product Heat Load: Sum of specific, latent, and respiration heat loads.

Propane: Volatile hydrocarbon used as a fuel or as a refrigerant.

Proportional: Being in the proper relative quantity or balance.

psi: Pounds per square inch.

psia: Pounds per square inch absolute. Absolute pressure equals gauge pressure plus atmospheric pressure.

psig: Pounds per square inch gauge.

Psychrometer: Instrument for measuring the relative humidity of atmospheric air. Also called *wet-bulb hygrometer*

Psychrometric Chart: Chart that shows relationship among the temperature, pressure, and moisture content of the air.

Puffback: The ignition of vaporized oil in the firepot.

Pulley: Flat wheel with a "V" groove. When attached to a drive and drive members, the pulley provides a means for driving the compressor.

Pulse: Term referring to one cycle of ignition and combustion of a gas–air mixture in a pulse combustion furnace.

Pulse Combustion Process: Repeated ignition of a gas and air mixture in a high-efficiency gas furnace.

Pulse Furnace: Furnace that has a "tuned" (resonant) combustion chamber. Part of the energy normally lost through the flue is returned to start next "pulse" of combustion.

Pump: Any one of various machines that force gas or liquid into—or draw it out of—something as by suction or pressure.

Pump, Centrifugal: Pump that produces fluid velocity and converts it to pressure head.

Pump Down: The act of using a compressor or a pump to reduce the pressure in a container or a system.

Pump, Fixed Displacement: A pump in which the displacement per cycle cannot be varied.

Pump, Reciprocating Single Piston: A pump having a single reciprocating (moving up and down or back and forth) piston.

Pump, Screw: Pump having two interlocking screws rotating in a housing.

Purging: The act of blowing out refrigerant gas from a refrigerant containing vessel usually for the purpose of removing noncondensables.

Pyrometer: Instrument for measuring high temperatures.

Q

Quenching: Submerging a hot object in cooling fluid.

Quick-Connect Coupling: A device that permits easy and fast connecting of two fluid lines.

R

R-11, Trichlorofluoromethane: Low-pressure, synthetic chemical refrigerant that is also used as a cleaning fluid.

R-12, Dichlorodifluoromethane: Popular refrigerant known as Freon-12.

R-22, Chlorodifluoromethane: Low temperature refrigerant with boiling point of −41°F (−40.5°C) at atmospheric pressure.

R-113, Trichlorotrifluoroethane: Synthetic chemical refrigerant that is nontoxic and nonflammable.

R-160, Ethyl Chloride: Toxic refrigerant now seldom used.

R-170, Ethane: Low-temperature refrigerant.

R-290, Propane: Low-temperature refrigerant.

R-500: Refrigerant that is an azeotropic mixture of R-12 and R-152a.

R-502: Refrigerant that is an azeotropic mixture of R-22 and R-115.

R-503: Refrigerant that is an azeotropic mixture of R-23 and R-13.

R-504: Refrigerant that is an azeotropic mixture of R-32 and R-115.

R-600, Butane: Low-temperature refrigerant; also used as a fuel.

R-611, Methyl Formate: Low-pressure refrigerant.

R-717, Ammonia: Popular refrigerant for industrial refrigerating systems; also a popular absorption system refrigerant.

Radiant Heating: Heating system in which warm or hot surfaces are used to radiate heat into the space to be conditioned.

Radiation: The passage of heat from one object to another without warming the space between. The heat is passed by wave motion similar to light.

Range: Pressure or temperature settings of a control; change within limits.

Rankine Scale: Name given the absolute (Fahrenheit) scale. Zero (0°R) on this scale is −460°F.

Receiver Drier: Cylinder (container) in a refrigerating system for storing liquid refrigerant and desiccant.

Receiver Heating Element: Electrical resistance heater mounted in or around liquid receiver. It is used to maintain head pressures when ambient temperature is low.

Reciprocal: Inverse.

Reciprocating: Back and forth motion in a straight line.

Reciprocating Compressor: A compressor driven by piston (positive displacement).

Reclaiming: Taking refrigerant that has been removed from a system and processing it in accordance with EPA rules.

Recording Thermometer: Temperature-measuring instrument that has a pen marking a moving chart.

Recovery: Removal of refrigerant from a system.

Rectifier, Electric: Electrical device for converting AC to DC.

Recuperative Coil: Secondary coil in glycol–water forced-air furnace that extracts latent heat from combustion gases.

Recycling: Passing of flue gases from combustion in a furnace to a secondary heat exchanger to remove latent heat.

Reed Valve: Compressor valve consisting of a thin, flat, high-carbon alloy steel.

Refractory Cement: A variety of mixtures used to line furnaces.

Refrigerant: Substance used in refrigerating mechanism. A substance that produces a refrigerating effect while expanding or vaporizing. It absorbs heat in the evaporator by change of state from a liquid to a gas, and releases its heat in a condenser as the substance returns from the gaseous state back to a liquid state.

Refrigerant Charge: Quantity of refrigerant in a system.

Refrigerant Control: Device that meters flow of refrigerant between two areas of a refrigerating system. It also maintains pressure difference between the high-pressure and low-pressure sides of the mechanical refrigerating system while unit is running.

Refrigerant Dye: Coloring agent that can be added to refrigerant to help locate leaks in a system.

Refrigerant Jets: A jet pump that sprays refrigerant into the condenser.

Refrigerant Management System: A refrigerant recovery/recycling unit.

Refrigerant Quality: Ratio of liquid refrigerant to refrigerant vapor.

Refrigerant Transfer Unit: Machine designed to safely remove refrigerant from a system.

Register: Combination grille and damper assembly covering an air opening or end of an air duct.

Relative Density: Ratio of the mass of a volume of gas compared to the mass of the same volume of hydrogen.

Relative Humidity: The ratio of the water-vapor pressure of air compared to the vapor pressure it would have if saturated at its dry-bulb temperature.

Relief Valve: A valve designed to open at excessively high pressures to allow refrigerant to escape.

Remedy: A procedure whereby refrigerants are prepared for reuse by returning them to new product specifications.

Remote System: Refrigerating system in which the condensing unit is away from the space to be cooled.

Remote Temperature-Sensing Element: Control device used to maintain desired temperature.

Resistance: An opposition to flow or movement. A coefficient of friction.

Restrictor: A device for producing a deliberate pressure drop or resistance in a line by reducing the flow area.

Retrofit: Term used in describing reworking an older installation to bring it up-to-date with modern equipment or to meet new code requirements.

Return Air: Air drawn into a heating unit after having been circulated from the heater's output supply to a room.

Reverse Cycle Defrost: Method of heating evaporator for defrosting. Valves move hot gas from the compressor into the evaporator.

Reversing Valve: Device used to reverse direction of the refrigerant flow, depending upon whether heating or cooling is desired.

Ringlemann Scale: Device for measuring smoke density.

Riser Valve: Device used to manually control flow of refrigerant in vertical piping.

Rotary Blade (Vanes) Compressor: Mechanism for

pumping fluid by revolving blades inside cylindrical housing.

Rotary Compressor: A compressor in which compression is attained in a cylinder by rotation of a semiradial member.

RSES: Refrigeration Service Engineers Society— http://www.rses.org/

Running Time: Amount of time a condensing unit is run per hour or per 24 hr.

R-Value: The thermal resistance of a given material.

S

Saddle Valve (Tap-a-Line): Valve body shaped so it may be silver-brazed or clamped onto a refrigerant tubing surface.

Safety Control: Device to stop refrigerating unit if unsafe pressure, temperatures, or dangerous conditions are reached.

Safety Interlock Switch: A switch that, when activated, prevents a piece of interlocked equipment from operating.

Safety Motor Control: Electrical device used to open the circuit to the motor if temperature, pressure, or current flow exceed safe conditions.

Safety Plug: Device that will release the contents of a container before rupture pressures are reached.

Safety Valve: Self-operated, quick-opening valve used for fast relief of excessive pressures.

Saturated Vapor: Vapor not superheated but of 100% quality, i.e., containing no unvaporized liquid.

Saturation: Condition existing when substance contains all of another substance it can hold.

Saturation Temperature: Also referred to as the boiling point or the condensing temperature. This is the temperature at which a refrigerant will change state from a liquid to a vapor or vice versa.

Scale: A coating of deposited material.

Scale-Free System: A system that eliminates deposits in condensers by picking up electrical energy from water, allowing deposits to be carried through the system and disposed.

Schrader Valve: Spring-loaded device that permits

fluid flow in one direction when a center pin is depressed and in the other direction when a pressure difference exists.

Scotch Yoke: Mechanism used to change reciprocating motion into rotary motion or vice versa. Used to connect crankshaft to piston in refrigeration compressor.

Screw Compressor: Compressor constructed of two mated revolving screws.

Scroll Compressor: A compressor that uses the interaction of two spiral coils (scrolls) to compress a vapor.

Sealed Unit: See Hermetically Sealed Unit.

Seal, Shaft: A mechanical system of parts for preventing gas leakage between a rotating shaft and a stationary crankcase.

Seat: That portion of a valve mechanism against which the valve presses to effect shutoff.

Secondary Air: Air added to a flame after ignition to maintain combustion.

Secondary Refrigerating System: Refrigerating system in which the condenser is cooled by the evaporator of another (primary) refrigerating system.

SEER (Seasonal Energy Efficiency Ratio): The amount of cooling your equipment delivers per every dollar spent on electricity. SEER applies to air conditioners and heat pumps. In the past, a unit with a SEER of 8.00 was considered standard efficiency, and a unit with a 10.00 SEER was considered high efficiency. After January 1, 1992, the minimum SEER required by the DOE is 10.00 and 15.00+ SEER is considered high efficiency. EER and SEER cannot be compared equally. Air source equipment is rated by SEER and geothermal equipment is rated by EER. The total cooling of a central unitary air conditioner or unitary heat pump in Btus during its normal annual usage period for cooling divided by the total electric energy input in watt-hours during the same period.

Selective Absorber Surface: Surface used to increase the temperature of a solar collector.

Semihermetic Compressor: Hermetic compressor with service valves.

Sensible Heat: Heat, that when added or removed, causes a change in temperature but not in state.

Sensor: Material or device that goes through physical or electronic change as surrounding conditions change.

Separator, Oil: Device to separate refrigerant oil from refrigerant gas and return the oil to the compressor crankcase.

Sequential Operating Control: A series of controls used in a preset order.

Serpentine Belt: Drive belt that assumes many winding forms eliminating other belts needed to drive accessories.

Serviceable Hermetic: Hermetic unit housing containing motor and compressor assembly by use of bolts or cap screws.

Service Valve: Manually operated valve mounted on refrigerating systems used for service operation.

Setpoint: The temperature to which a thermostat is set to result in a desired heated space temperature.

Shell-and-Coil Condenser: Condenser consisting of a coil of tubing housed in a shell. Similar to a *shell-and-tube condenser*.

Shell and Tube: Pertaining to heat exchangers in which a coil of tubing or pipe is contained in a shell or container.

Shell-and-Tube Flooded Evaporator: An evaporator that uses water flow (through tubes built into cylindrical vessels).

Short Cycling: Refrigerating system that starts and stops more frequently than it should.

Shroud: Housing over condenser, evaporator, or fan.

Sick Building Syndrome (SBS): In a building, conditions existing that may result in human illness.

Sight Glass: Glass tube or glass window in refrigerating mechanism. It shows amount of refrigerant or oil in system and indicates presence of gas bubbles in liquid line.

Silica Gel: A drier material having the formula SiO_2.

Single-Pass Recycling Machine: A recycling machine in which the refrigerant is passed through a filter drier once.

Single-Pipe System: System of steam heating in which a single pipe carries steam to radiator and is also used as a condensate return.

Single-Stage Compressor: Compressor having only one compressive step between inlet and outlet.

Sizing: Refers to the procedure a heating contractor goes through to determine how large a furnace (measured in Btu/h) is needed to heat a house efficiently. Sizing depends on the square footage of the home, the amount of ceiling and wall insulation, the window area, use of storm doors, storm windows, and more.

Skin Condenser: Condenser using the outer surface of the cabinet as the heat radiating medium.

Sleeve Covers: The top opening cover on an ice cream cabinet.

Sling-Psychrometer: Measuring device with wet- and dry-bulb thermometers. Moved rapidly in air, it measures relative humidity.

Slip Ring Lubricating Method: A lubricating method in which a brass ring lubricates the bearing.

Sludge: A decomposition product formed in a refrigerant due to impurities in the oil or due to moisture.

Slug: 1—Unit of mass equal to the weight of object (in pounds) divided by 32.2 (acceleration due to the force of gravity). 2—Detached mass of liquid or oil that causes an impact or hammer in a circulating system.

Slugging: Condition in which a mass of liquid enters the compressor, causing hammering.

Smoke Test: Test made to determine completeness of combustion.

Solar Energy Systems: Systems used to collect, convert, and distribute solar energy in forms useful within a business or residence. A passive system uses no additional energy from other sources for the distribution of the solar-generated heat. An active system may use blowers, supplementary coils, etc.

Soldering: Joining two metals by adhesion of a metal with a melting temperature of less than 800°F (427°C).

Solenoid Valve: A valve opened by a magnetic effect of an electric current through a solenoid coil.

Solid: The state of matter in which a force can be exerted in a downward direction only when not confined. As distinguished from fluids.

Solid Fuel Heating: The use of solid natural resources such as wood or coal to provide heat.

Solubility: The ability of one material to enter into solution with another.

Solution: The homogeneous mixture of two or more materials.

Specific Gravity: The weight of a volume of a material compared to the weight of the same volume of water.

Specific Heat: The quantity of heat required to raise the temperature of a definite mass of a material a definite amount compared to that required to raise the temperature of the same mass of water the same amount, expressed as Btu/per pound per degrees Fahrenheit.

Specific Volume: The volume of a definite weight of a material. Usually expressed in cubic feet per pound. The reciprocal of density.

Splash System, Oiling: Method of lubricating moving parts by agitating or splashing oil in the crankcase.

Split System: Refrigeration or air conditioning installation, which places condensing unit outside or away from evaporator. These units are connected together by supply and return refrigerant lines. Also applicable to heat pump installations.

Spray Cooling: Method of refrigerating by spraying expendable refrigerant or by spraying refrigerated water.

Squirrel Cage: Fan that has blades parallel to fan axis and moves air at right angles or perpendicular to fan axis.

SRN (Sound Rating Number): Sound is measured in bels (a bel equals 10 decibels). The SRN of a unit is based on ARI test, performed at ARI standard rating conditions. Average sound rating range from 7.0 to 8.0 decibels. The lower the SRN rating, the quieter the unit.

GLOSSARY (cont.)

Standard Air: Air having a mass density of 0.075 lb./ft.^{3} (1.204 kg/m^3), a temperature of 70°F (21°C), and a pressure of 30" Hg (760 mm Hg).

Standard Atmosphere: Condition when air is at 14.7 psia pressure, at 68°F (20°C) temperature and a relative humidity of 36%.

Standard Conditions: Used as a basis for air conditioning calculations: temperature of 68°F (20°C), pressure of 29.92" of mercury (Hg), and relative humidity of 30%.

Standing Pilot: Old system of furnace burner ignition, in which a pilot light is constantly burning.

Starve: Condition in which there is not enough refrigerant reaching the evaporator.

Static Head: Vertical piping run.

Static System: A system in which air is circulated through the condenser by natural convection.

Stationary Blade Compressor: Rotary pump that uses a nonrotating blade inside pump to separate intake chamber from exhaust chamber.

Steam: Water in vapor state.

Steam Heating: Heating system in which steam from a boiler is piped to radiators.

Steam Jet Refrigeration: Refrigerating system which uses a steam venturi to create high vacuum (low pressure) on a water container causing water to evaporate at low temperature.

Steam Trap: A device for allowing the passage of condensate, or air and condensate, and preventing the passage of steam.

Sterling Cycle: A refrigeration cycle that can produce temperatures down to −450°F (−268°C).

Strainer: Device such as a screen or filter used to retain solid particles while liquid passes through.

Stratification: Condition in which there is little or no air movement in room; air lies in temperature layers.

Stroke: The distance traveled by a piston.

Subcooled Liquid: Liquid refrigerant which is cooled below its saturation temperature.

Sublimation: The change from a solid to a vapor state without an intermediate liquid state.

Suction Line: The tube or pipe that carries refrigerant vapor from the evaporator to the compressor inlet.

Suction Pressure: Pressure on the suction side of the compressor.

Suction Pressure Control Valve: Device located in the suction line that maintains constant pressure in evaporator during the running portion of cycle.

Suction Service Valve: Two-way, manually operated valve located at the inlet to the compressor. It controls suction gas flow and is used to service the unit.

Suction Side: Low-pressure side of the system extending from the refrigerant control through the evaporator to the inlet valve of the compressor.

Superheat: 1—Temperature of vapor above its boiling temperature as a liquid at that pressure. 2—The difference between the temperature at the evaporator outlet and the lower temperature of the refrigerant evaporating in the evaporator.

Superheated Vapor: Refrigerant vapor which is heated above its saturation temperature. If a refrigerant is superheated, there is no liquid present.

Superheater: Heat exchanger arranged to take heat from liquid going to evaporator and use it to superheat vapor leaving the evaporator.

Surge: Regulating action of temperature or pressure before it reaches its final value or setting.

Surge Tank: Container connected to the low-pressure side of a refrigerating system that increases gas volume and reduces rate of pressure change.

Swaging: Enlarging one tube end so the end of another tube of the same size will fit within it.

Swamp Cooler: Evaporative type cooler in which air is drawn through porous mats soaked with water.

Swash Plate: Device used to change rotary motion to reciprocating motion. Used in some refrigeration compressors.

Sweating: Condensation of moisture from the air on surfaces below the dew-point temperature.

Switchover Valve: A device in a heat pump that reverses the flow of refrigerant as the system is switched from cooling to heating. Also called a reversing valve or four-way valve.

Synthetic Dust Weight Arrestance: Measurement of filter's ability to remove synthetic dust from test air.

T

Tail Pipe: Outlet pipe from the evaporator.

Tandem: A term used to identify a system that connects two motor compressors at the motor end.

Temperature: Heat level or pressure.

Temperature–Humidity Index: Actual temperature and humidity of air sample compared to air at standard conditions.

Temperature-Sensing Bulb: Bulb containing a volatile fluid and bellows or diaphragm. Temperature increase on the bulb causes the bellows or diaphragm to expand.

Test Light: Light provided with test leads. Used to test or probe electrical circuits to determine if they are working properly.

Therm: Another measurement of heat. One therm equals 100,000 Btu/h.

Thermal Conductivity: The ability of a material to conduct heat from one point to another. Indicated in terms of Btu per hour per square foot per inches of thickness per degrees Fahrenheit.

Thermal Precipitation: The collection of dirt around warm air grilles.

Thermal Relay (Hot Wire Relay): Heat-operated electrical control used to open or close a refrigeration system electrical circuit. This system uses a resistance wire to convert electrical energy into heat energy.

Thermal Resistance: See *R-Value.*

Thermistor: A semiconductor with electrical resistance that varies with temperature.

Thermocouple: A device consisting of two electrical conductors having two junctions—one at a point whose temperature is to be measured and the other at a known temperature.

Thermodisc Defrost Control: Electrical switch with bimetal disc controlled by temperature changes.

Thermodynamics: The science of the mechanics of heat.

Thermoelectric Refrigeration: Refrigerator mechanism that depends on the Peltier effect. Direct current flowing through an electrical junction between unlike metals provides a heating or cooling effect, depending on direction of current flow.

Thermometer: A device for indicating temperature.

Thermomodule: Number of thermocouples used in parallel to achieve low temperatures.

Thermopile: Number of thermocouples used in series to create a higher voltage.

Thermostat: A temperature sensitive switch for controlling the operation of a heater or furnance.

Thermostat Droop: Added heat in line-voltage thermostat produced by the thermostat itself.

Thermostatic Control: Device that operates system or part of system based on temperature change.

Thermostatic Expansion Valve: A device to regulate the flow of refrigerant into an evaporator so as to maintain an evaporation temperature in a definite relationship to the temperature of a thermostatic bulb.

Thermostatic Motor Control: Device used to control cycling of unit through use of a control bulb. The bulb reacts to temperature changes.

Thermostatic Switch: A switch controlled by temperature changes.

Thermostatic Valve: Valve controlled by temperature-change response elements.

Thermostatic Water Valve: Valve used to control flow of water, actuated (made to work) by temperature difference. Used in units such as water-cooled compressors and condensers.

Three-Way Valve: Flow control valve with three fluid flow openings.

Throttling: Expansion of gas through an orifice or controlled opening without gas performing any work as it expands.

Throw: The distance air travels from a grille before slowing to 50 ft./min.

Timed On–Off Control: Control needed when the existing differential is too great.

Timer-Thermostat: Thermostat control which includes a clock mechanism. Unit automatically controls room temperature and changes temperature range depending on time of day.

Ton: A cooling unit of measure. Each ton equals 12,000 Btu/h. Heat pumps and air conditioners are generally sized in tons. Typical sizes for single family residences are between two and five tons. It is important to note that actual capacity is not constant and will change based on outdoor or indoor temperatures. The published capacity rating of air conditioners and heat pumps is based on performance at the ARI standard temperature levels of 95°F outside, 80°F inside.

Ton of Refrigeration: Refrigeration equivalent to the melting of one ton of ice per 24 hr. 288,000 Btu per day, 12,000 Btu per hour, or 200 Btu per minute.

Ton Refrigeration Unit: Unit that removes same amount of heat in 24 hr. as melting of 1 ton of ice.

Torque: Turning or twisting force.

Torque, Starting: Amount of torque available to start and accelerate the load.

Torque Wrench: Wrench that may be used to measure torque or pressure applied to a nut or bolt.

Torr: A unit of pressure equal to 1/760 of an atmosphere (1 mm Hg), normally used for measuring vacuum pressure.

Total Air Balance (TAB): In an air circulation system, adjusting the system so that all rooms receive the proper amount of air.

Total Energy Management (TEM): Conservation concept where a building is looked at in terms of its total energy usage, rather than analyzing the requirements of separate systems.

Total Heat: Sum of both the sensible and latent heat.

Total Pressure: In fluid flow, the sum of static pressure and velocity pressure.

GLOSSARY (cont.)

Toxicity: A measure of the amount of poison in a substance or the amount of harm it can cause.

Transducer: Device turned on by a change of power from one source for the purpose of supplying power in another form to a second system.

Transformer: Electromagnetic device that transfers electrical energy from the primary circuit into variations of voltage in a secondary circuit.

Transformer-Rectifier: Combination transformer and rectifier in which input AC current may be varied and then rectified into DC current.

Transmission: Heat loss or gain from a building through exterior components such as windows, walls, or floors.

Trichlorotrifluoroethane: Complete name of R-113. Group A1 refrigerant in common use. Chemical compounds that make up this refrigerant are chlorine, fluorine, and ethane.

Triple Point: Pressure–temperature condition in which a substance is in equilibrium (balance) in solid, liquid, and vapor states.

Troposphere: Part of the atmosphere immediately above the earth's surface in which most weather changes occur.

Tube-within-a-Tube Condenser: Water-cooled condensing unit in which a small tube is placed inside a large unit. Refrigerant passes through the outer tube; water passes through the inner tube.

Tubing: Fluid-carrying thin-walled pipe.

Turbulent Flow: Fluid flow in which the fluid moves transversely as well as in the direction of the tube or pipe axis, as opposed to streamline or viscous flow.

Twin Parallel: Two or more units installed in line by piping.

Two-Pipe System: A heating system in which one pipe delivers steam to radiators and a second pipe is used to return condensate.

Two-Position On–Off Control: A control system in which the control device can only start and stop the equipment.

Two-Stage Vacuum Pump: A vacuum pump used to remove vapor and moisture from a system.

Two-Temperature Valve: Pressure-opened valve used in suction line on multiple refrigerator installations that maintains evaporators in system at different temperatures.

Two-Way Valve: Valve with one inlet port and one outlet port.

U

Ultraviolet: Invisible radiation waves with frequencies shorter than wavelengths of visible light and longer than X-rays.

Unitary System: A factory assembled heating/cooling system in one package and usually designed for conditioning one space or room.

Unit Heater: A direct-heating, factory-made, encased assembly including a heating element, fan, motor, and directional outlet.

Unit System: A system that can be removed from the user's premises without disconnecting refrigerant containing parts, water connection, or fixed electrical connections.

Universal Motor: Electric motor that will operate on either AC or DC.

Unloader: A device that allows for easier compressor start-up by temporarily reducing high-side pressure at the cylinder head.

Upflow Furnace: A furnace that pulls return air in from the bottom and expels warm air from the top.

Urethane Foam: Type of foam insulation that is placed between the inner and outer walls of a container.

U-Value: Represents the heat leakage from one side of a wall to the other.

V

Vacuum: A pressure below atmospheric, usually measured in inches of mercury below atmospheric pressure.

Vacuum Activators: Dampers and control valves used in automotive air conditioning system; controlled by the vacuum created by engine intake manifold vacuum.

Vacuum Control System: Intake manifold vacuum is used to operate dampers and controls in some automobile systems.

GLOSSARY (cont.)

Vacuum Pump: Device used for creating vacuums for testing or drying purposes.

Valve: In refrigeration, a device for regulation of a liquid, air, or gas.

Valve, Expansion: Type of refrigerant control that maintains constant pressure in the low side of the refrigerating mechanism. Valve is caused to operate by pressure in the low or suction side. Often referred to as an automatic expansion valve or *AEV.*

Valve Plate: Part of the compressor located between the top of compressor body and the head. It contains compressor valves and ports.

Valve, Service: Device used to check pressures and charge refrigerating systems.

Valve, Solenoid: Valve made to work by magnetic action through an electrically energized coil.

Valve, Suction: Valve in refrigeration compressor that allows vaporized refrigerant to enter cylinder from suction line and prevents its return.

Valve, Water: In most water cooling units, a valve that provides a flow of water to cool the system while it is running.

Vapor: A gas, particularly one near to equilibrium with the liquid phase of the substance, which does not follow the gas laws. Frequently used instead of gas for a refrigerant and, in general, for any gas below the critical temperature.

Vapor Barrier: Thin plastic or metal foil sheet used to prevent water vapor from penetrating insulating material.

Vaporization: Change of liquid into a gaseous state.

Vapor Lock: Condition where liquid is trapped in a line because of a bend or improper installation. Such vapor prevents liquid flow.

Vapor Pressure: Pressure imposed by a vapor.

Vapor Pressure Curve: Graphic presentation of various pressures produced by refrigerant under various temperatures.

Vapor Recovery Method: A refrigerant recovery system using relatively small equipment to remove refrigerant from residential, automobile, and light commercial units.

Vapor, Saturated: Vapor condition that will result in condensation into droplets of liquid if temperature is reduced.

Vapor Velocity: Speed or rate at which a gas moves.

Variable Air Volume (VAV) Controller: Device having electronic components used to regulate the volume of air in a distribution system.

Variable Control: A control that can make gradual adjustments to equipment, as opposed to a simple on–off control.

Variable Pitch Pulley: Pulley that can be adjusted to provide different pulley drive ratios.

V-Belt: Belt commonly used in refrigeration work with a contact surface and pulley in a V-shape.

Velocimeter: Instrument that measures air speeds.

Velocity: Quickness of motion; swiftness; speed. Change in position with respect to time.

Ventilation: Airflow from one area to another.

VFD (Variable Frequency Drive): Electronic speed control for motors.

Vibration Absorbers: Soft or flexible substance or device that will reduce the transmission of a vibration.

Viscosity: The property of a fluid to resist flow or change of shape.

Volatile: Easily vaporized; a liquid that changes to the gaseous state readily.

W

W (Watt): A watt is a unit of electricity.

WC (Water Column): Common measure of air pressure used in HVAC systems.

Wet-Bulb Depression: Difference between dry- and wet-bulb temperatures.

Wet Compression: A system of refrigeration in which some liquid refrigerant is mixed with vapor entering the compressor so as to cause discharge vapors from the compressor to tend to be saturated rather than superheated.

X

Xylene: A flammable solvent, similar to kerosene, used for dissolving or loosening sludges and for cleaning compressors and lines.

Z

Zero, Absolute, of Pressure: The pressure existing in a vessel that is entirely empty.